FRAMINGHAM STATE COLLEGE
3 3014 00055 2739
D1272379

THE COMMONWEALTH AND INTERNATIONAL LIBRARY
Joint Chairmen of the Honorary Editorial Advisory Board
SIR ROBERT ROBINSON, O.M., F.R.S., LONDON
DEAN ATHELSTAN SPILHAUS, MINNESOTA

SELECTED READINGS IN PHYSICS
General Editor: D. TER HAAR

GENERAL THEORY OF RELATIVITY

GENERAL THEORY OF RELATIVITY

by

C. W. KILMISTER

Professor of Mathematics, Kings College, London

PERGAMON PRESS

Oxford · New York · Toronto

Sydney · Braunschweig

Framingnam State College
Framingham, Massachusetts

PERGAMON PRESS LTD.
Headington Hill Hall, Oxford
PERGAMON PRESS INC.
Maxwell House, Fairview Park, Elmsford, New York 10523
PERGAMON OF CANADA LTD.
207 Queen's Quay West, Toronto 1
PERGAMON PRESS (AUST.) PTY. LTD.
19a Boundary Street, Rushcutters Bay, N.S.W. 2011, Australia
VIEWEG & SOHN GmbH
Burgplatz 1, Braunschweig

Copyright © 1973 Pergamon Press Ltd.
All Rights Reserved. No part of this publication may be reproduced, stored in a retrieval system, or transmitted, in any form or by any means, electronic, mechanical, photocopying, recording or otherwise, without the prior permission of Pergamon Press Ltd.

First edition 1973

Library of Congress Cataloging in Publication Data

Kilmister, Clive William.
General theory of relativity.

(The Commonwealth and international library.
Selected readings in physics)
Includes bibliographical references.
1. General relativity (Physics) I. Title.
QC173.6.K54 1973 530.1'1 73-7639
ISBN 0-08-017639-9
ISBN 0-08-017645-3 (flexicover)

Printed in Hungary

QC
173.6
K54
1973

Contents

INTRODUCTION vii

ACKNOWLEDGEMENTS ix

Part I

Chapter I *The Principle of Equivalence* 3

Chapter II *The Beginnings of General Relativity* 25

Chapter III *Modern Developments* 62

References

Part II

1. On the Hypotheses which Lie at the Bases of Geometry
   B. RIEMANN (translated by W. K. CLIFFORD) 101
2. On the Space-theory of Matter
   W. K. CLIFFORD 124
3. On the Effect of Gravitation on the Propagation of Light
   A. EINSTEIN 128
4. The Foundations of General Relativity Theory
   A. EINSTEIN 140
5. On the Motion of Particles in General Relativity Theory
   A. EINSTEIN and L. INFELD 174

6. Three Lectures on Relativity Theory
V. Fock 220

7. Invariant Formulation of Gravitational Radiation Theory
F. A. E. Pirani 228

8. Gravitational Waves in General Relativity: VII. Waves from Axisymmetric Isolated Systems
H. Bondi, M. G. J. van der Burg, and A. W. K. Metzner 258

9. On Continued Gravitational Contraction
J. R. Oppenheimer and H. Snyder 308

10. A Spinor Approach to General Relativity
R. Penrose 318

11. Gravitational Red-shift in Nuclear Resonance 358
R. V. Pound and G. A. Rebka, Jr.

Index 363

Introduction

THIS book must be considered a sequel to the earlier one on special relativity but the difficulties in writing it are of rather a different order from those described in the introduction to that book. General relativity, unlike the special theory, is not a general framework but a specific scientific theory, in fact the best theory of gravitation that we have to date. It is often said to suffer unduly from lack of experimental check or confirmation. This fact, however, is essentially due to the extremely good gravitational theory which we had before, which was that of Newton's. In the situation when one theory is very good it is difficult to distinguish between its predictions and those of an alternative. The main difficulty in writing the book, however, has been to know what to leave out. Every reader will find the choice of papers included extremely idiosyncratic, but this is inevitable when there has been such a sudden rapid advance as has occurred in general relativity since 1945. If anyone feels that their work has been passed over, or unfairly treated in any way, I can only offer my apologies.

C. W. KILMISTER

Acknowledgements

My thanks are due to the following bodies for permission to reprint the extracts in this volume:

For Extracts 1 and 2: Macmillan & Co., London.

For Extracts 3 and 4: Johann Ambrosius Barth.

For Extract 5: *The Canadian Journal of Mathematics.*

For Extracts 6, 7, 9, and 11: American Physical Society.

For Extract 8: The Royal Society.

For Extract 10: Academic Press.

PART I

CHAPTER I

The Principle of Equivalence

THE special theory of relativity, which was discussed in the previous volume, gives rise to a modification of Newtonian mechanics. In particular the mass, which was treated as constant in Newtonian mechanics, increases with the velocity. This modification is to the left-hand side of Newton's law of motion:

$$m \frac{d^2\mathbf{r}}{dt^2} = \mathbf{F}.$$

As far as the special theory of relativity is concerned the right-hand side, the force, can still be filled in arbitrarily. In the particular application of the equations of motion in electromagnetic theory there is, indeed, a definite form for the force on a charged particle in an electric and magnetic field

$$\mathbf{F} = e(\mathbf{E} + \mathbf{v} \wedge \mathbf{B}),$$

and this form turns out to be consistent with the requirements of special relativity. It can, therefore, be taken over into the modified mechanics. The next problem, however, arises with the gravitational field. This problem worried Einstein for the 10 years from the publication of the special theory in 1905 (Einstein, 1905) until the advent of the general theory of relativity in 1915. In this chapter we shall follow the general lines of Einstein's argument, but in order to tackle one problem at a time we shall do so first in Newtonian mechanics, i.e. we shall ask ourselves first what is the appropriate way of discussing the gravitational field in Newtonian mechanics, and when we have answered this question we can go on to extend this discussion to special relativity.

There is good reason for the gravitational field to be accorded a unique treatment amongst other fields because of the fact, which has been known for a long time, that the force which it exerts on a body is proportional to its mass. So, from theequation of motion, the acceleration produced in a body at a particular point of the gravitational field is the same whatever the mass of that body. To put it in an inexact but easily remembered form: "all bodies fall equally fast." This fact was certainly known to Galileo (1638) but before him Stevinus in 1586 emphasised that weight (i.e. gravitation) had a theory that was like geometry (though different) because it acts on everything. The reader who is familiar with the treatment of planetary orbits given in most textbooks will recall that here, and only here, the working is carried out in terms of force per unit mass, energy per unit mass, and so on. Thus the characteristic property of the gravitational field of producing a definite acceleration was already taken account of in the actual *technique* of Newtonian mechanics, but it was regarded as a technical trick and not as an important question of principle that this could be done.

Let us turn from this discussion for a moment to the question of the existence of inertial frames in Newtonian mechanics. An inertial frame is defined originally as a frame of reference in which Newton's laws hold. Although this definition has an air of unambiguousness, it is really less clear than it seems. According to Newton's laws a particle acted on by no forces moves uniformly in a straight line. One would therefore have thought that a frame of reference fixed to the surface of the Earth could not be considered as even approximately an inertial frame (even if the rotation of the Earth is ignored) since a particle released in it falls with a certain acceleration. However, Newtonian mechanics gets over this difficulty by the postulation of gravitational forces and says that, so long as gravitational forces are allowed as well as contact and electromagnetic forces, such a frame of reference is approximately inertial. The question of which frames are inertial can, then, be settled only when one has decided which are the forces which are to be allowed. An equally valid way of starting mechanics would be to use a frame of reference which is falling freely under gravity, so that a particle released from rest in this frame remains at rest in the frame, since, as we have said, all particles fall equally fast. Looked at from outside, the particle falls with just the same acceleration as the frame of reference. There are thus at least two ways of dealing with the gravitational field—either by postulating a certain frame to be inertial and treating the gravitational field as a field of force, or by refusing to admit the gravitational field as a field of force and describing it by means of the acceleration of the reference frame relative to the former choice of inertial frame. The possibility of these two equivalent but different descriptions depends essentially on the universal character of the gravitational force, i.e. on the fact that all bodies fall equally fast.

Now when we come to look at these two possibilities we see at once that one of them, although superficially attractive, has really only been chosen as a historical—or possibly geographical—accident. A frame of reference fixed to the surface of the Earth is certainly important for practical considerations, and so remains a useful way of doing mechanics for the engineer who is concerned with constructing buildings on the surface of the earth. In considering mechanics in general, however, there is nothing to be said for this method of description over the other one, and the other method of description has the great advantage that the universal character of the gravitational field is automatically incorporated as a matter of principle. It is this new method of description which is adopted by Einstein for his generalization of gravitation to make it consistent with special relativity. Let us look at this in some detail. It will be instructive to consider some particular problems in mechanics. Let us first take one of the simplest problems, that of the projectile on the surface of the earth supposed stationary. The orthodox treatment for this problem using as an inertial frame a frame fixed to the earth's surface, starts with the equation of motion,

$$\ddot{\mathbf{r}} = -\mathbf{g},$$

and integrates this twice to get a solution of the form

$$\mathbf{r} = \mathbf{V}t - \tfrac{1}{2}\mathbf{g}t^2,$$

which is then easily shown to represent a parabola. In order to find, for example, the range reached on a horizontal plane through the gun, one then chooses coordinates with the x-axis horizontal, so that $x = (V \cos \alpha)t$, $y = (V \sin \alpha)t - \frac{1}{2}gt^2$, where α is the angle of projection. Setting $y = 0$ gives $t = 2V \sin \alpha/g$ as the time of flight, and the horizontal range

$$R = (2V^2/g) \cos \alpha \sin \alpha.$$

From the new point of view the inertial frame is one which is freely falling, and so the equation of motion becomes

$$\ddot{\mathbf{r}} = 0,$$

that is to say, the path of the particle is a straight line. Of course, this simplification is balanced to some extent by the fact that the horizontal plane mentioned above will now no longer be a horizontal plane and the point of impact of the shell will accelerate upwards with acceleration g. In the corresponding figure in this frame of reference we find $R = Vt \cos \alpha$, where $\sin \alpha = (gt/2V)$. Eliminating t again gives the result for R.

This first example shows that it is, indeed, a practical proposition to do elementary problems in mechanics by the new method rather than by the old one. However, there is a considerable simplification in this particular problem because the gravitational field with which we are concerned is a uniform one. If the gravitational field is not uniform, as, for example, when we deal with the orbits of the planets, then evidently the problem of coordinate transformations will be much more difficult. In fact for a non-uniform field all that we can do is to transform away the gravitational field in the immediate neighbourhood of one point. As a practical way of solving problems in mechanics this turns out to be much less valuable, but as a matter of principle this possibility of transforming away a field at one point is extremely important because it still incorporates the essential universality of the gravitational field. Now once we decide to carry out transformations of this sort—which reduce the gravitational field at a pre-assigned point to zero—we are evidently involved with accelerated

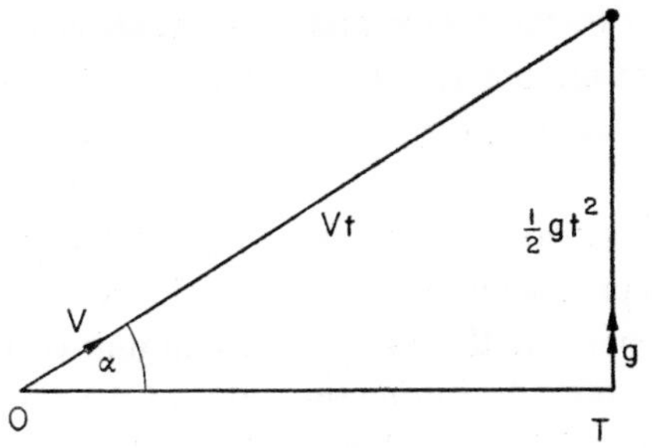

FIG. 1.

transformations, indeed, even in the simple case of the example just considered the transformation from the first way of looking at it to the second is by means of the transformation

$$\mathbf{r}_1 \rightarrow \mathbf{r}_2 = \mathbf{r}_1 + \tfrac{1}{2}\mathbf{g}t^2.$$

There is one instance in mechanics in which extensive use is made of such accelerated frames, and this will serve as our second example. This is in the case when one frame of reference is rotated relative to another. Suppose that a frame of reference is defined by three orthogonal unit vectors $\mathbf{e}_1'$, $\mathbf{e}_2'$, $\mathbf{e}_3'$, and suppose that this frame of reference for our present purpose can be considered as an inertial frame. Consider now another frame of refererence $\mathbf{e}_1$, $\mathbf{e}_2$, $\mathbf{e}_3$ with the same origin. This frame of reference is moving relative to the first one and the first step in the problem is to show that this movement consists of an angular velocity about a certain axis. This is easily done as follows. Since the three unit vectors remain always at right angles, we have the equation

$$\mathbf{e}_i \cdot \mathbf{e}_j = \delta_{ij},$$

where δ_{ij} is the Kronecker delta symbol which is 1 if $i = j$ and 0 otherwise. On differentiating this with respect to the time we get

$$\dot{\mathbf{e}}_i \cdot \mathbf{e}_j + \mathbf{e}_i \cdot \dot{\mathbf{e}}_j = 0.$$

This constraint on the derivatives of the unit vectors splits into two parts. Firstly, if $i = j$ it shows that the derivative of each unit vector is at right angles to the vector and accordingly it follows that there exist three vectors $\boldsymbol{\omega}_i$ such that

$$\dot{\mathbf{e}}_i = \boldsymbol{\omega}_i \wedge \mathbf{e}_i.$$

The vectors $\boldsymbol{\omega}_i$ are not completely defined by this since they only enter in a vector product with one of the unit vectors, and accordingly each one's component in the direction of that unit vector is completely at our disposal. Now substituting this value of the derivative back into the constraint for different values of i and j gives us, after a little manipulation,

$$(\boldsymbol{\omega}_i - \boldsymbol{\omega}_j)\cdot(\mathbf{e}_i \wedge \mathbf{e}_j) = 0.$$

When we look at this equation for different pairs of values of i and j it shows us that $\boldsymbol{\omega}_i$ and $\boldsymbol{\omega}_j$ have the same $\mathbf{e}_k$ component, where i, j, k is a permutation of 1, 2, 3. Let us now write out the possible values of the three vectors $\boldsymbol{\omega}_i$:

$$\begin{aligned}
\boldsymbol{\omega}_1 &= \cdot\mathbf{e}_1 + y\mathbf{e}_2 + z\mathbf{e}_3 \\
\boldsymbol{\omega}_2 &= x\mathbf{e}_1 + \mathbf{e}_2 + z\mathbf{e}_3 \\
\boldsymbol{\omega}_3 &= x\mathbf{e}_1 + y\mathbf{e}_2 + \mathbf{e}_3.
\end{aligned}$$

Using the arbitrariness of one component of each vector (signified here by its being omitted) it is clear that we may choose all three of the vectors to have the same value $\boldsymbol{\omega}$, say, and accordingly write

$$\dot{\mathbf{e}}_i = \boldsymbol{\omega} \wedge \mathbf{e}_i.$$

Consider now the position of a point in the moving coordinate system in the form

$$\mathbf{r} = \Sigma x_i \mathbf{e}_i.$$

The velocity of such a point will be

$$\dot{\mathbf{r}} = \Sigma(\dot{x}_i \mathbf{e}_i + x_i \dot{\mathbf{e}}_i).$$

The first term in this expression corresponds to differentiating the components of the vector in the usual way as if it were referred to a fixed coordinate system. We could adopt the notion $\dot{\mathbf{r}}$ for this. The second term arises from the rotation of the coordinate system but, because of the expression which we have for the derivatives of the unit vectors, can be written $\boldsymbol{\omega} \wedge \mathbf{r}$. When we want to write down equations of motion, however, we need the acceleration and accordingly we have to apply the rule which we have found to the velocity vector rather than to the position vector, and so we derive for the acceleration measured relative to an inertial frame the expression

$$\frac{d^2\mathbf{r}}{dt^2} = \ddot{\mathbf{r}} + 2\boldsymbol{\omega} \wedge \dot{\mathbf{r}} + \dot{\boldsymbol{\omega}} \wedge \mathbf{r} + \boldsymbol{\omega} \wedge (\boldsymbol{\omega} \wedge \mathbf{r}).$$

When we consider this expression in detail we see that in addition to the usual term which we expect in an expression for acceleration there are terms involving the position of the particle and also a term involving its velocity. This latter term, the so-called Coriolis acceleration, is the one which gives rise to numerous interesting phenomena on the rotating earth, playing an important part in the circulation of the atmosphere, for example.

In order to show the importance of this transformation to an accelerated reference frame, it will be instructive to work out a detailed example. So long as we allow gravitational forces, a frame of reference fixed to the surface of the earth is not a bad approximation to an inertial frame, but a much better one is one with its origin at the earth's centre and with axes in fixed directions. If we assume this one is inertial, and consider (Fig. 2) the transformation to one fixed to the earth's surface, in latitude λ, we shall have (using an equally good origin O′ for the fixed axes on the earth's axis)

$$\mathbf{r}' = \mathbf{a} + \mathbf{r}$$

so that

$$\frac{d^2\mathbf{r}'}{dt^2} = \frac{d^2\mathbf{a}}{dt^2} + \frac{d^2\mathbf{r}}{dt^2} \quad \text{with} \quad \frac{d^2\mathbf{a}}{dt^2} = -\Omega^2\mathbf{a},$$

so that, altogether, for motion under a gravitational field $-\mathbf{g}$ (assumed constant),

$$\ddot{\mathbf{r}} + 2\boldsymbol{\Omega} \wedge \dot{\mathbf{r}} + \boldsymbol{\Omega} \wedge (\boldsymbol{\Omega} \wedge \mathbf{r}) = -\mathbf{g} + \mathbf{a}\Omega^2.$$

Here we have disregarded the time derivative of $\boldsymbol{\Omega}$; if we also neglect

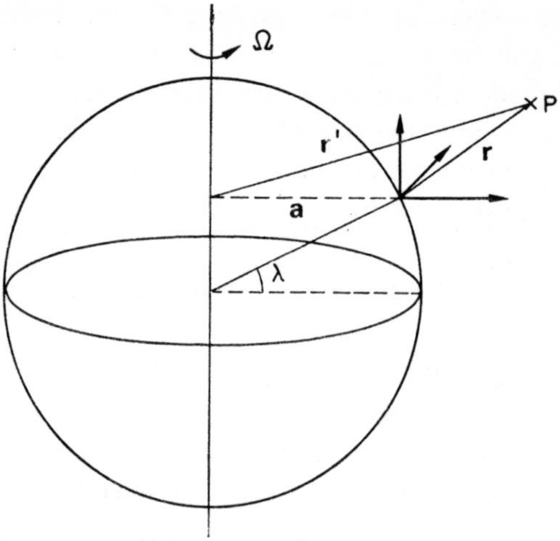

FIG. 2.

terms in Ω^2, since the earth's rotation is only once per day, we are left with

$$\ddot{\mathbf{r}}+2\boldsymbol{\Omega}\wedge\dot{\mathbf{r}} = -\mathbf{g}.$$

Let us apply this approximate result to the motion of a simple pendulum hanging from the origin O; if there were no rotation the equation of motion would be

$$m\ddot{\mathbf{r}} = -mg\mathbf{e}_3 - T\mathbf{r}/l,$$

where l is the length of the string and m the mass, since $\mathbf{g} = g\mathbf{e}_3$. Here T is the tension in the string. But since the axes are rotating the correct equation becomes

$$\ddot{\mathbf{r}}+2\boldsymbol{\Omega}\wedge\dot{\mathbf{r}} = -g\mathbf{e}_3 - \frac{T\mathbf{r}}{lm}.$$

Using Cartesian components, in which

$$\boldsymbol{\Omega} = \Omega(-\mathbf{e}_1\cos\lambda+\mathbf{e}_3\sin\lambda),$$

this becomes

$$\begin{aligned} \ddot{x}-2\Omega(\sin\lambda)\dot{y} \qquad\qquad\qquad &= -Tx/lm, \\ \ddot{y}+2\Omega(\sin\lambda)\dot{x}+2\Omega(\cos\lambda)\dot{z} &= -Ty/lm, \\ \ddot{z}-2\Omega(\cos\lambda)\dot{y} \qquad\qquad\qquad &= -g-Tz/lm. \end{aligned}$$

These are the equations for a general motion of the pendulum; but consider now small oscillations for which $z \simeq l$. In fact, then

$$x^2+y^2+z^2 = l^2$$

so that

$$z = l[1-(x^2+y^2)/l^2]^{\frac{1}{2}}$$

so that z differs from l only by *second*-order terms in the displacements x and y. Accordingly, neglecting $\dot{z}$ and $\ddot{z}$, the equations become

$$\begin{aligned} \ddot{x}-2\Omega\sin\lambda\dot{y} &= -Tx/lm, \\ \ddot{y}+2\Omega\sin\lambda\dot{x} &= -Ty/lm, \\ -2\Omega\cos\lambda\dot{y} &= -g-T/m. \end{aligned}$$

The last equation may be used to calculate the tension T when the motion is known; the first two may be integrated by writing $w = x+iy$, since then

$$\ddot{w}+2(i\Omega\sin\lambda)\dot{w} = -(T/lm)w,$$

and if $w = ve^{-i\Omega t\sin\lambda}$, $lmp^2 = T$,
then (neglecting powers of Ω again)

$$\ddot{v}+p^2v = 0,$$

with solution

$$v = Ae^{ipt}+Be^{-ipt},$$

giving

$$w = x+iy = e^{-i\Omega t\sin\lambda}[Ae^{ipt}+Be^{-ipt}].$$

In general this is an ellipse, rotating with angular speed $\Omega\sin\lambda$. It is to be noted that (as with most effects of the Coriolis acceleration) the rotation will be zero for points on the equator, and in opposite directions in the northern and southern hemispheres. The theory just described is that of the *Foucault pendulum*, and the effect may be observed by using any simple pendulum for which l is sufficiently great and the bob sufficiently heavy for air resistance not to reduce the swing too greatly. But the fact that this can be observed is really very surprising, because the theory started with the assumption that the frame of reference fixed to the earth was rotating relative to some *fixed* set of axes (in fact those axes with respect to which the rotation of the earth is measured). These axes are determined by the distant background of stars; but equally this theory shows that they could be found by means of a Foucault pendulum, an instrument which can be conducted entirely with the laboratory. Thus the local inertial frames on the surface of the earth are related in some way to the (position and motion) of the distant matter. Assuming that the connection is a causal one, the only reasonable conclusion is that the distribution of matter in the universe determines the local inertial frames, an assumption attributed to Mach, and so called *Mach's principle* (Mach, 1960). This principle had a profound effect on Einstein's thought, and was part of the motivation for his re-expressing gravitational theory. For no explanation of this causal effect has ever been found; but Einstein hoped that general relativity would "incorporate Mach's principle" (Holton, 1965).

The terms like the Coriolis acceleration arise, it is to be noted, even for the case of a rigid Newtonian frame rotating with uniform angular velocity. We can therefore expect that, if we are to employ generally accelerated coordinate systems, considerably greater complexity will result. The criterion of an accelerated coordinate system is that, whereas the time is measured in the same way as in the old coordinate system as always in Newtonian mechanics, the space variables depend not only on the old space variables but on the time as well and, indeed, do so in a non-linear fashion (since a linear dependence on time comes in the transformation between inertial frames and corresponds to motion with uniform speed). Accordingly the transformations which we shall have to allow will be of the form

$$x^{\alpha} \rightarrow x^{\alpha} = f^{\alpha'}(x^{\alpha}, t) \quad (\alpha = 1 \ldots 3),$$
$$t' = t,$$

which may be written with the convention

$$x^i = (x^\alpha, t), \quad x^4 = t$$

in an abbreviated notation as

$$x^i \to x^{i'} = f^{i'}(x^j);$$

that is, we make the rule that the Greek suffixes run from 1 to 3 and Latin suffixes from 1 to 4, the fourth coordinate being the time, as it is in special relativity. The difference here, of course, is that, since we are now concerned with Newtonian mechanics, the time is entirely unconnected with the space variables, unlike the way which it was in special relativity.

The group of transformations which we are concerned with, as defined by

$$x^\alpha \to x^{\alpha'} = f^{\alpha'}(x^\alpha, t) \quad (\alpha = 1 \ldots 3)$$
$$t' = t,$$

gives rise to quantities transforming under representations of the group, just as the Lorentz group gives rise to such quantities in special relativity. In fact, all our experimental observations must be expressed in terms of such quantities, and they can be derived in very much the same way as was sketched out in the earlier book. For example, the differentials of coordinates give rise to one representation, and a general quantity transforming according to this representation is a contravariant vector A^i. Since, however, the coordinates in this particular case are Newtonian ones, the contravariant vector splits up into the three-dimensional vector and an invariant part

$$A^i = (A^\alpha, A^4)$$

This is not the case, on the other hand, with the covariant vector A_i, which is defined as transforming under the same representation as the gradient of an invariant quantity. This asymmetry between the two kinds of vectors disappears again when we get to the full statement of general relativity. But it is emphasized here because it does affect the working in this detailed example of the corresponding Newtonian theory.

The velocity vector

$$v^i = (v^\alpha, 1) = \frac{dx^i}{dt}$$

can be defined in the obvious way, and it splits up into the usual three-vector part, the usual velocity, together with a numerical constant. When the velocity vector is transformed it will have as its values in a new coordinate system

$$v \;\; = x_i{}'v^i,$$

where we employ the abbreviation

$$x_i^{i'} = \frac{\partial x^{i'}}{\partial x^i}$$

for the array of differential coefficients. Notice here that we employ dashes on the suffixes for the new coordinate system, a convenient convention when a number of coordinate transformations have to be carried out. Splitting up the transformation of the velocity into the three-dimensional part, it becomes

$$v^{\alpha'} = x_\alpha^{\alpha'} v^\alpha + x_4^{\alpha'},$$

and since, from the form of the group of transformations, the following conditions hold for the partial derivatives:

$$x_\alpha^{4'} = x_{\alpha'}^4 = 0,$$
$$x_4^{4'} = x_{4'}^4 = 1;$$

the remaining part of the transformation of the velocity is trivial. If we confine our attention to a transformation in which the axes are not rotating, we can rewrite the three-vector part in vector notation as

$$\mathbf{v}' = \mathbf{v} + \mathbf{V},$$

which is the usual relative velocity formula.

So far everything is as expected, but certain difficulties enter when we consider the acceleration, i.e. the derivative of the velocity. If we compute, in the dashed coordinate system, the derivative of the velocity, we get

$$\begin{aligned}\frac{dv^{i'}}{dt} &= \frac{d}{dt}(x_i^{i'}v^i)\\ &= x_i^{i'}\frac{dv^i}{dt}+v^i x_{i4}^{i'}\\ &= x_i^{i'}\frac{dv^i}{dt}+v^\alpha x_{\alpha 4}^{i'}+x_{44}^{i'},\end{aligned}$$

and we observe here terms linear in the velocity entering on one side but not on the other. In other words, if the derivative of the velocity is taken as the acceleration in one coordinate system this coordinate system will be privileged over all those others in which this derivative has certain terms linear in the velocity added to it. There is a well-known trick for getting out of this difficulty of a privileged coordinate system; it is to modify the definition of the acceleration so that the linear terms appear in all coordinate systems. In other words one defines in general an expression for the acceleration:

$$f^i = \frac{dv^i}{dt}+\Omega_j^i v^j = v^j(v_{i,j}+\Omega_j^i).$$

In order to carry this approach through, however, the nature of the coefficients Ω_j^i has to be investigated a little further. At present all we know is that we are no longer to assume the existence of a coordinate system in which $\Omega_j^i = 0$ holds.

It is at this point that we must make an appeal to rather deeper considerations that have not been evident up to now. These considerations arise from the paper of Riemann which appears, translated by W. K. Clifford (Clifford, 1873), as Extract 1 of the present volume. In this paper Riemann undertakes to consider the general concepts from which geometry is built, generalizing his investigations to the case of an n-dimensional space and to what he calls a general notion of magnitude. His starting point is really the investigation of Gauss about curved surfaces, a summary of which will be found in the introduction to Extract 1 of the present volume. He concludes from Gauss's results that it is necessary to study the infinitesimal geometry of the space and that the simplest form to assume for the displacement between two points is a quadratic one in the coordinate differentials. He mentions as a further possibility the question of a biquadratic expression (cf. Eddington, 1924) but disregards this as it would throw little new light on the idea of space. A most important consideration for later developments in relativity enters at this point, where he remarks on the fact that the number $\frac{1}{2}n(n+1)$ of independent coefficients in the metrical form is reduced to $\frac{1}{2}n(n-1)$ by being allowed to take n new variables. He now goes on to consider a flat space, i.e. an immediate generalization of a Euclidean one and asks the question: What other spaces may there be? By drawing geodesics through a point to get a geodesic surface he adopts, as the measure of curvature of the space, the Gaussian curvature of this surface. Finally, he considers spaces of constant curvature and applies all these ideas to three dimensions where, he suggests, an empirical decision is necessary. In 1876, that is 3 years after this translation, Clifford takes up the idea enthusiastically (in Extract 2), and wants to regard space as flat on the average but with small hills, whose motion is determined by the motion of matter, indeed, *is* the motion of matter.

At this point it is not convenient for us to take over the whole of Riemann's ideas, although they do apply very closely when we come to general relativity. But what is important is the idea, tentatively put forward by Riemann and seized upon by Clifford, of a geometrical background being used to describe physical phenomena. What we have to do in the present instance in mechanics is to inquire what appropriate geometrical structure to assume in our space.

It is a little more convenient, because of the general differential geometry character of the investigation, to imagine not a single particle but a fluid consisting of many particles flowing in the coordinate system. When we do this the ordinary expression of the acceleration, excluding any extra terms, becomes

$$\frac{dv^i}{dt} = v^j v^i{}_{,j},$$

and if linear terms are to be added in order to get the correct transformation of the acceleration, then this must have the form

$$f^i = v^j v^i_{;j},$$

where the new quantities, denoted by the semicolon, are defined in some way in terms of the ordinary derivatives. The new kind of derivative introduced like this is known as the covariant derivative because our requirement that f^i should have the same form in all coordinate systems implies that f^i is a contravariant vector and, therefore, that the covariant derivative is a mixed covariant tensor of rank 2. This covariant derivative is usually assumed to fulfil the following conditions:

(1) That the covariant derivative of a sum of vectors is the sum of their covariant derivatives.
(2) That the covariant derivative of a scalar is its ordinary derivative.
(3) That the usual rule holds for differentiating a product.

If these assumptions are made it is clear that the most general form possible for a covariant derivative of any vector v^i is

$$v^i_{;j} = v^i_{,j} + \Gamma^i_{pj} v^p,$$

and when this assumption is made the acceleration can be written as

$$f^i = \frac{dv^i}{dt} + \Gamma^i_{pj} v^p v^j.$$

We notice that in order to avoid the uniqueness of one coordinate system we have been forced to import a certain geometrical structure into the space just as Riemann did. Ours is defined by an array of sixty-four coefficients Γ^i_{pj} which may, in general, be functions of position. In the definition of the acceleration, however, these sixty-four quantities do not enter independently but multiply by a symmetric product of the velocities. It is therefore only the part of these quantities symmetric in the lower suffixes which can arise and it is convenient, and economizes in assumptions, to suppose that the quantities themselves are symmetric in the lower indices. It is now a simple exercise to calculate how these coefficients, the so-called coefficients of affine connection, transform from one coordinate system to any other. We have

$$v^{i'}{}_{;j'} = v^{i'}{}_{,j'} + \Gamma^{i'}_{p'j'} v^{p'} = x^{j}_{j'}(x^{i'}_{i} v^{i})_{,j} + \Gamma^{i'}_{p'j'} v^{p'}$$
$$= x^{j}_{j'} x^{i'}_{i} (v^{i'}{}_{,j} + \Gamma^{i}_{pj} v^{p}) + x^{j}_{j'} x^{i'}_{ij} v^{i} + (\Gamma^{i'}_{p'j'} - x^{j}_{j'} x^{i'}_{i} \Gamma^{i}_{pj} x^{p}_{p'}) v^{p'}.$$

In order that the covariant derivative transforms like a tensor, then

$$\Gamma^{i'}_{p'j'} = x^{i'}_{i} x^{j}_{j'} x^{p}_{p'} \Gamma^{i}_{pj} - x^{j}_{j'} x^{i'}_{ij} x^{i}_{p'} = x^{i'}_{i} x^{j'}_{j} x^{p}_{p'} \Gamma^{i}_{pj} + x^{i}_{p',j'} x^{i'}_{i}$$
$$(\text{since } (x^{i'}_{i} x^{i}_{p'}),_{j} = (\delta^{i'}_{p'}),_{j} = 0).$$

Let us make sure that what we have done is quite clear by rewriting the earlier examples in this new notation. More precisely, let us assume that we begin with a coordinate system in which the coefficients of affine connection are zero, $\Gamma^{i}_{pj} = 0$ (since in the two earlier examples there *was* such a preferred coordinate system) and let us transform to a dashed coordinate system, computing the consequent coefficients of affine connection. In the first case the transformation was written

$$\mathbf{r}' = \mathbf{r} + \tfrac{1}{2}\mathbf{g}t^2$$

or, in our present notation: $x^{\alpha'} = x^{\alpha} + \frac{1}{2} g^{\alpha} t^2$. We have now to work out the various partial derivatives, and the first partial derivatives are obviously $x^{\alpha'}_{\alpha} = \delta^{\alpha'}_{\alpha}$, $x^{\alpha'}_{4} = g^{\alpha} t$, since g^{α} is a constant. Now as far as working out the affine connection in the new coordinate system is concerned, the main interest lies in the second partial derivatives, and it is clear from these formulae that the only three of these which are non-zero are given by $x^{\alpha'}_{44} = g^{\alpha}$. Inserting this value into the expression for the transformed affine connection, and noting that the original affine connection is zero, we get

$$\Gamma^{\alpha'}_{4'4'} = g^{\alpha},$$

giving, as the expression of the generalized acceleration,

$$f^{\alpha} = \frac{dv^{\alpha}}{dt} + g^{\alpha},$$
$$f^{4} = 0,$$

This agrees entirely with our earlier result for this example, i.e. that if the freely falling coordinate system is used there is no acceleration because the first term in the expression for the acceleration given by this formula will cancel out with the second.

Let us now turn to the case of a rotating coordinate system and it will simplify the working a little to confine ourselves to the rotation about the z-axis so that the transformation equations have the form

$$x' = x \cos \theta + y \sin \theta,$$
$$y' = -x \sin \theta + y \cos \theta,$$

where $\theta = \theta(t)$.

The first partial derivatives of interest are given by

$$x_1^{1'} = \cos \theta, \quad x_2^{1'} = \sin \theta, \quad x_1^{2'} = -\sin \theta, \quad x_2^{2'} = \cos \theta,$$

and

$$x_4^{1'} = \dot{\theta} y', \quad x_4^{2'} = -\dot{\theta} x'.$$

Hence the non-zero second derivatives will be

$$x_{14}^{1'} = -\dot{\theta} \sin \theta, \quad x_{24}^{1'} = \dot{\theta} \cos \theta, \quad x_{14}^{2'} = -\dot{\theta} \cos \theta, \quad x_{24}^{2'} = -\dot{\theta} \sin \theta,$$
$$x_{44}^{1'} = \ddot{\theta} y' - \dot{\theta}^2 x',$$
$$x_{44}^{2'} = -\ddot{\theta} x' - \dot{\theta}^2 y'.$$

It will be sufficient for our purposes to look at the effect at *any* instant of the rotation, and so to calculate the transformation at that moment when the axes momentarily coincide. Hence we can take $\theta = 0$, $x = x'$, $y = y'$, and so have

$$x_1^{1'} = 1, \quad x_2^{2'} = 1, \quad x_4^{1'} = \dot{\theta} y, \quad x_4^{2'} = -\dot{\theta} x,$$
$$x_{24}^{1'} = \dot{\theta}, \quad x_{14}^{2'} = -\dot{\theta}, \quad x_{44}^{1'} = \ddot{\theta} y - \dot{\theta}^2 x,$$
$$x_{44}^{2'} = -\ddot{\theta} x - \dot{\theta}^2 y.$$

Using the formula for the transformation of the affine connection and noting that the undashed connection is zero (and also that, in the partial derivatives just quoted, the interchange of dashed and undashed suffixes is obviously accomplished by changing the sign of θ) it is at once clear that

$$\Gamma^{\alpha'}_{\beta'\gamma'} = 0, \quad \Gamma^{4'}_{\beta'\gamma'} = 0,$$

whilst

$$\Gamma^{1'}_{4'2'} = \Gamma^{1'}_{2'4'} = -\dot{\theta}, \quad \Gamma^{2'}_{4'1'} = \Gamma^{2'}_{1'4'} = \dot{\theta},$$

$$\Gamma^{1'}_{4'4'} = -\ddot{\theta}y - \dot{\theta}^2 x, \qquad \Gamma^{2'}_{4,4'} = \ddot{\theta}x - \dot{\theta}^2 y.$$

Inserting these values into the expression for the acceleration, it has the x, y components

$$\ddot{x} - 2\dot{\theta}\dot{y} - \ddot{\theta}y - \dot{\theta}^2 x,$$

$$\ddot{y} + 2\dot{\theta}\dot{x} + \ddot{\theta}x - \dot{\theta}^2 y$$

(the factor 2 arising because, for example in the first line, the terms $\Gamma^1_{24}\dot{y}$ and $\Gamma^1_{42}\dot{y}$ must both occur). The reader can verify that these are just the expressions worked out earlier (in the particular case chosen here).

Having reassured ourselves that we are on the right lines, let us now try and see how the gravitational field is to be described with these new techniques. In one way, of course, the answer is very simple: we adopt a freely falling coordinate system so that at any one point the field vanishes. But this is only a very partial answer, except in very special cases such as the uniform field, since this transformation will not have the effect of removing the field on neighbouring points. It would be useful then to see to what extent the field *cannot* be removed at neighbouring points by imagining ourselves fixed to a freely falling particle and observing the relative motions of nearby particles of a cloud. We could do this, of course, entirely in the Newtonian theory where the equation of motion is

$$m\ddot{\mathbf{r}} = \nabla\phi = m\mathbf{g} \text{ (say)},$$

where $\mathbf{g}$ is not to be thought of as constant and ϕ is the gravitationa potential, satisfying Laplace's equation, $\nabla^2\phi = 0$, in free space. When we look at a nearby particle to the one which we have just considered, say at the nearby point $\mathbf{r}+\mathbf{s}$, the equation of motion becomes (to the first order in $\mathbf{s}$)

$$m(\ddot{\mathbf{r}} + \ddot{\mathbf{s}}) = m\mathbf{g} + m\mathbf{s}\cdot\nabla\mathbf{g}.$$

For the relative motion we have, therefore,

$$m\ddot{\mathbf{s}} = m\mathbf{s}\cdot\nabla\mathbf{g}.$$

Now let us write this out in terms of Cartesian coordinates x^1, x^2, x^3; it becomes

$$\ddot{s}^\alpha = s^\beta g^\alpha{}_{,\beta}.$$

Evidently in the equation for the relative motion it is the derivatives of the old gravitational field which enter, as we expected. If the relative motion is specified in general (not only with the special coordinate system given here) by an equation of the form

$$\ddot{s}^\alpha = \Phi^\alpha_\beta s^\beta,$$

we can expect the equivalent equation to that of Laplace (which implies that the original gravitational field is the inverse square law field) to take the form

$$\Phi^\alpha_\alpha = 0,$$

It remains to rewrite this rather simple derivation in terms of the more general framework which we have constructed. We are to consider a freely falling particle as our origin of coordinates, i.e. one whose acceleration is zero. That means that the motion of the particle satisfies the condition

$$f^i = \frac{dv^i}{dt} + \Gamma^i_{jk}v^j v^k = 0.$$

Imagine now a cloud of such particles moving in roughly the same direction, so that the particle O, which we are looking upon as the origin, has the velocity shown in the diagram (Fig. 3) and the nearby particle has also at the same instant of time the velocity shown. The quantity which we are interested in is the variation of y^i as the particles move along their paths. If we examine the change from the small parallelogram in Fig. 3, we have

$$v^i\,\delta t + y^i + \delta y^i = y^i + (v^i + y^j v^i{}_{,j})\,\delta t,$$

so that

$$\frac{dy^i}{dt} = y^j v^i{}_{,j}.$$

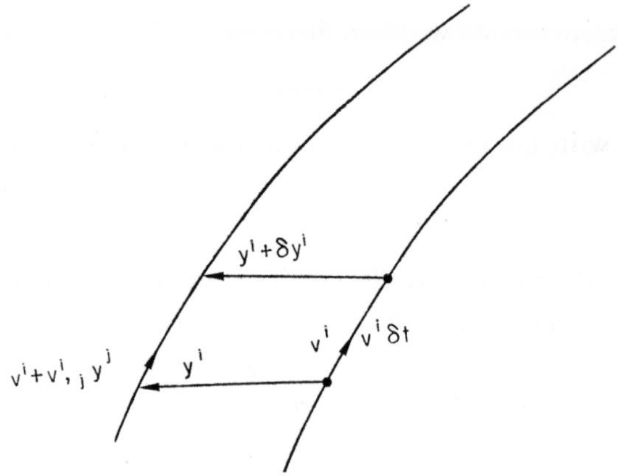

FIG. 3.

This appears to be four equations, but in fact it is only three, since y is a vector joining two points at the same instant of time, so that $y^4 = 0$, and since also $v^4 = 1$, the fourth equation is an identity.

We are failing at this point, however, to carry out our own precepts. The equation which we have just written must evidently be true in any coordinate system since no use has been made of the special coordinate system (i.e. freely falling origin) devised in finding it. But the two sides of this equation transform in most complicated ways, because on the left-hand side we have an ordinary derivative of the vector quantity y^i and this is given as an ordinary derivative of another vector v^i. We can, however, put matters right very conveniently by the following trick. It is always possible at any one point to use freely falling coordinates so that the gravitational field vanishes. Since the gravitational field is represented here by the coefficients of affine connection, this means that we must be able to choose coordinates at one point so that

$$\Gamma^i_{jk} = 0.$$

In fact it is easy to see that this is possible since the coordinate transformation

$$x^i \to x^{i'} = x^i - \tfrac{1}{2}\Gamma^i_{jk}x^j x^k,$$

where the coefficients of the squared terms are those of the affine connection at the origin, evidently reduces this connection to zero at the origin because of the transformation equations which we have had before. Now it is evidently such a coordinate system that we have chosen at the point O under consideration. The equation for the rate of change of y^i can therefore be written

$$\frac{\delta y^i}{\delta t} = y^j v^i_{;j},$$

where the left-hand side of this equation is the *absolute derivative*, defined by

$$\frac{\delta y^i}{\delta t} = v^j y^i_{;j}.$$

This new equation is also true in every coordinate system, *in exactly this form*, because both sides of it transform as contravariant vectors.

We are interested, however, not in the first derivative of the vector y^i but in its second derivative, and this can be calculated by repeating the process:

$$\begin{aligned}\frac{\delta^2 y^i}{\delta t} &= v^j_{;k} y^k v^i_{;j} + y^j v^k v^i_{;jk} = y^k(v^j v^i_{;j})_{;k} + y^j v^k v^i_{;jk} - y^k v^j v^i_{;jk} \\ &= y^j v^k (v^i_{;jk} - v^i_{;kj})\end{aligned}$$

after using the fact that $\delta v^i/\delta t = 0$ from our assumption (freely falling origin). Now the quantity in brackets in the above formula may be found as follows:

$$\begin{aligned}v^i_{;jk} - v^i_{;kj} &= (v^i_{,j} + \Gamma^i_{pj} v^p)_{;k} - (v^i_{,k} + \Gamma^i_{pk} v^p)_{;j} \\ &= v^p(\Gamma^i_{pj,k} - \Gamma^i_{pk,j} + \Gamma^i_{lk}\Gamma^l_{pj} - \Gamma^i_{lj}\Gamma^l_{pk}).\end{aligned}$$

It is noteworthy that this quantity does not depend on the derivatives of v^i but may be written in the form

$$R^i_{pkj} v^p,$$

where the array of numbers R^i_{pkj} is called the *Riemann–Christoffel* tensor, or *curvature tensor*, of the connection. The second rate of change of the vector y^i therefore takes the form

$$\frac{\delta^2 y^i}{\delta t^2} = R^i_{pkj} v^p v^k y^j,$$

the so-called equation of geodesic deviation (Levi-Civita, 1926; Pirani, 1956).

Let us go back to the inertial frame in which we started the discussion. We could choose a freely falling frame attached to the first particle so that

$$v^p = (0, 0, 0; 1).$$

When we do this the equation of geodesic deviation reduces to

$$\frac{\delta^2 y^i}{\delta t^2} = R^i_{44j} y^j$$

which is of the same form as the equation we had in the elementary discussion (there is, of course, the question of the fourth component of this equation, but since y has been drawn to join two points at the same time its fourth component is zero and both sides of the equation vanish identically then). Laplace's equation therefore has the form, *in this particular coordinate system*,

$$R^i_{44i} = 0.$$

However, if we make a transformation of coordinates, Laplace's equation will only remain true in all coordinate systems if, in addition to the equation which we have just quoted, the remaining equations of the set

$$R_{jk} \equiv R^i_{jki} = 0$$

hold as well. At this point our Newtonian analogy begins to break down because this set of equations is not sufficiently numerous to form a set of equations for the forty coefficients of affine connection. It is possible to get out of this difficulty, which is essentially connected with the need to assume that the gravitational field is a conservative field, so that it is derivable from a scalar potential. However, since our purpose here is merely to provide a simple introduction to general relativity, it is not profitable to pursue this argument further, and instead we shall turn to the corresponding theory in general relativity.

CHAPTER II

The Beginnings of General Relativity

THE first intimations of the experimental results of this theory (as distinct from the theory itself) are in Einstein's 1911 paper (Einstein, (1911) (Extract 3 of the present volume) in which he considers, by means of a thought-experiment, the effect on a light-ray of a uniform gravitational field. He shows that this effect arises because, firstly, the energy of the light is related to the potential energy of the gravitational field. Secondly, Einstein is proposing what he calls an important interpretation of the fact that all bodies accelerate equally in the uniform field; this fact, he says, is one of the most general that nature has given us, and yet it has no fundamental place in our physical constructions. His new interpretation is to assume that a coordinate system in which there is a uniform gravitational field, and one which is freely falling in the field, are physically equivalent. These ideas must have been constantly with Einstein in the succeeding 4 years, finally emerging in triumphantly clear form in 1915 (Einstein, 1916) (most of which forms Extract 4 of the present volume). We can use them by returning to our earlier discussion. The first part of the argument which we have been discussing was carried out in Newtonian theory with an absolute time. Now in special relativity, as we saw in the companion book in this series, there is no absolute time, and instead one considers the proper time defined by the equation

$$ds^2 = \eta_{ij}\, dx^i\, dx^j = dt^2 - \frac{1}{c^2}(dx^2 + dy^2 + dz^2),$$

where $\quad x^1 = \dfrac{x}{c}, \quad x^2 = \dfrac{y}{c}, \quad x^3 = \dfrac{z}{c}, \quad x^4 = t.$

In this case the appropriate definition for the acceleration is no longer in terms of the coordinate time t but in terms of the proper time, and so one defines the acceleration by

$$f^i = \frac{d^2x^i}{ds^2}.$$

Now these assumptions correspond in our analogy to starting with inertial frames in which there is a zero affine connection. In the analogy we then considered the effect of transforming to the accelerated frames needed, and this left one frame, the original inertial frame, in a privileged position. We were able to remove this privilege by introducing an arbitrary set of coefficients of affine connection. Here we might as well short-cut this process by going to the general situation straight away. If one uses any different coordinate system the metrical form must still be a quadratic one and so we write quite generally, as Einstein did,

$$ds^2 = g_{ij}\, dx^i\, dx^j$$

without making the assumption that there is a coordinate system in which g_{ij} reduces to the original form. That is to say, we are employing a *Riemannian* geometry. The corresponding form for the acceleration of a particle will then be

$$f^j = \frac{\delta^2 x^j}{\delta s^2} = v^i v^j{}_{;i},$$

where

$$v^i = \frac{dx^i}{ds}.$$

The interesting thing, which is not to be expected from our analogy, is that with certain plausible assumptions the coefficients of the metric then give all the geometrical structure that we need for discussing the theory, the affine connection which enters being determined in a particularly natural way in terms of them. For purposes of manipulation it is convenient to define a related set of metric coefficients by the equation

$$g^{ij} = \frac{\text{cofactor of } g_{ij} \text{ in } \det(g_{ij})}{\det(g_{ij}),},$$

so that, by the usual rule for expanding a determinant,

$$g^{ij}g_{jk} = \delta^i_k.$$

It is also convenient in the exposition to introduce the idea of parallel displacement; we shall say that a vector is parallelly displaced if in the course of its movement from one point of the manifold to another it satisfies

$$A^i_{;j} = 0.$$

This corresponds to the idea, when a Cartesian coordinate system is possible, of parallel displacement, meaning displacement without change of components. In the general coordinate system it means that the change in the component of the vector during a small displacement dx^j is given by

$$dA^i = -\Gamma^i_{kj}A^k\,dx^j.$$

In such a parallel displacement one would expect there to be no change of length of the vector. Since the covariant derivative of a scalar is its covariant derivative, this has the effect of requiring

$$(g_{ij}A^iA^j)_{,k} = (g_{ij}A^iA^j)_{;k} = 0,$$

and since the coefficients of the metric are a symmetric array this implies

$$g_{ij;k} = 0.$$

These equations may be regarded as linear ones for the affine connection and they may easily be solved uniquely to give (Christoffel, 1869)

$$\Gamma^i_{jk} = \left\{ \begin{matrix} i \\ jk \end{matrix} \right\} = g^{ip}[p, jk] = \frac{1}{2}g^{ip}(g_{pj,k} - g_{jk,p} + g_{kp,j}).$$

Thus in the extension of our theory to special relativity the coefficients of affine connection are determined by certain very natural assumptions in terms of the metric, and it is only necessary to find field equations which will determine the metric, i.e. to find ten equations rather than forty. In this respect the situation in general relativity is paradoxically simpler than that in the Newtonian theory.

The field equations themselves may be derived by glancing back at the analogy and noting the way in which Laplace's equation is related to them by geodesic variation. In exactly the same way as the equation

$$R_{jk} = R^i_{jki} = 0,$$

was derived there as field equations, so the corresponding equation may be found here. In this case, however, the Riemann–Christoffel tensor has certain symmetry properties, and as a result this equation represents ten scalar equations. That is just the right number to determine the ten coefficients of the metric. The best-known solution of these equations and the first solution to be found is that corresponding to spherical symmetry. The metrical form for a static spherically symmetric solution cannot be more general than

$$ds^2 = e^{2\lambda}\,dt^2 - e^{2\mu}\,dr^2 - r^2 e^{2\nu}(d\theta^2 + \sin^2\theta\,d\phi^2),$$

where λ, μ, ν are functions of r. However, we must remember that we are allowed arbitrary coordinate transformations, and it is therefore possible to replace the radius vector by another r in such a way as to ensure $\nu = 0$ (at least this was what was assumed in the early days of the theory, though it is now realized (Robinson and Trautmann, 1962) that there is one exceptional case in which this is impossible, i.e. when $\nu = -\log r$). The non-zero coefficients of the metric are therefore

$$g_{44} = e^{2\lambda}, \quad g_{11} = -e^{2\mu}, \quad g_{22} = -r^2, \quad g_{33} = -r^2 \sin^2\theta,$$

and the corresponding symbols with raised suffixes are

$$g^{44} = e^{-2\lambda}, \quad g^{11} = -e^{-2\mu}, \quad g^{22} = -\frac{1}{r^2}, \quad g^{33} = -\frac{1}{r^2\sin^2\theta}.$$

The only non-zero derivatives of the original coefficients are then

$$g_{44,1} = 2\lambda' e^{2\lambda}, \quad g_{11,1} = -2\mu' e^{2\mu}, \quad g_{22,1} = -2r,$$
$$g_{33,1} = -2r\sin^2\theta, \quad g_{33,2} = -2r^2\sin\theta\cos\theta,$$

where primes denote differentiation with respect to r, and a glance at the formulae for the coefficients of affine connection shows that the only non-zero ones which can arise will be those derived from

$$
\begin{aligned}
[1,44] &= -\lambda' e^{2\lambda}, \\
[4,14] &= \lambda' e^{2\lambda}, \\
[1,11] &= -\mu' e^{2\mu}, \\
[1,22] &= r, \\
[2,12] &= -r, \\
[1,33] &= r \sin^2 \theta, \\
[3,13] &= -r \sin^2 \theta, \\
[2,33] &= r^2 \sin \theta \cos \theta. \\
[3,23] &= -r^2 \sin \theta \cos \theta.
\end{aligned}
$$

These will therefore have the nine values

$$
\begin{aligned}
\left\{ {1 \atop 44} \right\} &= \lambda' e^{2(\lambda-\mu)}, \\
\left\{ {4 \atop 14} \right\} &= \lambda', \\
\left\{ {1 \atop 11} \right\} &= \mu', \\
\left\{ {1 \atop 22} \right\} &= -re^{-2\mu}, \\
\left\{ {2 \atop 12} \right\} &= 1/r. \\
\left\{ {1 \atop 33} \right\} &= -e^{-2\mu} r \sin^2 \theta, \\
\left\{ {3 \atop 13} \right\} &= 1/r, \\
\left\{ {2 \atop 33} \right\} &= -\sin \theta \cos \theta, \\
\left\{ {3 \atop 23} \right\} &= \cot \theta.
\end{aligned}
$$

The next step is to calculate the contracted curvature tensor or Ricci tensor as it is known. The formulae for this can be written slightly differently in the form

$$R_{ij} = \phi_{,ij} - \Gamma^k_{ij,k} + \Gamma^k_{jp}\Gamma^p_{ik} - \Gamma^p_{ij}\phi_{,p},$$

where $\phi = \log \sqrt{(-g)}$. This is so, since

$$\Gamma^k_{pk} = \tfrac{1}{2} g^{kl}[l, pk] = \tfrac{1}{2} g^{kl}(g_{lp,k} - g_{pk,l} + g_{kl,p}).$$

In this expression the first two terms are anti-symmetric in k, l and so give nothing when summed with g^{kl}. The remaining term is

$$\frac{1}{2}g^{kl}g_{kl,\,p} = \frac{1}{2}\frac{g_{,\,p}}{g},$$

where g is now written for the determinant of the g_{ij}. Since, however, g is negative (in a locally Cartesian system it is -1) this is best written in the form $\phi_{,\,p}$. It is now a slightly tedious but straightforward matter to calculate all those components which may be non-zero:

$$\begin{aligned}
R_{11} &= \lambda'' - \mu'\lambda' + \lambda'^2 - 2\mu'/r,\\
R_{22} &= e^{-2\mu}(1 + r(\lambda' - \mu')) - 1,\\
R_{33} &= R_{22}\sin^2\theta,\\
R_{44} &= e^{2(\lambda-\mu)}(-\lambda'' + \lambda'\mu' - \lambda'^2 - 2\lambda'/r),
\end{aligned}$$

and R_{12} which in fact comes out to be zero.

The field equations are now $R_{ij} = 0$. From

$$R_{11} + e^{2(\mu-\lambda)}R_{44} = -2(\mu' + \lambda')/r,$$

it follows that

$$\mu' + \lambda' = 0.$$

At this point in the theory it begins to prove very inconvenient to display the velocity of light c whenever it occurs. It is therefore more convenient to adopt such a unit of *length* that the unit of time is one second, and $c = 1$. (That is, we adopt as the unit of length the distance travelled by light in one second.) The special relativity metric then has the form

$$ds^2 = dt^2 - (dx^2 + dy^2 + dz^2).$$

We shall later have cause to fix the units of mass as well, so that the final situation will be that only one standard, that of time, is needed. If we take flat space at infinity as a *boundary condition*, so that we can fix the coordinate system by the requirement

$$\mu \to 0, \qquad \lambda \to 0$$

as $r \to \infty$, this gives $\lambda + \mu = 0$. Substituting in $R_{22} = 0$ then gives

$$e^{2\lambda}(1 + 2\lambda' r) = 1,$$

i.e.

$$\frac{d}{dr}(re^{2\lambda}) = 1,$$

so that

$$e^{2\lambda} = 1-\frac{2m}{r}, \quad \text{(say)},$$

where a certain constant of integration has been called $-2m$, for reasons which will be clear shortly.

The solution which we have found, which was originally due to Schwarzschild (1916), then has the form

$$ds^2 = \left(1-\frac{2m}{r}\right) dt^2 - \frac{dr^2}{1-(2m/r)} - r^2(d\theta^2 + \sin^2\theta\, d\phi^2).$$

We can at once appreciate one feature of this solution without any further calculation if we consider the weak static field, slow motion, approximation to the theory. For such a weak field when the motions of the particles are small compared with that of light we get for the equation of a free particle, i.e. freely falling in the gravitational field,

$$\frac{d^2x^i}{ds^2} = \frac{dv^i}{ds} = -\left\{{i \atop jk}\right\} v^j v^k.$$

Since the motion is *slow*, however, and since

$$v^1, v^2, v^3, v^4 = \frac{\mathbf{v}}{\sqrt{(1-\mathbf{v}^2)}}, \frac{1}{\sqrt{(1-\mathbf{v}^2)}},$$

where $\mathbf{v}$ is the usual Newtonian velocity, it follows (bearing in mind the choice of units) that v^1, v^2, v^3 are all very small compared with v^4. The only surviving term is therefore that for which $j = k = 4$, so that

$$\frac{dv^i}{ds} = -\left\{{i \atop 44}\right\}(v^4)^2.$$

Now $$\left\{{i \atop 44}\right\} = \tfrac{1}{2}g^{ij}(g_{j4,4} - g_{44,j} + g_{4j,4}) = -\tfrac{1}{2}g^{ij}g_{44,j}$$

since the field is static. Choosing a locally Cartesian coordinate system, the equation of motion is approximately (remembering that $g^{ji} \simeq -1$)

$$\frac{d\mathbf{v}}{dt} = -\frac{1}{2}\nabla g_{44},$$

so that (apart from a possible additive constant) $\frac{1}{2}g_{44}$ is the Newtonian gravitational potential. In other words, reverting to the Schwarzschild solution in which $\frac{1}{2}g_{44} = \frac{1}{2}-(m/r)$ the Newtonian theory is the first approximation to the theory which we have found, as it should be, considering our choice of field equations. The constant m is accordingly identified with the constant GM that occurs in the expression $-GM/r$ for the gravitational potential at distance r from a spherically symmetric mass M, G being the constant of gravitation. It is convenient to fix the unit of mass (as we fixed the unit of length earlier), this time so that $G = 1$; m is then the mass of the central gravitating particle in these units.

We have now seen that the theory agrees closely with Newtonian mechanics at a first approximation. Reaction to the existence of this theory from 1915 onwards right up to 1939 was to seek the small differences between the theory and Newtonian mechanics and look into experimental results of these. Looking back on it, this can be seen straight away to be a rather unprofitable exercise, since the Newtonian theory of gravitation was an exceptionally good one and the differences are therefore bound to be very difficult to observe. To some extent this was not so clear at the time, principally because astronomy is a subject in which such exceedingly refined observations are possible. The first solution of the field equations, that of Schwarzschild, is enough to provide three experimental checks on the theory. It is first necessary to find the orbit of a test particle moving round the central gravitating mass described by this solution. For this purpose one writes down the equation for a particle moving under no forces, as before, and specifies the particular value of the coefficients of affine connection. The equation for θ is then

$$\frac{d^2\theta}{ds^2}+\frac{2}{r}\frac{dr}{ds}\frac{d\theta}{ds}-\cos\theta\sin\theta\left(\frac{d\phi}{ds}\right)^2 = 0.$$

It is to be noticed from this equation that if originally θ is chosen to be $\frac{1}{2}\pi$ and its derivative is chosen zero, i.e. if the particle moves initially in the plane $\theta = \frac{1}{2}\pi$, then its second derivative vanishes, and by differentiating, so do all its derivatives, and the particle continues always to move in this plane. That is to say, the orbit of a planet is a plane curve, just as it is in Newtonian mechanics. It is therefore possible to simplify all the equations by making $\theta = \frac{1}{2}\pi$ everywhere. The equations in ϕ and t then become

$$\frac{d^2\phi}{ds^2}+\frac{2}{r}\frac{dr}{ds}\frac{d\phi}{ds} = 0 \quad \text{and} \quad \frac{d^2t}{ds^2}+2\lambda'\frac{dr}{ds}\frac{dt}{ds} = 0,$$

and these two equations are immediately integrable to the form

$$r^2\frac{d\phi}{ds} = h \quad \text{and} \quad \frac{dt}{ds} = ce^{-2\lambda} = \frac{c}{\gamma} \quad \text{(say)},$$

where h, c are introduced as constants of integration.

It is possible now to tackle the remaining equation, that for r, but, as usual with a set of equations derived from a variational principle, it is not necessary to integrate all of them. It is often more convenient to use the easy integral which corresponds to what would be energy in the Newtonian mechanical problem. In other words we use the fact that the metric is, indeed, given by these coefficients, so that

$$\gamma^{-1}\left(\frac{dr}{ds}\right)^2+r^2\left(\frac{d\phi}{ds}\right)^2-\gamma\left(\frac{dt}{ds}\right)^2 = -1.$$

Now in order to compare with the Newtonian theory we must get rid of both the time and the relativistic proper time, since the Newtonian theory always considers first the shape of the orbit and then determines the time by a second integration. Performing this, we get

$$\frac{1}{\gamma}\left(\frac{h}{r^2}\frac{dr}{d\phi}\right)^2+\frac{h^2}{r^2}-\frac{c^2}{\gamma} = -1,$$

and by making the familiar assumption $r = (1/u)$ this becomes

$$\left(\frac{du}{d\phi}\right)^2+u^2 = \frac{c^2-1}{h^2}+\frac{2m}{h^2}u+2mu^3,$$

and by differentiation we get

$$\frac{d^2u}{d\phi^2}+u = \frac{m}{h^2}+3mu^2.$$

Comparing this with the corresponding Newtonian equation

$$\frac{d^2u}{d\phi^2}+u = \frac{m}{h^2}$$

it is clear that there is an extra term on the right-hand side. The constant h in these two equations is also not exactly the same, since the first one is defined by

$$r^2\frac{d\phi}{ds} = h$$

and the second one by

$$r^2\frac{d\phi}{dt} = h.$$

Moreover, there is a certain amount of ambiguity in the definition of r in the relativistic case; but it is clear that it agrees with the Newtonian r a long way from the central mass, so we can disregard this difference.

Taking the equation we may solve it by successive approximation by first neglecting the extra term on the right-hand side, which must evidently be very small since we know that the Newtonian theory is a very good first approximation. (The reader may verify, by inserting the units explicitly, that it is indeed very small.) If we neglect it the solution of the equation is well known to be

$$u = \frac{m}{h^2}(1+e\cos\phi)$$

for a suitable choice of the initial line, $\phi = 0$. This is exactly as in the Newtonian theory, but we now insert this value for u in the very small term and integrate again. On the right-hand side there are several terms which produce small differences in the value of u in the second integration, but there is one term which produces a continually increasing effect. This is because it has the same period in ϕ as the terms on the left-hand side. Keeping only this term, so that the equation is really

$$\frac{d^2u}{d\phi^2}+u = \frac{m}{h^2}+\frac{6m^3}{h^4}e\cos\phi,$$

the new solution will be

$$u = \frac{m}{h^2}\left[1+e\cos\phi+\frac{3m^2}{h^2}e\phi\sin\phi\right],$$

and this may be written in the form

$$u = \frac{m}{h^2}[1+e\cos(\phi-\varepsilon)],$$

where $\varepsilon = (3m^2/h^2)\phi$. That is to say, for each revolution of the planet the orbit advances by the fraction of a revolution equal to

$$\frac{\varepsilon}{\phi} = \frac{3m^2}{h^2} = \frac{3m}{a(1-e^2)}.$$

Now the orbits of the planets all rotate by considerable amounts, but this had been accounted for in all cases, except that of Mercury, by the effect of the other planets on the orbits. Only in the case of Mercury was there a discrepancy between a Newtonian correction of about 450 seconds of arc per century and the observed rotation of about 500 seconds of arc per century. The difference between these two values was exactly that predicted by the general theory of relativity (Clemence, 1947). Of course there is a slight doubt whether it is appropriate to add to the Newtonian correction the relativistic connection. Strictly speaking this is working in a combined theory which does not exist, but since the Newtonian approximation is such a good one there can be little doubt of the correctness of the process.

Returning again to the equations of the orbit one may ask for the path of a ray of light. If we think of light as consisting of photons, i.e. particles of zero rest mass, the constant h will be infinite for these since such a particle moves among a path which has $ds = 0$. For this limiting case the equation of the orbit has *only* the "correction term" and takes the form

$$\frac{d^2u}{d\phi^2}+u = 3mu^2.$$

Carrying out the same process as before, the first approximation, neglecting the correcting term, is a straight line, having the form in these unusual coordinates,

$$u = \frac{\cos\phi}{R},$$

where R is evidently the closest approach of the straight line to the origin. Substituting in the correction term then gives

$$\frac{d^2u}{d\phi^2}+u = \frac{3m}{R^2}\cos^2\phi.$$

Integrating again, the second approximation is

$$u = \frac{\cos\phi}{R}+\frac{m}{R^2}(\cos^2\phi+2\sin^2\phi),$$

which may conveniently be put into Cartesian coordinates in the form

$$x = R-\frac{m}{R}\frac{x^2+2y^2}{\sqrt{(x^2+y^2)}}.$$

In this equation the second term is the measure of deviation from the straight line. At a great distance in either direction the light is moving in a straight line, but it is not in the same one in each direction, and the angle between these two lines is $4m/R$.

The way in which such an effect could be measured is by observing the stars whose light reaches the earth passing near to the sun. Such stars are invisible, however, except at the time of a total solar eclipse, so that the observations can only be carried out infrequently. The deflection of a ray which just grazes the sun's rim should be $1\frac{3}{4}$ seconds of arc; although these observations have been attempted many times it cannot be said that the confirmation of general relativity by them is completely certain (Beisbroek, 1950, 1953).

The third prediction of the Schwarzschild solution, which relies on a small additional assumption, is concerned with the red shift of light in the gravitational field. If one supposes that the atoms on the earth have the same frequency of vibration no matter what gravitational field they find themselves in, then it is clear from the Schwarzschild solution that the atoms of the sun will be *seen* to have a different period of vibration when viewed from the earth from those on the earth, because, since they are at rest, $ds^2 = \gamma dt^2$ giving different times of vibration inversely proportional to $\sqrt{\gamma}$. Here again the effect is a very small one; in fact the difference between wavelengths of radiation from atoms on the sun and those on the earth amounts to two parts in 10^6. The confirmation of the theory by this method has been very unsatisfactory because it is necessary to take account of the motion of the sun, and the corresponding Doppler shift of the light appears to be much larger (Adams, 1952, 1955, 1958, 1959).

However, this particular test of the theory has been carried much further in more recent years, notably by Pound and Rebka (1959), and their first paper is reproduced as Extract 11 of the present volume. The possibility of checking the difference of frequency was made possible by the discovery of the Mössbauer effect. Mössbauer (1958) discovered that some gamma rays emitted from solids come out without the nucleus recoiling individually, the recoil being taken up by the whole of the crystal lattice, so there is a negligible Doppler shift. In the Pound–Rebka experiment this possibility of extremely accurate frequency gamma rays with the corresponding method of detection was used to measure change in frequency in fall down a tower. Such an experiment (which gave results in close agreement with the theory) is probably even more significant as a beginning in laboratory experiment in the theory.

In this situation in which so little experimental check was possible between the theory and Newtonian mechanics, it is not surprising that interest wained amongst a large body of scientists during the twenties and thirties. Most interest which did survive was in the field of cosmology, but here the contribution of the theory proved very disappointing. It does not seem worth reproducing any of the papers on this aspect of general relativity, but for the sake of completeness we give here a brief summary of the applications to cosmology. Essentially the problem of studying the universe as a whole was bedevilled in Newtonian mechanics by the appearance of infinities of all kinds. The hope was that these difficulties would be removed by discussing the problem with general relativity. Nor was this unreasonable. If we seek a solution of Poisson's equation, in Newtonian mechanics,

$$\nabla^2\phi = 4\pi G\varrho$$

in a situation corresponding to a homogeneous universe (i.e. a constant value for ϱ, the smoothed-out density), we get in spherical polars (and assuming that the universe is isotropic, i.e. the same in all directions)

$$\frac{1}{r^2}\frac{d}{dr}\left(r^2\frac{d\phi}{dr}\right) = 4\pi G\varrho,$$

with unique solution, so long as there is no singularity at the origin,

$$\phi = \tfrac{2}{3}\pi G\varrho r^2.$$

In such a universe there is at each point r a force towards the origin $-\frac{4}{3}\pi G\varrho r$. As a result the matter will not stay at rest, but will collapse. A static universe can only be provided if one modifies Poisson's equation in some way, say by writing

$$\nabla^2\phi = 4\pi G\varrho - \lambda$$

in general, and supposing that the actual universe has $\varrho = \lambda/4\pi G$. But the modified Poisson equation corresponds to the situation, for the field round a point mass, of

$$\frac{1}{r^2}\frac{d}{dr}\left(r^2\frac{d\phi}{dr}\right) = \nabla^2\phi = -\lambda,$$

so that

$$\phi = -\frac{m}{r} - \frac{1}{6}\lambda r^2,$$

corresponding to a force of attraction towards *o* of

$$\frac{m}{r^2}-\frac{1}{3}\lambda r.$$

The modification has therefore produced a field of repulsion, proportional to distance, around each gravitating mass. Since the number of such masses is infinite, the total contribution of their repulsive fields to the potential at the origin (since these fields *increase* with distance) will be infinite. And such an infinite contribution results from building in a repulsive force to overcome the gravitational attraction. But since general relativity succeeds in describing gravitation without explicitly introducing a gravitational field, it is imaginable that it will also remove the infinity.

Einstein was also particularly concerned with another such problem of the effect of distant matter, i.e. the one associated with the name of Mach. It is a well-known observation that the local dynamical behaviour of bodies, as exemplified, for example, by the Foucault pendulum, is describable simply only in the local inertial frames of Newtonian mechanics, which are determined by the distant matter in the universe. But what is lacking completely is the mechanism for this determination, and Einstein hoped that general relativity would provide such a mechanism. Looking back on it it is clear even from the Schwarzschild solution that this could not be because in the Schwarzschild case the very accurate determination of the orbits of the planets proves them to be rotating ellipses, and this rotation is verified by observations made in a frame of reference defined relative to the distant stars, although no such stars enter into the calculation. But this was not realized at the time, and Einstein hoped in any case to be able to modify general relativity slightly in order to incorporate Mach's principle.

While it is true that most workers in the subject view Einstein's approach as essentially correct, it should be noted that the very well-informed and notable voice of Fock is on the opposing side. For Fock (as can be seen in Extract 6 of the present volume) the mistake of the general theory is the extreme emphasis on general covariance. Fock

(1957) notes that it is perfectly possible to express special relativity in a generally covariant form; accordingly it is not the general covariance, but the non-uniformity of the space–time that distinguishes the gravitational situation. Accordingly Fock believes that physical systems may correspond to a particular class of coordinate systems; not only in the sense that the solution of the physical problem is easier in the preferred coordinate systems (there would be general agreement about that) but (as far as can be judged from Fock's writings) because it is only in these coordinate systems that the boundary conditions can be adequately expressed, and the transformation properties (Lorentz-like group of transformations) incorporated. Having noted Fock's demur, we must let him speak for himself and return to Einstein's attempt to modify general relativity to incorporate Mach's principle.

His approach to this was by seeking to modify the field equations. The equation $R_{ij} = 0$ was derived above by generalizing Laplace's equation, but this approach leans rather heavily on Newtonian mechanics. The first step from this point of view was evidently to find some other way to derive field equations, and Einstein turned to the idea of a variational principle. The use of such variational principles has already been discussed at some length in the volume on special relativity. When we come to the general theory and we are using arbitrary coordinate systems a little more needs to be said however. A variational principle in the form

$$\delta \int L\, d^4x = \delta \int L\, dx^1\, dx^2\, dx^3\, dx^4 = 0$$

will not be invariant under this coordinate transformation if L is a scalar, because the integration is with respect to a *coordinate* volume element. In order to add up a scalar over a region of four-dimensional space one must multiply by a "genuine" volume element. Such an element can be found as follows. We know that if we transform the coordinate system to a new one, the integral is multiplied by the Jacobian of the coordinate transformation. We therefore need to have a factor in the integrand which will account for this multiplication by the Jacobian. Consider now the transformation of a tensor of rank 2, e.g. the metric tensor

$$g'_{ab} = \frac{\partial x^c \, \partial x^d}{\partial x'^a \, \partial x'^b} g_{cd}.$$

This could be written in an obvious matrix notation as

$$g' = TgT^t,$$

where
$$T = (T_{ab}) = \left(\frac{\partial x^b}{\partial x'^a} \right)$$

and T^t denotes the transposed matrix. By taking the determinant on each side we see that $g' = J^2 g$, where J is the Jacobian of the transformation, so that the square root of this determinant will be a factor of the type required. Remembering, however, that the determinant in a flat space will have the value -1 rather than $+1$, it is more convenient to use $\sqrt{-g}$.

Accordingly we are to seek for a variational principle of the form

$$\delta \int L \sqrt{(-g)} \, d^4x = 0,$$

where L is a scalar. Now this scalar has to be derived from the field quantities, i.e. the coefficients of the metric, so there is very little choice in its construction, and by far the most obvious variational principle to choose is

$$\delta \int R \sqrt{(-g)} \, d^4x = 0,$$

where $R = g^{ab} R_{ab}$ and $R_{ab} = R^c_{abc}$ as usual. As a matter of fact, as will be seen below, this does, indeed, give the Einstein field equations, but there is rather more to be said about it than that. Firstly, the scalar curvature R involves second derivatives of the metric tensor, and so by carrying out the usual process of the calculus of variation we would expect to get field equations involving fourth derivatives of these quantities. The Einstein field equations, however, involve only second derivatives. It is clear that we are confronted with a rather special kind of variational principle in which the higher derivatives cancel out, for some reason. Secondly, there will be a conservation theorem associated with such a variational principle which is akin to Noether's theorem (Noether, 1918), which we discussed in the volume on special relativity. The group of transformations involved here is that of general functions of four variables so that the result is somewhat different from the one found there, but there must be some theorem of this kind, and it is the extension of this theorem which gave Einstein the clue to generalizing his field equations.

Let us first carry out the variation mentioned and show that this variational principle does, indeed, give the field equations required. From the point of view of carrying out the calculations it is best to suppose the curvature tensor to be that of an affine connection which initially is given independently of the metric tensor. We can then work out the variation of the integral and at a suitable stage put on the condition that the affine connection is, indeed, that of the Christoffel brackets. Performing the variation we get

$$\delta \int R\sqrt{(-g)}\, d^4x = \int [\delta R \sqrt{(-g)} + R\delta \sqrt{(-g)}]\, d^4x.$$

Now the curvature tensor is given by

$$R^a_{bcd} = \Gamma^a_{bd,\, c} - \Gamma^a_{bc,\, d} + \Gamma^r_{bd}\Gamma^a_{cr} - \Gamma^r_{bc}\Gamma^a_{dr},$$

and it follows from this that

$$\delta R^a_{bcd} = \delta\Gamma^a_{bd;\, c} - \delta\Gamma^a_{bc;\, d}.$$

(The simplest way of establishing this is to remark that the variation in the affine connection is, from the law of transformation of affine connections, a tensor of rank 3, and also that the equation stated can be derived at once from the definition of the curvature tensor at any particular point in the coordinate system which is freely falling there, i.e. such that at that point the affine connection vanishes. But since both sides of the equation are tensors their equality in one coordinate system involves their equality in all, and therefore the theorem is established in general.) By contraction it follows that

$$\delta R_{bc} = \delta \Gamma^a_{ba;\,c} - \delta \Gamma^a_{bc;\,a}$$

and therefore that

$$\delta R = \delta(g^{ab}R_{ab}) = \delta g^{ab}R_{ab} + (g^{bc}\delta\Gamma^a_{ba})_{;c} - (g^{bc}\delta\Gamma^a_{bc})_{;\,a}.$$

At this point we assume that the affine connection is the one usually used in general relativity both before and after the variation. This means that the covariant derivative of the metric tensors will be zero. It is a slightly more subtle question to ask whether

$$(\sqrt{-g})_{;\alpha} = 0;$$

the quantity here being differentiated is not a scalar, as we remarked above, but is what is known as a density, because it is analogous to mass density measured in the given coordinate system. Since we have not given any rules for the covariant derivatives of densities it will be in order to *define* $(\sqrt{-g})_{;\alpha} = 0$. This will lead to a general formula for the covariant derivatives of densities, which need not concern us for the moment. It will have the effect, moreover, that for any vector A^c we shall have

$$\int (\sqrt{-g}A^c)_{;\,c}\, d^4x = \int \sqrt{-g}A^c_{;\,c}\, d^4x;$$

On the other hand, it is easy to see that, since

$$A^c_{;\,c} = A^c_{,\,c} + \Gamma^a_{ca}A^c,$$

where the affine connection term is given by

$$\Gamma^a_{ca} = \tfrac{1}{2}g^{am}(g_{mc,\,a} - g_{ca,\,m} + g_{am,\,c})$$
$$= \tfrac{1}{2}g^{am}g_{am,\,c}$$

(taking account of the antisymmetry of the first 2 terms in the expression for the affine connection), we can use the law for differentiating a determinant to write this as

$$\Gamma^a_{ca} = \frac{1}{\sqrt{-g}}(\sqrt{-g})_{,c}\,.$$

Accordingly the left-hand integral may be written in the form

$$\int (\sqrt{-g}A^c)_{,c}\, d^4x,$$

where the divergence which enters is an ordinary divergence and so, by Gauss's theorem, may be converted into a three-dimensional surface integral. In the usual way, so long as the field quantities tend to zero appropriately at infinity, this may be disregarded.

This enables us to integrate certain expressions by parts even when the derivatives are covariant derivatives, and we apply this to the last two terms in the expression for the variation of R. Disregarding the surface integrals we get

$$\delta \int R\sqrt{(-g)}\, d^4x = \int [\sqrt{(-g)}\, \delta g^{ab} R_{ab} + R\delta\sqrt{-g}]\, d^4x.$$

However, by a similar calculation to the one just carried out it follows that

$$\delta g = \delta g_{ab} \cdot g g^{ab}$$

or

$$\frac{\delta g}{g} = g^{ab}\delta g_{ab} = -g_{ab}\delta g^{ab}$$

since

$$g^{ab}g_{ab} = 4.$$

Finally, collecting everything together we have

$$\delta \int R\sqrt{(-g)}\, d^4x = \int \sqrt{(-g)}\, \delta g^{ab}(R_{ab} - \tfrac{1}{2}g_{ab}R)\, d^4x.$$

This does, indeed, give a variational principle for the field equations, because since the variations of the metric tensor are arbitrary, we conclude that the quantity in brackets must vanish, i.e.

$$R_{ab} = \tfrac{1}{2}g_{ab}R.$$

By contracting this quantity over its two suffixes it follows that

$$g^{ab}R_{ab} = R = \tfrac{1}{2}g^{ab}g_{ab}R = 2R,$$

so that $R = 0$, which then leads back to the field equations in their usual form. Einstein's approach in generalizing this was, however, a little different. The quantity in brackets satisfies the identity

$$(R^{ab} - \tfrac{1}{2}g^{ab}R)_{;\,b} = 0.$$

This may be verified directly most simply by the reader by observing that it is a vector equation, and that it can be verified straightforwardly at a point in which the coordinate system has been chosen so that the first derivatives of the metric vanish and therefore the coefficients of affine connection vanish. But it is also a general rule that a quantity derived in the way in which we have derived this one will always have vanishing covariant divergences.

This is really an application of Noether's theorem which was discussed at some length in the book on special relativity, but since the generalization of Noether's theorem to the present case is a little difficult we shall derive the result directly for any scalar function K of the metric tensor. Let us assume, then, that

$$\delta \int K\sqrt{(-g)}\, d^4x \equiv \int P^{ab}\, \delta g_{ab}\sqrt{(-g)}\, d^4x.$$

There is no question here of a variational *principle*. On the other hand, we could consider special variations which were simply due to a coordinate transformation. In that case, since the left-hand side is an invariant it cannot be changed by this transformation, and therefore the result will be zero, very much as in the case of the variational principle. Consider, then, the infinitesimal coordinate transformation

$$x^a \to x'^a = x^a + \varepsilon^a$$

for which the differential coefficients which enter into the transformation equations can be written

$$\frac{\partial x'^a}{\partial x^b} = \delta^a_b + \varepsilon^a{}_{,b}$$

with the corresponding inverse transformation. The transformation of the metric tensor will then take the form

$$g'_{ab}(x'^r) = \frac{\partial x^c}{\partial x'^a}\frac{\partial x^d}{\partial x'^b}\, g_{cd}(x^r)$$

$$= g_{ab} - \varepsilon^c_{,a} g_{cb} - \varepsilon^d_{,b} g_{ad}.$$

Notice, however, that by definition this gives the new metric tensor at the point with transformed coordinates, whereas the difference employed in the variational principle is a difference keeping the coordinates fixed, that is to say we need

$$g'_{ab}(x^r) = g'_{ab}(x'^r - \varepsilon^r)$$

$$= g'_{ab}(x'^r) - \varepsilon^r g_{ab,\,r}.$$

Inserting all these it is clear that

$$0 = \int P^{mn}[\varepsilon^c_{,m} g_{cn} + \varepsilon^c_{,n} g_{mc} + \varepsilon^r g_{mn,r}]\sqrt{(-g)}\, d^4x$$

$$= -\int \varepsilon^c[(P^{mn} g_{cn}\sqrt{-g})_{,m} + (P^{mn} g_{nc}\sqrt{-g})_{,m} + P^{mn}\sqrt{(-g)}g_{mn,\,c}]\, d^4x,$$

and it follows that since the infinitesimal coordinate transformation is an arbitrary one the contents of the square bracket must vanish. This easily reduces to

$$(P^{mn}\sqrt{-g})_{,\,n} = 0$$

which has as its consequence

$$P^{mn}_{;n} = 0.$$

Einstein's approach was the following. He interpreted the tensor which arose in this way as the energy momentum tensor for the gravitational field. The vanishing of its covariant divergence he viewed as in some sense representing the conservation of total energy and momentum. The equation derived by putting the tensor equal to zero is then the condition for free space, that is to say, the field equations. The generalization is to ask whether any more general energy tensor can be derived. It is necessary to put some restrictions on this, and Einstein looked for the most general energy tensor which did not involve covariant derivatives of the metric above the second and which was linear in those derivatives. By choosing again a special coordinate system in which the first derivatives of the metric vanish at one particular point, it is not difficult to prove that the most general tensor is a multiple of

$$R_{mn}-\tfrac{1}{2}g_{mn}(R-2\lambda),$$

where λ is a constant. The same process of equating this to zero now gives as field equations

$$R_{mn} = \lambda g_{mn}.$$

Before solving these equations there are two small investigations which we may make. Firstly, we shall discuss what they correspond to in the Newtonian approximation, and, secondly, we shall consider a recent new approach to this problem by Lovelock (1970). We are already aware of the Newtonian equations to which Einstein's field equations correspond, i.e. Laplace's equation. Accordingly, in the static solution, or at any rate in one in which the matter does not move very quickly, taking account of the fact that the metric tensor is then related to the gravitational potential ϕ by the equation

$$g_{44} = 1+2\phi,$$

it follows that, to the first order in the gravitational potential, the Newtonian approximation will be exactly that modification of the Newtonian theory discussed above,

$$\nabla^2\phi = \lambda.$$

(The interest of general relativity in this connection is in the way this turns up as virtually the *only* possibility. This fact is also true in the Newtonian theory in the case of cosmology because of the additional assumptions of homogeneity and isotropy. But it follows already from the general covariance in general relativity.) The right-hand side of the equation is now a constant, since the cosmical constant λ

is very small, and therefore the terms involving a gravitational potential on the right-hand side may be disregarded. To see the significance of this equation we may look again at the static spherically symmetric solution; the equation becomes, as we saw,

$$\frac{1}{r^2}\frac{\partial}{\partial r}\left(r^2\frac{\partial \phi}{\partial r}\right) = \lambda,$$

with the solution

$$\frac{\partial \phi}{\partial r} = \frac{1}{3}\lambda r - \frac{m}{r^2}$$

for the gravitational force. In the Newtonian approximation, then, the equation corresponds, just as in the modification of the Newtonian theory, to the well-known inverse square law, to which is added a direct distance law. With a suitable sign of the cosmical constant this direct distance law can cancel out the gravitational attraction and accordingly give rise to a stable universe.

Moreover, the direct distance law, being a force which increases with distance instead of decreasing, has a superficial attraction because its value in our immediate locality is almost entirely determined by the most distant matter, for there is more of this matter, and it is also at a greater distance away. This at once suggests the fulfilment of Einstein's desire to incorporate Mach's principle into the theory. Unfortunately, however, this attractive feature vanishes on closer investigation. The trouble now is not the same as it was in the Newtonian theory, that of an infinite potential at the origin. It is possible, as will be clear below, to solve the field equations for a finite distribution of matter, which is, none the less, unbounded in extent; the obvious analogy is with two-dimensional spaces, amongst which the surface of a sphere has the properties of isotropy and homogeneity, and is of finite total area, yet without a boundary. But the direct distance law is not actually attached to any sources at all, the constant in front of it is an absolute constant, bearing no relation to the distribution or motion of masses. Indeed, it is hard to see how it could be otherwise, since conventional laws of force which decrease with distance like the inverse square law, are provided with sources at the singular points of the field where it has increased without limit. A law of force of this kind has no singularities of that sort and is therefore not connected to its sources in the way that we are used to. It is not surprising, therefore, to find again that the introduction of this additional term into the equations of motion fails to achieve its result.

Turning now to Lovelock's work, he considers the curious fact that although a variational principle exists for general relativity in terms of a Lagrangian involving the metric tensor and its first and second derivatives, the field equations which result are not of the fourth order, as would be expected from the usual theory, but only of the second. Perhaps an analogy to Lovelock's investigation in a simpler case would be of help. Consider the well-known result, in special relativity, of a scalar field ψ derived from a Lagrangian $L = L(\psi, \psi_{,i})$. The usual field equations are, of course, of the second order, having the form

$$\frac{\partial L}{\partial \psi} - \frac{\partial}{\partial x^i}\left(\frac{\partial L}{\partial \psi_{,i}}\right) = 0$$

We may ask the question: Can these equations be of the first order? If the theory is a Lorentz invariant one, the Lagrangian must be a function of invariants under the Lorentz group and so can only depend upon ψ and

$$\varrho = \tfrac{1}{2}\psi^i\psi_i = \tfrac{1}{2}\eta^{ij}\psi_{,i}\psi_{,j}, \quad \text{say.}$$

Writing, therefore,

$$L = f(\psi, \varrho),$$

we have

$$\frac{\partial L}{\partial \psi} = f_1, \quad \frac{\partial L}{\partial \psi_i} = f_2\psi^i,$$

which then produces the field equations in the form

$$f_1 - (f_2\psi^i)_{,i} = 0.$$

These field equations will certainly contain second derivatives of the scalar field unless

$$f_2 = 0, \quad \text{i.e.} \quad \frac{\partial f}{\partial \varrho} = 0.$$

In that case the Lagrangian is a function of the scalar field only not involving its derivatives, and so, in fact, the field equation must be algebraic, i.e. of zero order. We have therefore shown, in this particularly simple case, that it is possible to have field equations of lower order than the second but that they must then be of zero order.

The investigation in the case of general relativity is very much more complicated for two reasons. Firstly, it is obviously going to be more difficult to deal with a set of ten independent field variables instead of one. Secondly, the theory has to be invariant under the general group of transformations rather than merely the Lorentz group, and the implications of this for the form of the Lagrangian and its derivatives take quite a lot of working out. It will suffice merely to summarize Lovelock's result here. In a four-dimensional space the only second-order field equations which can be derived from a Lagrangian invariant depending on the metric tensor and its first and second derivatives are Einstein's equations with the cosmical term. The result is false in spaces of higher dimension (and Lovelock also shows that there is only one possible third-order equation in the four-dimensional case). Lovelock's investigation goes as far as one can hope to settle the form of the field equations so long as you believe them to be of the second order and derived from a variational principle.

It is now quite straightforward to carry out the calculations for the cosmological theory. The early investigators were looking for a static universe, and so they again assumed for the metric

$$ds^2 = e^{\nu}\, dt^2 - e^{\lambda}\, dr^2 - r^2(d\theta^2 + \sin^2\theta\, d\phi^2).$$

It is convenient to continue using the same notation as before, involving the function $\lambda = \lambda(r)$, and so it is necessary to change the letter chosen for the cosmical constant. We shall now use Λ for the cosmical constant. Exactly the same calculation as before now gives

$$\begin{aligned}
R_{11} &= \tfrac{1}{2}\nu'' - \tfrac{1}{4}\lambda'\nu' + \tfrac{1}{4}\nu'^2 - \lambda'/r,\\
R_{22} &= e^{-\lambda}[1 + \tfrac{1}{2}r(\nu' - \lambda')] - 1,\\
R_{33} &= R_{22}\sin^2\theta,\\
R_{44} &= e^{\nu-\lambda}(-\tfrac{1}{2}\nu'' + \tfrac{1}{4}\lambda'\nu' - \tfrac{1}{4}\nu'^2 - \nu'/r),
\end{aligned}$$

but these quantities are not now to be equated to zero. Instead we have to find the tensor

$$R_{\mu\nu}-\tfrac{1}{2}g_{\mu\nu}(R-2\Lambda).$$

This is to represent the distribution of energy and momentum, and we want to impose upon it certain symmetry conditions. It obviously makes a difference, however, whether we impose these conditions on the form of the tensor given or on the contravariant or mixed forms. We shall have to consider which of these, if any, is the physically significant one, for which we can assume that it is the same at all points and that the components for different directions are the same. That is to say, we have to make precise in which *way* we can put in the assumption that the universe is homogeneous and isotropic.

This at once raises a severe problem, the importance of which was not realized in the early days of relativity theory, although it was clearly foreshadowed in Einstein's original paper. This is the fact that the coordinates, since they can be changed by perfectly arbitrary transformations, can have no physical significance until some has been found for them. We have, *after* solving a problem and using a particular coordinate system, to investigate the circumstances of the problem and relate the important features of the coordinate system with what is actually measured. The phrase "actually measured" has a comfortable sound, but it is necessary to be more precise about what it means. The seductive apparatus of general relativity makes us feel that we understand the physical significance of tensors of various ranks, but in fact the way in which we carry out measurements in the theory is that an observer sets up a local Cartesian coordinate system. Such an observer may be freely falling, in which case his coordinate system locally forms an inertial frame without any gravitational field, or it may be convenient in some problems to use an observer fixed (e.g. on the surface of the earth), in which case a gravitational field has also to be introduced. In either case a general feature is that a set of four unit vectors in the four coordinate directions are set up. Of course any other set derived from this by a Lorentz transformation will be equally suitable. Once these vectors are set up, however, it is

Framingham State College
Framingham, Massachusetts

the components of the other quantities with respect to this coordinate system that we are able to measure.

By a component is meant here the inner product. Suppose that the four unit vectors along the axes are written as

$$\underset{\alpha}{h_i} \quad \text{or} \quad \underset{\alpha}{h^i},$$

where the α labels the individual vectors. The condition that they are unit vectors at right angles in the coordinate system which they define, can be written in the form

$$\underset{\alpha}{h_i}\,\underset{\beta}{h^i} = \eta_{\alpha\beta},$$

where the occurrence of $\eta_{\alpha\beta}$ on the right-hand side is due to the fact that one of the four unit vectors must be time-like (and has been chosen here as $\underset{4}{h_i}$), whilst the other three will then be space-like. Given a tensor of rank 2, e.g. the energy tensor T_{ij}, the quantities which can actually be measured are the components

$$T_{\alpha\beta} = \underset{\alpha}{h^i}\,\underset{\beta}{h^j}\,T_{ij}.$$

Now a convenient set of unit vectors for the coordinate systems which we are using here will be those in the direction of the coordinates t, r, θ, Φ increasing. In that case the conditions of orthogonality and being a unit vector reduces to a much simpler form. It is first necessary to rearrange the conditions in the form

$$\eta^{\alpha\beta}\,\underset{\alpha}{h_i}\,\underset{\beta}{h_j} = g_{ij}.$$

This may be most easily proved by modifying the original condition slightly, writing

$$\underset{4}{h^i} = A_{4i},$$

$$\underset{1}{h_i} = iA_{1i}, \quad \underset{2}{h_i} = iA_{2i}, \quad \underset{3}{h_i} = iA_{3i},$$

and, similarly,

$$\underset{4}{h^i} = B_{i4}, \quad \text{and so on.}$$

The original condition then has the matrix form

$$AB = I,$$

where I is the unit matrix, and this is well known in matrix algebra to have the consequence

$$BA = I,$$

which is easily seen to be the new form of the conditions.

The vectors now have the form, then,

$$\underset{1}{h_i} = e^{\lambda/2}, \quad \underset{2}{h_i} = r, \quad \underset{3}{h_i} = r\sin\theta, \quad \underset{4}{h_i} = e^{\nu/2}.$$

Accordingly, the actual measured quantities are derived by multiplying the covariant components by the quantities

$$e^{-\lambda/2} \quad \text{and so on,}$$

or the contravariant components by the quantities

$$e^{\lambda/2} \quad \text{and so on.}$$

Finally, if we take the mixed form of the tensor we find that the measured quantities are *equal* to the mixed components. Accordingly, the conditions which we want to impose are

$$\frac{\partial}{\partial r} T_j^i = 0, \quad T_1^1 = T_2^2 = T_3^3$$

(of which the last equality is an identity, i.e. we have already been able to assume it by our choice of metric.)

If we use the expressions for the components of the Ricci tensor

$$R_1^1 = -e^{-\lambda}\left(\frac{1}{2}\nu'' - \frac{1}{4}\lambda'\nu' + \frac{1}{4}\nu'^2 - \frac{\lambda'}{r}\right),$$

$$R_2^2 = -\frac{1}{r^2}\left\{e^{-\lambda}\left[1 + r(\nu' - \lambda')\right] - 1\right\} = R_3^3,$$

$$R_4^4 = e^{-\lambda}\left(-\frac{1}{2}\nu'' + \frac{1}{4}\lambda'\nu' - \frac{1}{4}\nu'^2 - \frac{\nu'}{r}\right),$$

which accordingly give

$$R = e^{-\lambda}\left(-\nu''+\frac{1}{2}\lambda'\nu'-\frac{1}{2}\nu'^2-\frac{2(\nu'-\lambda')}{r}-\frac{2}{r^2}\right)+\frac{2}{r^2}.$$

We derive for the radial and the time components of the energy tensor the values

$$-T_1^1 = e^{-\lambda}\left(\frac{\nu'}{r}+\frac{1}{r^2}\right)-\frac{1}{r^2}+\Lambda,$$

$$-T_4^4 = e^{-\lambda}\left(-\frac{\lambda'}{r}+\frac{1}{r^2}\right)-\frac{1}{r^2}+\Lambda.$$

Here the cosmical constant has been inserted, and the negative sign on the left-hand side is necessary since, firstly, we have only proved proportionality between the energy tensor and the expression

$$R_{ij}-\tfrac{1}{2}g_{ij}(R-2\Lambda),$$

and a short calculation with the Newtonian approximation shows, in fact, that the constant of proportionality must be negative. In identifying the expressions with the energy tensor instead of with the multiple of it, we are, of course, making a certain special choice of units for energy, but it does not matter for our purposes exactly what this choice is. The other two non-zero components of the energy tensor have complicated forms.

From the time component, which must be a constant, it follows that

$$\frac{1}{r^2}\frac{\partial}{\partial r}(re^{-\lambda})-\frac{1}{r^2} = A,$$

which integrates to the form

$$re^{-\lambda} = \tfrac{1}{3}Ar^3+r+B,$$

where A and B are constants of integration. However, since the solution cannot have a singularity at the origin, because this would mark the origin off as different from other points, the arbitrary constant B

must be zero, and, accordingly, the solution is

$$e^{-\lambda} = 1+\tfrac{1}{3}Ar^2.$$

Substituting this solution back into the radial component and equating this to a constant, gives

$$\left(1+\frac{1}{3}Ar^2\right)\left(\frac{\nu'}{r}+\frac{1}{r^2}\right)-\frac{1}{r^2} = C,$$

and this easily simplifies to

$$\nu'\left(\frac{1}{r}+\frac{1}{3}Ar\right) = D,$$

where D is a new constant. Again this is straightforward to integrate, giving

$$e^{\nu} = (1+\tfrac{1}{3}Ar^2)^E,$$

where $2AE = 3D$. The form of the metric so far is, therefore,

$$ds^2 = \left(1+\frac{1}{3}Ar^2\right)^E dt^2-\frac{dr^2}{1+\frac{1}{3}Ar^2}-r^2(d\theta^2+\sin^2\theta\, d\phi^2).$$

We must now take account of the other components of the energy tensor, but this is simpler in that we have definite expressions for the coefficients of the metric. In fact the transverse component of the contracted curvature is given by

$$R_2^2 = -\tfrac{1}{2}(D+\tfrac{4}{3}A),$$

whilst the radial component reduces to

$$R_1^1 = -\frac{1}{1+\frac{1}{3}Ar^2}\left[\frac{1}{2}\left(D+\frac{4}{3}A\right)+\left(\frac{1}{4}D^2+\frac{2}{9}A^2\right)r^2\right].$$

Since these two must be equal if the corresponding mixed components of the energy tensor are to be equal, it follows that

$$\frac{1}{2}\left(D+\frac{4}{3}A\right) = \frac{\frac{1}{4}D^2+\frac{2}{9}A^2}{\frac{1}{3}A}.$$

Simplifying, either $D = 0$, which implies that $E = 0$, or $D = \frac{2}{3}A$, so that $E = 1$. We must now examine these two possibilities separately. Before we begin, however, we may notice that as well as the two cases described by this choice of constants there is also the one obtained by taking $A = 0$ in the original integration. This third choice is simply the flat space–time of special relativity and need not concern us any more in the description of a universe supposed to contain matter.

Let us consider first the possibility that $E = 0$ so that the time component of the metric tensor is constant and the mixed components of the energy are given by

$$-T_1^1 = \tfrac{1}{3}A + \Lambda,$$
$$-T_4^4 = A + \Lambda.$$

The usual choice in this case is to make $A = -3\Lambda$ so that there is no pressure in the universe but only matter at rest, with density (in suitable units) given by $\varrho = 2\Lambda$. This was the choice originally made by Einstein (1917). Since we can rewrite the metric coefficient in the form

$$e^{-\lambda} = 1 - \Lambda r^2 = 1 - r^2/R^2 \quad \text{(say)},$$

so long as Λ is positive it follows that

$$\Lambda = 1/R^2, \quad \varrho = 2/R^2,$$

which relates the density of matter, which evidently must refer to a smoothed-out average density, to a length in the universe corresponding roughly speaking to the greatest possible length. For the kind of values of the length constant thought to be applicable in astronomical situations, the corresponding density turns out to be very much higher than that observed, a fact which has been picturesquely described by Eddington in the words "the Einstein universe has as much matter as it can possibly hold". He envisaged the process of increasing the matter in the universe as at the same time increasing the amount of curvature, since the gravitational field becomes stronger, until a point is reached at which the universe ceases to be open at infinity in the sense that flat space is open, and becomes closed. And Einstein's universe can, indeed, be considered closed in this way, as we can see by considering a radial geodesic. By making a transformation of coordinates akin to that from Cartesian to polars, i.e. by putting $r = R \sin x$, the metric becomes

$$ds^2 = dt^2 - R^2[d\chi^2 + \sin^2\chi(d\theta^2 + \sin^2\theta\, d\phi^2)].$$

Length measured radially ($dt = d\theta = d\phi = 0$) (which is easily seen to be along a geodesic, by using the geodesic equations) is then $R\chi$. But as χ increases from 0 to π, the metric coefficient $R \sin \chi$ increases to a maximum value R and then decreases to zero. The vanishing of a coefficient of the metric in a diagonal metric like this involves the vanishing of the determinant, and this means that it is impossible to get beyond $\chi = \pi$. (The singularity at $\chi = 0$ is to be expected, as it always arises in polar coordinates.) It is helpful to look at this from a geometrical point of view. Writing

$$x_1 = R\cos\chi, \quad x_2 = R\sin\chi\cos\theta,$$
$$x_3 = R\sin\chi\sin\theta\cos\phi, \quad x_4 = R\sin\chi\sin\theta\sin\partial\phi,$$

it follows that

$$x_1^2 + x_2^2 + x_3^2 + x_4^2 = R^2$$

so that (x_1, x_2, x_3, x_4) are the Cartesian coordinates of a point in an (imaginary) four-dimensional space which lies on a "sphere" of radius R. Moreover,

$$dx_2^2 + dx_3^2 + dx_4^2 = \{d(R\sin\chi)\}^2 + R^2\sin^2\chi(d\theta^2 + \sin^2\theta\, d\phi^2),$$

so that the original spatial metric is given by

$$ds^2 = dx_1^2 + dx_2^2 + dx_3^2 + dx_4^2.$$

The point $\chi = 0$, at which there is the expected singularity, is at (R, 0, 0, 0), whilst the other one is at ($-R$, 0, 0, 0). The radial geodesic is like a "great circle" on the "sphere" and returns to its original point after a (measured) distance $2\pi R$.

It should be mentioned that this global interpretation of the Einstein metric (which is only given locally), is by no means unique. Another obvious possibility is that the sphere just mentioned could be considered to have its opposite points identified, so that the point $(-R, 0, 0, 0)$ is to be taken as representing the same one as $(R, 0, 0, 0)$, and the greatest length is then πR. An infinite range of other global possibilities is also available, all consistent with the original local metric.

The analysis of the other case when $E = 1$ is a little more complicated. This solution was discovered by de Sitter (1917), and one of the interesting features is that it is possible to choose the constants in such a way that there is no matter present at all. This would also have been possible in the case of the Newtonian universe if we had chosen $A = -\Lambda$, but the result would then have been a negative pressure at all points. The de Sitter universe manages to provide a zero-density, zero-pressure solution, i.e. the choice $A = -\Lambda$ gives

$$-T^i_j = R^i_j - \tfrac{1}{2}\delta^i_j(R - 2\Lambda) = 0$$

everywhere, although the space–time is certainly *not* the flat space–time of special relativity. This is very disquieting for those who believe in Mach's principle and who hope that it is incorporated in general relativity.

However, the de Sitter universe has another feature which makes it very interesting from the point of view of the development of general relativity and cosmology. Let us consider the equation for a particle moving radially outwards along a geodesic. This has the form

$$\frac{d^2r}{ds^2} + \begin{Bmatrix} 1 \\ ij \end{Bmatrix} \frac{dx^i}{ds}\frac{dx^j}{ds} = 0,$$

and substituting the values for the metric tensor it is easy to calculate that its derivatives, which are non-zero, are

$$\begin{aligned} g_{44,1} &= -2r/R^2, \quad & g_{11,1} &= -(2r/R^2)/(1-r^2/R^2)^2, \\ g_{22,1} &= -2r, & g_{33,1} &= -2r\sin^2\theta, \\ g_{33,2} &= -2r^2\sin\theta\cos\theta, \end{aligned}$$

giving rise to the following three-index symbols:

$$\left\{ {1 \atop 44} \right\} = -(r/R^2)(1-r^2/R^2),$$
$$\left\{ {1 \atop 11} \right\} = (r/R^2)/(1-r^2/R^2),$$
$$\left\{ {1 \atop 22} \right\} = -r(1-r^2/R^2),$$
$$\left\{ {1 \atop 33} \right\} = -r\sin^2\theta(1-r^2/R^2).$$

Hence we have, for $d\theta = d\phi = 0$,

$$\frac{\dfrac{d^2r}{ds^2}+\dfrac{r}{R^2}\left(\dfrac{dr}{ds}\right)^2}{1-r^2/R^2}-\left(\frac{r}{R^2}\right)\left(\frac{dt}{ds}\right)^2(1-r^2/R^2) = 0.$$

This equation (together with the t-equation) is hard to integrate exactly. For a particle which is initially at rest, however, the equation will simplify to

$$\frac{d^2r}{ds^2}-\frac{r}{R^2}\left(1-\frac{r^2}{R^2}\right)\left(\frac{dt}{ds}\right)^2 = 0.$$

Moreover, for such a particle, using the original form of the metric, i.e.

$$ds^2 = \left(1-\frac{r^2}{R^2}\right)dt^2,$$

it follows that when the particle is at rest

$$\left(\frac{dt}{ds}\right)^2 = \frac{1}{1-r^2/R^2}.$$

Accordingly, the initial motion of the particle is determined by the equation

$$\frac{d^2r}{ds^2} = \frac{r}{R^2}.$$

The constant $1/R^2$ in this equation is certainly non-zero, since otherwise we would have a flat space–time. Accordingly, the universe must either be collapsing or expanding. We noticed above that the constant

A turned out to be negative in the Einstein case if the density is positive, and accordingly we introduced $-\frac{1}{3}A = 1/R^2$. By working out the components of the energy tensor here it will be found that the same situation arises even for zero density, and accordingly the de Sitter universe has the following property: there is no uniform distribution of matter in it, but if a test particle is introduced at a point, at rest, it begins to accelerate away from the origin. The analysis which we have given is not enough to show how this acceleration continues, but the original equation for radial motion can be written

$$\frac{d^2r}{ds^2} = \frac{r}{R^2}\left[\left(1-\frac{r^2}{R^2}\right)dt^2 - \frac{dr^2}{1-r^2/R^2}\right]\frac{1}{ds^2} = \frac{r}{R^2},$$

so that we see that the acceleration has the same value at a point even for particles not at rest. In this way the de Sitter metric suggested that astronomers should look for a systematic recession of distant matter.

It was already known (Wirtz, 1922; Hubble, 1929), that the spiral galaxies tended to have velocities away from the earth, as shown by their red shifts, and this was regarded at the time as a curious phenomenon since it suggested a total untenable view of the earth as the centre of the universe. In fact the de Sitter universe provides a model for such a recession; i.e. one in which the recession velocity is proportional to the distance from the earth. But, of course, once we have come to this point in our theoretical analysis it strongly suggests that we should consider models of the universe containing matter, which is actually expanding, and the Einstein universe which we discussed above becomes an obsolete picture. It was usually considered that the Einstein universe was unstable, so that the slightest disturbance of it would send it into an expanding phase, finishing ultimately in the de Sitter form. This fact, however, is not quite so straightforward as was formerly supposed, since, as Bonnor (1954) has shown, this instability involves a discontinuity in the pressure; in other words, the Einstein universe is only unstable for very extreme disturbances.

Be that as it may, there arose in the years following the discovery of the static models of the universe, on the one hand, a great deal of experimental information of the velocities of recession of the galaxies, and, on the other, an exhaustive theoretical description of expanding universes. No final conclusions emerged from all this work. There are a number of reasons for this. Perhaps the most important is that the whole problem of cosmology is to describe the unique universe in which we find ourselves. In cosmology a theory which gives rise to a number of possibilities, with choices of arbitrary parameters, fails to do this. The situation with the static universes of general relativity was not so unsatisfactory in this respect since only one of them contained matter. But once one considers expanding models there is such a wide range that the subject becomes a purely scholastic one.

There is another deeper reason which also contributes to the difficulties of cosmology. That is, that if the universe is expanding and has expanded from a much more concentrated state, then since some of the light we see from the stars comes from them at a time when they were in this state very different from local conditions, we have no systematic theory of what the laws of physics would be in the very different circumstances in which the light was emitted. These laws are formulated on the earth for a very small range of temperature, pressure, and density. Until we have a theory which explains how they should be changed in different circumstances it is highly speculative to try to produce a cosmology. None the less, a great deal of the effort expended on general relativity between the wars went in this direction.

CHAPTER III

Modern Developments

AFTER a disappointing period in the late twenties and thirties, general relativity became a really active subject again immediately before and following the Second World War. The problems which have led to its various further developments are so interlinked that it is necessary to describe them to a large extent together. But before entering into details, we ought to try to characterize these problems in some way. General relativity is undoubtedly a complicated theory, and when dealing with such a theory some simple analogue is useful as a guide. Of course, to understand the whole of a theory will need more than one such analogue (for otherwise the theory itself is really equivalent to one of the simpler ones). The problems of general relativity before 1938 are all immediately comprehensible in terms of Newtonian gravitation. The gravitational field considered in them is measured relative to some given coordinate system, and so can be represented by the right-hand side of the geodesic equation

$$\frac{d^2x^p}{ds^2} = -\begin{Bmatrix} p \\ qr \end{Bmatrix} \frac{dx^q}{ds} \frac{dx^r}{ds},$$

notwithstanding the non-covariant nature of this procedure. But, of course, a corollary of this is that the predictions of the theory are of small variations from the Newtonian theory, and since Newtonian gravitation happens to be an extremely exact theory, these variations are in fact very small indeed, and correspondingly difficult to measure.

The analogue that has been the mainspring of developments since 1938 has been Maxwell's electrodynamics; correspondingly, since the field there, in the Lorentz invariant formulation, is a skew-symmetric tensor of rank 2, F_{ij}, it is natural that the Riemann tensor R_{ijkl} should enter as the gravitational field in which the theory is now interested. For this tensor is doubly antisymmetric; so that those features of electromagnetic theory which are connected with the antisymmetry of F_{ij} reappear in general relativity (only twice over). It is to be expected that the problems suggested by this analogue will be quite unlike those in Newtonian theory, and so the experimental distinction between the theories becomes more possible. We shall make considerable use of the electromagnetic analogue to clarify the recent developments of the theory.

Historically the first of these problems was already tackled by Einstein in the twenties, and yet it is one of the most mysterious features of the theory. In a paper with Grommer (Einstein and Grommer, 1927), and then in a further one by himself, Einstein (1927) posed the problem: If the gravitating masses are represented by the singularities (as in the Newtonian theory of point masses), is it possible to prescribe the motion of these singularities arbitrarily or will this violate the field equations? In fact Einstein was at that time only able to answer the question negatively to a first approximation. But the calculations were carried further, and Einstein, Infeld, and Hoffman (1938) verified that the usual formulation of general relativity, in which one first postulates the geodesic equation for a particle, in this way assuming that the metric of the space can be so chosen that this equation properly represents the particle's motion and *then* postulates field equations to determine this metric, is not, after all, a formulation consisting of independent parts. In fact, the original geodesic assumption need not be made at all if one agrees to represent a particle, as in Newtonian mechanics, by a singularity in the solution of the field equations. For Einstein, Infeld, and Hoffman were able to show that such a singularity moved along a geodesic in the limit when the strength of the singularity was vanishingly small.

The electromagnetic analogue is useful here. Maxwell's equations in special relativity, in free space, can be written in the form

$$F^{ij}_{\ ,j} = J^i, \quad F_{ij,k}+F_{jk,i}+F_{ki,j} = 0,$$

where J^i is the four-vector of charge and current density. Now in the Maxwell theory an important part is played by the energy tensor

$$E^j_k = F^{ij}F_{ik} - \tfrac{1}{4}\,\delta^j_k F^{pq}F_{pq}.$$

(One is led to consider this tensor, either by generalizing Maxwell's original treatment of the "stresses in the medium" (Maxwell, 1892)— following Faraday (1852)— to a Lorentz invariant form, or by considering that the original field equations are invariant, if $J^i = 0$, under the inhomogeneous Lorentz group, applying Noether's theorem, and making the resultant tensor symmetric.) This tensor has the divergence

$$E^j_{k,j} = F^{ij}_{\ \ ,j}F_{ik} + F^{ij}F_{ik,j} - \tfrac{1}{2}\,\delta^j_k F^{pq}F_{pq,j}.$$

The first term is J^iF_{ik}, and the second is

$$-F^{ij}(F_{kj,i} + F_{ji,k}) = -F^{ij}F_{ik,j} + F^{pq}F_{pq,k}.$$

Hence, in all

$$E^i_{k,j} = -F_{ki}J^i = -G_k,$$

where G_k is the Lorentz force on the current J^i. We are not therefore at liberty to postulate *any* law of force in the theory. We have good reasons for thinking that $E^j_{k,j}$ will just balance the forces on the currents, and this determines the Lorentz force.

The differences which arise in general relativity do so because there

$$-T^i_j = R^i_j - \tfrac{1}{2}\,\delta^i_j R$$

has its (covariant) divergence *identically* zero. The existence of such a quantity is, in fact, a consequence (as we saw above) of the general covariance of the theory. Before going on to the actual derivation, which is represented in Extract 5 of this volume by a later paper of Einstein and Infeld (1949) in which the ideas have been more clearly presented, we can give a general idea of the physical principles behind the method by considering the field equations in matter which, as we saw, were adopted by Einstein in the form

$$-T_{ij} = R_{ij} - \tfrac{1}{2}g_{ij}R.$$

(We are now no longer concerned with problems of cosmology, and so we shall ignore the possibility of the cosmical term λg_{ij} being added to $-T_{ij}$.)

The energy density here may be taken in the form

$$T^{ij} = \varrho u^i u^j,$$

where u^i is the velocity dx^i/ds, and ϱ is the rest density (i.e. the local density of matter measured in a frame of reference locally chosen so that $u^i = (0, 0, 0, 1)$), if we are considering a cloud of dust. (If we were to consider a fluid there would, of course, have to be additional terms representing the pressure.) The energy tensor reduces to the form

$$T^{ij} = \begin{bmatrix} \beta^2 \varrho v^\alpha v^\beta, & \beta^2 \varrho v^\alpha \\ \beta^2 \varrho v^\alpha & \beta^2 \varrho \end{bmatrix} \quad (\alpha, \beta = 1, 2, 3),$$

where $v^\alpha = dx^\alpha/dt$, and at low velocities, when the β-factors can be ignored, this corresponds very well with the expressions which one would expect from the Newtonian theory. There is one component giving the energy or mass, three components giving the momentum, and the other six components are very small compared with these. As to the β-factors, the only remark that needs to be made is that they occur squared because we are considering a *density* distribution of *energy and momentum*. One of the factors is necessary because of the contraction of length, the other because of the increase of mass with velocity.

Now having postulated the field equations in matter this involves the following identity for the energy tensor:

$$T^{ij}_{;j} = \varrho u^i_{;j} u^j + (\varrho u^j)_{;j} u^i = 0.$$

However, the vector u^i, being the velocity of the particle in four dimensions, is a unit vector, so that

$$u_i u^i = 1,$$

and this gives

$$u_i u^i_{;j} = \tfrac{1}{2}(u_i u^i)_{;j} = 0.$$

By taking the inner product of the original identity with the velocity vector we can therefore deduce

$$(\varrho u^j)_{;j} = 0,$$

which is simply the equation of continuity; for example, in the Newtonian approximation it becomes

$$\frac{\partial \varrho}{\partial t} + \text{div}\,(\varrho \mathbf{v}) = 0.$$

This results in the second term of the identity vanishing, and as a consequence the first term must also vanish, which is essentially the geodesic equation

$$\frac{\delta u^i}{\delta s} = 0.$$

In this treatment we have used the field equations in matter but, on the other hand, no mention has been made of singularities. We have considered a cloud of dust instead of a test particle. The Einstein, Infeld, Hoffman technique was to use the field equations in vacuum except, of course, that there would be a singularity present acting as a source of the field, and so to find the motion of test particles without any special assumption about the form of the energy tensor. Since the field equations have a singularity at which they are not satisfied, Einstein, Infeld, and Hoffman enclose this singularity in a small sphere and write the field equations in such a form that they can be turned in part into surface integrals over this sphere. The treatment which they give is said in the paper quoted to be possible as a consequence of the non-linearity of the theory. This is not really true, however (Tulczyjew, 1965). It has been shown that it is possible for a linear theory to prescribe the equations of motion of its singularities. The real difference is that in a linear theory, since solutions can be superposed, there can be no interaction between these singularities; each of them may move independently of the others. Thus the non-linearity, although not essential for the calculation, is essential for it to be of any physical interest. Essentially, however, the calculation is simply a means of avoiding postulating any particular form for the energy tensor as we have done above. What is needed for it to succeed is a theory which is invariant under a general group of transformations. As we saw in an earlier chapter, such an invariance leads to conserved quantities, in this case the energy tensor, and whether we take the short cut given here of assuming a particular form for this energy tensor or whether we refuse to make an additional assumption of this kind, is to some extent a matter of taste. The calculations given by Einstein and Infeld have been repeatedly simplified on later occasions by using more and more refined analytical techniques, but essentially the step forward which they took in 1938 is one of those results which have to be shown once and for all, and must then be accepted as part of the theory and need not be re-worked.

It is worth while examining their methods a little more closely, however, since certain assumptions repay investigation. For example, in order to derive their results they are bound to use an approximation procedure for solutions of the field equations. The quantities entering, such as the metric tensor, are expanded in infinite series, and the various terms are then found one after the other in the usual way. It is necessary to make certain assumptions about the degrees of the parameter (which enters into the solution) in the various expansions, and Einstein and Infeld remarked that these assumptions are equivalent to those in the Maxwell theory which choose for the potential the average of the advanced and retarded potentials. The fact that any assumption of this kind is to be made early in the investigation is of extreme importance. The whole question really comes down to how one avoids considering a radiating system, where radiation is here meant to signify gravitational radiation (whatever the meaning of that term).

A system which radiates may well be expected to lose energy, and correspondingly the motion of the test particle may be expected no longer to be along a geodesic. Accordingly, Einstein and Infeld have somehow managed at the beginning of their investigation to rule out the possibility of radiation. In order to complete this investigation, then, we must certainly have a clear idea of what is meant by gravitational radiation and also, we would hope, to investigate the transport of energy by it. Now this whole idea of radiation comes from electromagnetic theory and it is there that we must first look for the definition. We can see the general idea if we write Maxwell's equations in their four-dimensional form

$$F^{\mu\nu}_{\ ,\nu} = 0, \quad F_{(\mu\nu,\,\sigma)} = 0,$$

where $F_{(\mu\nu,\,\sigma)} = F_{\mu\nu,\,\sigma}+F_{\nu\sigma,\,\mu}+F_{\sigma\mu,\,\nu}$, and where the various components of the field tensor are related to the three-dimensional description by

$$(F_{23}, F_{31}, F_{12}) = \mathbf{H},$$
$$(F_{14}, F_{24}, F_{34}) = \mathbf{E},$$

in suitable units. The following treatment is a perfectly standard one and has as its aim the derivation of the field around an oscillating dipole. Firstly, in free space the second of the Maxwell equations is simply the condition that the field tensor can be written in the form

$$F_{\mu\nu} = A_{\mu,\,\nu}-A_{\nu,\,\mu},$$

where the vector potential A_μ can with advantage be made to satisfy the identity $A^{\mu}_{\ ,\mu} = 0$ (Lorentz condition). Here we can rewrite this in three-dimensional form as

$$\mathbf{H} = \text{curl } \mathbf{A},$$
$$\mathbf{E} = (A_{1,4}-A_{4,1}) = -\nabla\phi-\frac{\partial\mathbf{A}}{\partial t},$$

where

$$A^\mu = (\phi, \mathbf{A}).$$

The Lorentz condition in this form becomes

$$A^{\mu}_{\ ,\mu} = \frac{\partial\phi}{\partial t}+\text{div } \mathbf{A} = 0.$$

We are concerned with the equations in free space which take the simple form

$$\text{curl } \mathbf{E} = -\frac{\partial\mathbf{H}}{\partial t}, \quad \text{curl } \mathbf{H} = \frac{\partial\mathbf{E}}{\partial t}.$$

With a view to integrating the second one with respect to time, a substitution which suggests itself is

$$\mathbf{A} = \frac{\partial}{\partial t}\mathbf{\Pi}.$$

When we insert this into the Lorentz condition we derive

$$\frac{\partial}{\partial t}(\phi + \text{div}\,\mathbf{\Pi}) = 0,$$

and one particular solution of this, admittedly a very special one, will be

$$\phi = -\text{div}\,\mathbf{\Pi}.$$

Let us substitute this into the second of the Maxwell equations. The result simplifies to the vector wave equation

$$\nabla^2\mathbf{\Pi} = \frac{\partial^2\mathbf{\Pi}}{\partial t^2}.$$

Further, let us simplify matters by considering fields which are derived from a particular vector field always in one fixed direction which we may choose as the z-axis so that

$$\mathbf{\Pi} = \psi\mathbf{e}_3,$$

the wave equation then becoming the scalar wave equation

$$\nabla^2\psi = \frac{\partial^2\psi}{\partial t^2}.$$

It is well known that for spherical waves a solution of this equation is

$$\psi = \frac{1}{r}\chi(t-r),$$

the occurrence of the term $t-r$ in the function corresponding to the retarded time. There is, of course, another solution, $(1/r)\mu(t+r)$, with the advanced time, and equally one may take averages involving both solutions. Although both of these are formal solutions of the equation, they do not equally agree with our ideas of causality. The question of how we should exclude one solution is a subtle one which we do not wish to take up here, so we will accept the retarded solution alone. Substituting we get for the components of the potential

$$\mathbf{A} = \mathbf{e}_3\chi'/r,$$
$$\phi = \mathbf{e}_3\cdot\mathbf{r}(\chi/r^2+\chi'/r^3).$$

Differentiating again, the electric field is given by

$$\mathbf{E} = -\mathbf{e}_3\chi''/r+\mathbf{e}_3\cdot\hat{\mathbf{r}}\hat{\mathbf{r}}\chi''/r+\boldsymbol{\alpha}/r^2+\boldsymbol{\beta}/r^3,$$

where $\boldsymbol{\alpha}$, $\boldsymbol{\beta}$ are two expressions involving the various functions which enter but not depending on the distance. We can rewrite this in a shortened form as

$$\mathbf{E} = \frac{\hat{\mathbf{r}}\wedge(\mathbf{e}_3\wedge\hat{\mathbf{r}})\chi''}{r}+\frac{\boldsymbol{\alpha}}{r^2}+\frac{\boldsymbol{\beta}}{r^3}.$$

Similarly, the magnetic field comes to

$$\mathbf{H} = \frac{(\mathbf{e}_3\wedge\hat{\mathbf{r}})\chi''}{r}+\frac{\boldsymbol{\gamma}}{r^2}.$$

The interpretation of these fields is well known. The larger that r becomes the more important are the first terms in each expression. These are the so-called distant field, a radiation field. At a great distance from the source, which is evidently the origin because of the singularity, we have a field which is propagating in a radial direction, such that the electric and magnetic vectors are transverse to the direction of the propagation. The other terms represent the so-called near field which are only important in the neighbourhood of the source and, as a matter of fact, it is only the radiation field which carries away the energy. In the near field the energy is transmitted away from the oscillating source during one part of the cycle and back to it again during the other part. It is this splitting of the field of the oscillator into radiation carrying energy, and a local field, which gives rise to the idea of radiation in electromagnetic theory.

We notice that the electric and magnetic vectors of the radiation field, as well as being at right angles to the direction of propagation, are also at right angles to each other, so that

$$\mathbf{E}\cdot\mathbf{H} = 0,$$

and, moreover,

$$\mathbf{E}^2 - \mathbf{H}^2 = 0.$$

At first sight these two identities do not appear to be invariant under Lorentz transformations, but as a matter of fact we see that they are equivalent to

$$I = F^{\mu\nu}F_{\mu\nu} = 0,$$
$$J = F^{\mu\nu}\tilde{F}_{\mu\nu} = 0,$$

where $\tilde{F}_{\mu\nu} = \frac{1}{2}\varepsilon_{\mu\nu\sigma\varrho}F^{\sigma\varrho}$, which are, indeed, invariants.

Let us pause for a moment to consider the difference between the electromagnetic case and the gravitational one. In the first place the electromagnetic radiation comes from a dipole. This fact was clear to Hertz when he found the solution which we have just quoted because the nearest field of all comes from the term β/r^3, that is to say, the term which arises from that part of the scalar potential of the form

$$\frac{\mathbf{e}_3\cdot\chi}{r^2} = \frac{\chi\sin\theta}{r^2},$$

corresponding to a dipole of varying moment χ. A corresponding investigation of the gravitational field of a moving system would finish with quadrupole radiation. Thus, apart from anything else, the gravitational case is bound to be considerably more complicated than the electromagnetic. But also Maxwell's equations are linear and solutions of them can be written down in terms of certain potentials which in turn can be simplified in particular problems by substitutions of the sort we have used above. No such simplification arises in gravitational theory. The equations are non-linear and, indeed, it is at first slightly mysterious what one should treat as the gravitational field at all. Originally people thought in terms of the Christoffel bracket connection as the gravitational field, the metric tensor being a potential from which it was derived. But this evidently only represents the field relative to some previously determined coordinate system since, when one writes the theory in an invariant form, this field can always be reduced to zero at any point. The field must then really be constituted by the Riemann tensor, and algebraically this suggests that the analogy with electromagnetic theory is not so far-fetched as it might have seemed. For just as in the electromagnetic case a skew-symmetric tensor of rank 2 arises, the generalization to the gravitational case is to a tensor of rank 4, skew-symmetric in each pair of indices. We have then somehow to characterize the idea of radiation in terms of the Riemann tensor. There have been a number of treatments of this, but the earliest attempt which really provides the basis for all the later ones is that of Pirani described in Extract 7.

The essential idea of Pirani's paper (1957) can be seen in terms of electromagnetic theory. The radiation part of the field is characterized by the vanishing of the two invariants mentioned before. Let us consider more closely the role of these invariants in the classification of the electromagnetic field. Given a pair of vectors (**E**, **H**) in three-dimensional space we can certainly rotate the axes so that they take the form

$$\mathbf{E} = (0, 0, E),$$
$$\mathbf{H} = (0, H_2, H_3).$$

Consider now the effect of a Lorentz transformation in the x-direction. These two vectors then become, if we suppose them to constitute parts of a skew-symmetric tensor of rank 2, as the notation suggests,

$$\mathbf{E} = (0, \beta V H_3, \quad \beta(E - VH_2)),$$
$$\mathbf{H} = (0, \beta(H_2 - VE), \quad \beta H_3).$$

A number of consequences follow, dependent upon whether or not the third component of H vanishes, that is whether or not $J = 0$. In the case when this invariant does vanish so that $H_3 = 0$, we have three cases to consider:

(i) If $I > 0$, so that $H_2^2 > E^2$, we can choose $V = E/H_2$, and in the new coordinate system $E = 0$, so that the field is purely magnetic.

(ii) If $I < 0$, so that $H_2^2 > E^2$, we choose $V = H_2/E$, and then the new $H = 0$, so that the field is purely electric.

(iii) If $I = 0$, neither of these reductions is possible.

Before discussing (iii) further, we need to consider the corresponding cases when $H_3 \neq 0$, i.e. $J \neq 0$. Certainly if $J \neq 0$, both fields are present in all Lorentz frames, but another reduction is possible, in which the fields are parallel.

For if

$$\frac{H_2 - VE}{VH_3} = \frac{H_3}{E - VH_2},$$

then

$$\frac{V}{1+V^2} = \frac{EH_2}{H_2^2 + H_3^2 + E^2}.$$

Since

$$\frac{V}{1+V^2} = \left(V + \frac{1}{V}\right)^{-1} \quad \text{and} \quad V + \frac{1}{V} > 2 \quad \text{if} \quad V < 1,$$

the only constraint on whether or not there is a solution is

$$H_2^2 + H_3^2 + E^2 > 2EH_2,$$

i.e.

$$H_3^2 + (H_2 - E)^2 > 0,$$

which is certainly fulfilled if $J \neq 0$. The only case in which no reduction is possible is the so-called null field in which $I = J = 0$, i.e.

$$\mathbf{E}^2 - \mathbf{H}^2 = 0 \quad \text{and} \quad \mathbf{E} \cdot \mathbf{H} = 0,$$

the radiation case.

Another way of looking at these results will also be instructive in the future account of relativity. The classification which has already been given is in terms of the two invariants I, J of the field tensor. One may well ask whether there are other invariants so that the classifications can be carried out more finely. That there are not is clear when one writes out the transformation of the electric and magnetic fields in three-dimensional form:

$$H_1' = H_1, \quad H_2' = \beta(H_2 - VE_3), \quad H_3' = \beta(H_3 + VE_2),$$
$$E_1' = E_1, \quad E_2' = \beta(E_2 + VH_3), \quad E_3' = \beta(E_3 - VH_2),$$

If we introduce the complex quantity

$$\mathbf{Z} = \mathbf{E}+i\mathbf{H},$$

these transformations assume the form

$$Z_1' = Z_1, \quad Z_2' = \beta(Z_2 - iVZ_3) = Z_2 \cos\theta - Z_3 \sin\theta,$$
$$Z_3' = \beta(Z_3 + iVZ_2) = Z_3 \cos\theta + Z_2 \sin\theta,$$

that is to say, a rotation about the x-axis with complex angle of rotation defined by $\tan\theta = iV$.

Thus the Lorentz transformations of the field vectors are in a one-to-one correspondence with a certain group of transformations of complex three-dimensional space. Now the only invariant of a single vector in three-dimensional space is its magnitude, given by

$$|\mathbf{Z}|^2 = (\mathbf{E}+i\mathbf{H})^2 = \mathbf{E}^2 - \mathbf{H}^2 + 2i\mathbf{E}\cdot\mathbf{H},$$

and the two invariants are seen here as the real and imaginary parts of this. This reduction of the complexity of the algebra by going over to a complex three-dimensional space will prove to be of value again in considering the Riemann tensor.

Yet another approach is also of value in the electromagnetic case, and can then be carried over to general relativity by analogy. This is to consider electromagnetic tensors as a linear transformation of the vectors of the four-dimensional space–time into themselves and to ask whether there are any such vectors v^j; which are left unchanged in direction by this transformation, i.e. we search for vectors satisfying

$$F^i_j v^j = \lambda v^i.$$

Because of the skew-symmetry of the electromagnetic tensor it follows by taking an inner product that

$$\lambda v^i v_i = 0,$$

so that either the vector concerned is a null vector or else the eigenvalue concerned is zero. Let us confine attention to the null vectors since this is obviously the general case. We can then write the vector in the form

$$v^i = (\mathbf{v},\ v),$$

where $\mathbf{v}^2 = v^2$.

Noticing, however, that the electromagnetic tensor corresponds to a matrix of the form

$$F^i_j = \begin{bmatrix} 0 & -H_3 & H_2 & -E_1 \\ H & 0 & -H_1 & -E_2 \\ -H_2 & H_1 & 0 & -E_3 \\ -E_1 & -E_2 & -E_3 & 0 \end{bmatrix},$$

the conditions become

$$\mathbf{H} \wedge \mathbf{v} - \mathbf{E}v = \lambda \mathbf{v},$$
$$-\mathbf{E} \cdot \mathbf{v} = \lambda v,$$

of which the second can be derived from the first, as one would expect, since we have already used the non-independence of these four conditions in assuming the vector to be a null vector. The classification of the field can now be carried out by asking the questions (i) whether the eigenvalues are different or coincident, real, zero, or imaginary, (ii) whether the corresponding vectors are different. It is easiest, in order to do this, first to reduce the field by a rotation of axes to the standard form

$$\mathbf{H} = (0,\ H_2,\ H_3), \quad \mathbf{E} = (0,\ 0,\ E).$$

The equation for the eigenvalue then becomes

$$\begin{vmatrix} -\lambda & -H_3 & H_2 & 0 \\ H_3 & -\lambda & 0 & 0 \\ -H_2 & 0 & -\lambda & -E \\ 0 & 0 & -E & -\lambda \end{vmatrix} = 0,$$

and it is easy to expand this into the form

$$\lambda^4 + I\lambda^2 - J^2 = 0.$$

The following possibilities arise about the roots of the equation:

(a) If $I = J = 0$, there are four equal zero roots which may be described symbolically as (0000).
(b) If $J = 0$ but $I > 0$, there are two zero roots and a pair of conjugate imaginary ones $(00i, -i)$.
(c) If $J = 0$ but $I < 0$, there are two zero roots and two equal and opposite real ones $(0\,0\,1, -1)$.
(d) If $J \neq 0$ the two roots for λ^2 are

$$\lambda^2 = \tfrac{1}{2}(-I \pm (I^2+4J^2)^{\frac{1}{2}}),$$

one of which is positive and one negative, so that the four roots for λ have the pattern $(1, -1, i, -i)$.

These four cases are exactly the ones noticed earlier, so that the classification suggested by determining the null eigenvectors of the electromagnetic tensor parallels closely the classification derived before by considering whether or not the field can be made purely electric or purely magnetic, etc. The radiation case corresponds to the existence of four equal zero roots of the eigenvalue equation. The corresponding results in general relativity may now be discussed.

We are concerned in this case with the curvature tensor, with the symmetry properties

$$R_{abcd} = -R_{bacd} = -R_{abdc} = R_{cdab},$$
$$R_{abcd} + R_{acdb} + R_{adbc} = 0.$$

The tensor has therefore twenty independent components. There are a number of ways of discussing the special cases which this tensor may take, and the earliest of these is that due to Petrov (1954) described at some length in the paper by Pirani (Extract 7). Pirani's motive in this will be clear from the electromagnetic case discussed above. The algebraically most special case in electromagnetism corresponds to what is known to be the radiative field. The same situation in gravitation might then, one hopes, be defined as the radiative one there. The following classification, which is slightly more elegant than Petrov's is essentially due to Debever (1958), but the version given here is that of Sachs (1961). In the electromagnetic case the classification by looking for eigenvectors falls into two cases as it leads to zero roots of the eigenvalue equation or else to non-zero ones. The corresponding situations could be described by Table 1. To construct this, let us first notice that, if

$$F_{ab}k^b = \lambda k_a,$$

then we must have

$$F_{ab}k_ck^a - F_{ac}k_bk^a = -\lambda(k_bk_c - k_ck_b) = 0,$$

which may be written (using square brackets to denote anti-symmetrization)

$$F_{a[b}k_{c]}k^a = 0.$$

This equation is always satisfied by a null eigenvector k^a, but, of course, it may be replaced by a simpler one (if the eigenvalue is zero),

$$F_{ab}k^b = 0.$$

The equation found for the eigenvalue λ was, in a special coordinate system,

$$\lambda^4 + \lambda^2(H_2^2 - E^2) - E^2H_3^2 = 0,$$

which, as we saw, gives rise to the following cases:

TABLE 1

(1)	(2)	(3)
N	$F_{ab}k^b = 0$	4
II	$F_{ab}k^b = 0 \quad F_{a[b}m^am_{c]} = 0$	211
I	$F_{a[b}k^ak_{c]} = 0$	1111

The first column labels the type of field (N denoting "null"). The second column gives the equations satisfied by the null directions, and the third gives the number of coincidences (i.e. 4 denotes 4 coincident directions, 211 a pair of coincident and 2 others, 1111 4 directions all different).

In the same way for the curvature tensor, so long as the field equations are satisfied, that is to say, in empty space, Debever proved that there exists at least 1 and at most 4 null directions such that

$$k_{[a}R_{b]ij[c}k_{d]}k^i k^j = 0.$$

TABLE 2

(1)	(2)	(3)
N	$R_{abcd}k^d = 0$	4
III	$R_{abc[d}k^c k_{e]} = 0$	3, 1
D	$\begin{cases} R_{abc[d}k_{e]}k^a k^c = 0 \\ R_{abc[d}m_{e]}m^a m^c = 0 \end{cases}$	2, 2
II	$R_{abc[d}k_{e]}k^a k^c = 0$	2, 11
I	$k_{[a}R_{b]ij[c}k_{d]}k^i k^j = 0$	1111

The null directions satisfying equations of this kind are said to define rays and by knowing these rays the field may be classified. The Petrov classification cited by Pirani becomes in the form shown in Table 2.

In Table 2 the first column gives the so-called Petrov type of the metric, the second gives the equation satisfied by 1 or possibly more of the rays, and the number of coincidences amongst the rays is shown in the third column. For example in the null case, which is evidently the one of radiation, there is only one distinct solution, which therefore satisfies the equation stated. In the case of type D, however, there are two distinct solutions, the corresponding case where only one ray actually satisfies the equation falling into type II. The Table 2 has been constructed so that as one goes up from the bottom line the metrics become more and more specialized. (The term *algebraically special* is used to describe all those not of type I.)

The fact that the classification is in terms of the rays led to a more intensive study of the so-called gravitational ray optics by Sachs. He considered the way in which such a family of rays, if they were imagined to be light rays, would expand, rotate, and distort (shear) the shadow of an object, and it proved possible to relate the Petrov type of the metric to these geometrical properties. There is, for example, the well-known Goldberg–Sachs (1962) theorem that an empty space–time is algebraically special if and only if it contains a shear-free null geodesic family of rays.

Although the discussion given here is a very elegant one, it is a little remote from Petrov's original one, as expounded by Pirani, and it seems useful to give an outline of a more pedestrian development, which needs less technical knowledge. In the course of this development, we need to use the technique of the *dual* of a bivector F_{ab}, defined by

$$\tilde{F}_{ab} = \tfrac{1}{2}\varepsilon_{abcd}F^{cd},$$

where $\varepsilon_{abcd} = 1$ if *abcd* is an even permutation of

$$1\,2\,3\,4$$

$= -1$ if *abcd* is an odd permutation of

$$1\,2\,3\,4$$

$= 0$ otherwise.

Thus a moment's calculation will show that

$$\tilde{F}_{14} = \varepsilon_{1234}F^{23} = F_{23},$$
$$\tilde{F}_{23} = \varepsilon_{1234}F^{14} = -F_{14},$$

and so on. It is at once apparent that

$$\tilde{\tilde{F}}_{ab} = -F_{ab}.$$

Another useful result is the following: if both F_{ab} and G_{ab} are bivectors, then

$$\begin{aligned} F_{ab}G^{ab} &= 2(F_{23}G^{23}+\ \dots\ +F_{14}G^{14}+\ \dots) \\ &= 2(-\tilde{F}_{14}\tilde{G}^{14}-\ \dots\ -\tilde{F}_{23}\tilde{G}^{23}-\ \dots) = -\tilde{F}_{ab}\tilde{G}^{ab}. \end{aligned}$$

We have already seen that, in any Lorentz frame, a bivector splits into a pair of three-vectors, which we called **E**, **H**. Let us now combine these to make a complex three-vector $\mathbf{Z} = \mathbf{E}+i\mathbf{H}$. Thus

$$Z_1 = E_1+iH_1 = F_{14}+iF_{23}, \quad \text{and so on,}$$

and so, if $\tilde{\mathbf{Z}}$ denotes the vector corresponding to the dual bivector

$$\tilde{Z}_1 = F_{23}-iF_{14} = H_1-iE_1 = -i(E_1+iH_1) = -iZ_1,$$

i.e. $$\tilde{\mathbf{Z}} = -i\mathbf{Z}.$$

Now R_{abcd} defines a linear mapping of all bivectors into bivectors:

$$F_{ab} \rightarrow G_{ab} = R_{abcd}F^{cd}.$$

So it also defines a mapping of complex three-vectors $\mathbf{Z}$ into themselves, only this mapping *is no longer linear* (since all the six real components of $\mathbf{Z}$ must be allowed to enter separately). In fact the mapping can obviously be written

$$\mathbf{Z} \rightarrow \mathbf{Z}' = A\mathbf{Z}+B\mathbf{Z}^*,$$

where A, B are linear vector functions of vectors with complex values (i.e. 3×3 complex matrices) and $\mathbf{Z}^*$ is the complex conjugate of $\mathbf{Z}$.

Next let us consider the effect of dual operations on R_{abcd}. Of course, there are two pairs of antisymmetric suffixes to which the operation can be applied. We shall define $\tilde{\tilde{R}}_{abcd}$ to be the result of performing both of these, so that

$$\tilde{\tilde{R}}_{abcd} = \tfrac{1}{4}\varepsilon_{abij}\varepsilon_{cdkl}R^{ijkl}$$

and so, now, $\tilde{\tilde{\tilde{\tilde{R}}}}_{abcd} = R_{abcd}$.

If $$G_{ab} = R_{abcd}\,F^{cd},$$

it is obvious, using the result above about duals, that

$$\tilde{G}_{ab} = -\tilde{\tilde{R}}_{abcd}\tilde{F}^{cd}.$$

Suppose we translate this into the complex three-space notation. If $\mathbf{W}$ corresponds to G_{ab}, and if the terms A and B become (say) $\tilde{A}$ and $\tilde{B}$

when we consider $\tilde{R}_{abcd}$, then we have

$$\mathbf{W} = A\mathbf{Z}+B\mathbf{Z}^*$$

and $$\tilde{\mathbf{W}} = i\mathbf{W} = -i(A\mathbf{Z}+B\mathbf{Z}^*) = A\tilde{\mathbf{Z}}-B\tilde{\mathbf{Z}}^*.$$

But $$\tilde{\mathbf{W}} = -(\tilde{A}\tilde{\mathbf{Z}}+\tilde{B}\tilde{\mathbf{Z}}^*)$$

by definition and the result above, so that

$$A = -\tilde{A} \quad \text{and} \quad B = \tilde{B}.$$

In other words, the obvious decomposition of R_{abcd} into self-dual and anti-self-dual parts,

$$R_{abcd} = \tfrac{1}{2}(R_{abcd}+\tilde{R}_{abcd})+\tfrac{1}{2}(R_{abcd}-\tilde{R}_{abcd})$$

is the split into B and A respectively.

The next step, which gives the key to the Petrov classification, is to prove the subsidiary result that the self-dual part $\frac{1}{2}(R_{abcd}+\tilde{R}_{abcd})$ is an algebraic function of R_{ab} only. This, then, shows that, when the field equations $R_{ab} = 0$ are satisfied, $B = 0$. To prove this result it is sufficient to note that there are essentially only four cases to consider:

(a) $R_{2323}+R_{1414}$, (c) $R_{2314}-R_{1423}$,

(b) $R_{2331}+R_{1424}$, (d) $R_{2324}-R_{1431}$.

Of these only case (a) is of any difficulty. For case (b), adopting for convenience a coordinate system at the point under consideration for which $g_{ab} = \eta_{ab}$,

$$\begin{aligned} R_{2331}+R_{1424} &= -R^{3}_{123}-R^{4}_{124} \\ &= -R_{12}, \end{aligned}$$

and for (d) similarly

$$R_{2324}-R_{1431} = R^{2}_{342}+R^{1}_{341} = R_{34}.$$

Case (c) is identically zero, whilst in case (a)

$$\begin{aligned} R_{2323}+R_{1414} &= (R^{2}_{332}+R^{1}_{331}+R^{1}_{441}+R^{2}_{442}) \\ &\quad -(R^{1}_{331}+R^{2}_{442}) \\ &= R_{33}+R_{44}-(R_{1313}+R_{2424}). \end{aligned}$$

Hence, if we write for a moment

$$S_1 = R_{2323}+R_{1414},$$

we have

$$S_1+S_2 = R_{33}+R_{44},$$

so that, by symmetry,

$$S_2+S_3 = R_{11}+R_{44},$$
$$S_3+S_1 = R_{22}+R_{44},$$

whence

$$2S_1 = R_{11}+R_{22}+R_{33}+R_{44},$$

and the result is proved.

Accordingly, in free space the problem of classifying the curvature tensor is equivalent to that of classifying the 3×3 complex matrix A. We will not enter further into the details of this here, since they can be found in books on algebra, but our investigation suffices to show why the Petrov classification is so easy and straightforward for empty space, but complicated and difficult when the field equations are not satisfied and so two 3×3 complex matrices A and B have to be discussed.

The initial impetus given to gravitational radiation theory by Pirani's investigation was very considerable, but like many other developments in general relativity it runs into trouble from the non-linearity of the theory. When we discussed the electromagnetic case we found the radiation field followed for a Hertzian oscillator by finding the complete field and picking out those terms which fall away most slowly at great distances. These terms by themselves would not be solutions to the physical problem, but it is possible to consider them separately in electrodynamics because the complete field is derived by adding together the partial fields. Of course, if one asked whether a given electromagnetic field were a radiation field or not, and used the criteria developed above, the answer would almost certainly be no; for whatever the radiation properties of the field there would in any physical situation always be local near fields present as well, and the criterion only distinguishes the case when there is nothing but radiation from all other cases. The situation is unfortunately much the same in gravitation. If one tries to apply the Petrov classification to the Riemann tensor of the metric the result is almost bound to be of type I. It is necessary to develop the metric in terms of distance from the physical objects involved and the criterion for radiation must then be applied to the terms in $1/r$ in this development.

This brings us to another aspect of the problem of radiation which, it will be recalled, we have taken up because of the necessity of excluding radiation in determining the equations of motion. This new problem is that of relating the field to its sources, rather in the way in which the electromagnetic radiation field was identified as coming from an oscillating dipole. A number of attempts were made by means of successive approximations to determine whether or not freely gravitating particles radiate when moving in each other's field. There are difficulties of two kinds. In the first place the solutions are so complicated that it is difficult in practice, and in principle, to say whether radiation is taking place or not. Secondly, because of this complication, the mere calculations are often intolerable. A step forward was taken by Bonnor (1959) when he considered the much simpler problem of whether a constant isolated system could lose mass by radiation. The advantage here was that he was able to require the system to vibrate in an arbitrary way. For example, he could consider a system initially at rest which then oscillated for a finite time before returning to rest. Since the initial and final states are those of rest, in which the concept of the mass of the system is a well understood one, the question of whether the mass has the same value afterwards or not is a well-defined one. Bonnor was able to find results which suggested that there were genuine gravitational waves which carried energy away from the source, but there was still the problem of the convergence of his method, and the calculations were extremely long. It would have been interesting to know a higher approximation to the result, but this would have been virtually impossible because of the complexity.

The whole situation in this field was transformed by the introduction by Bondi and his fellow workers of a special coordinate system. Bondi's work is described at length in the paper by Bondi, van der Burg, and Metzner (1962), Extract 8 of the present volume. Instead of using the ordinary polar coordinates, Bondi imagines a source of light placed at the origin O and surrounded by a small sphere so that the ordinary angles θ and ϕ can be defined. There is also a time coordinate, but this is now called u and the $u\theta\phi$ coordinates of any event are defined to be those at which the light ray from O to the event cuts the sphere. Taking r as a radial variable, i.e. the distance along this light ray, the relationship with the usual t in flat space–time is that

$$u = t - r$$

and the corresponding special relativistic metric becomes

$$ds^2 = du^2 + 2du\,dr - r^2\,(d\theta^2 + \sin^2\theta\,d\phi^2).$$

It is very striking that, whereas in most calculations in both special and general relativity a diagonal form of the metric is preferred, Bondi finds that physically it is most convenient to transform to this non-diagonal form. For the case of general relativity he generalizes this to the metric

$$ds^2 = g_{44}\,du^2 + 2e^{2\beta}\,du\,dr + 2g_{42}\,du\,d\theta - r^2(e^{2\gamma}\,d\theta^2 + e^{-2\gamma}\sin^2\theta\,d\phi^2),$$

where g_{44}, β, g_{42}, and γ are functions of u, r, and θ.

This generalization simplifies tremendously the calculations needed for the higher approximation in Bonnor's technique. In this way Bondi and his collaborators are able, in the paper mentioned, to discuss outgoing radiation, and they find it possible to expand the metric in a series of powers of $1/r$ very much like the result for electromagnetic theory. If the expansion is of the form

$$\gamma = \frac{c_1(u, \theta)}{r} + \frac{c_2(u, \theta)}{r^2} + \ldots,$$

and this is substituted into the field equations, relations arise between the coefficients. It is proved in the paper that only one function

$c_1(u, \theta)$ is left undetermined, and the derivative of this function with respect to u is what is called by Bondi the *news function*. According to his view, gravitational radiation is simply the transmission of "news" through space.

Important as these theoretical ideas of radiation have been in the development of the subject, deepening our knowledge of the structure of the gravitational field, they cannot yet be said to have had very great influence in the experimental confirmation of the theory. It is true that, in recent months, Weber (1961) in Maryland has reported definite positive results in his attempt to measure gravitational radiation. The apparatus which he employs consists of a very heavy aluminium cylinder to which quartz transducers are fixed. The passage of a gravitational wave causes elastic waves in the cylinder and this in turn produces electric oscillation on the quartz transducers which are amplified by a very high gain d.c. amplifier. The need for shielding from other perturbations leads to a very poor signal to noise ratio (of about 1:30). In order to improve on this very poor ratio, Weber employed two sets of apparatus at a considerable distance apart and tried to correlate the output from the two of them. It is too soon to assess the results of his experiments but, according to his own interpretation, there appears to be a strong source of gravitational radiation which is probably situated somewhere near the centre of our galaxy (Carmeli, Fickler, and Witten, 1970). A number of other experiments are being carried out in different parts of the world to confirm or contradict Weber's results, and in the next year or so we should have much more definite information. This is because, owing to the way in which gravitational waves pass through matter almost without modification, all of these observatories should receive the radiation at the same time and, therefore, any lack of consistency between their results will at once show Weber's to have been due to some other influence than the gravitational.

A more exciting and exotic possibility of confirmation of general relativity has, however, come up in more recent years from the consideration of gravitational collapse. The early work on this subject, due to Oppenheimer and Snyder (1939), is described in the paper forming

Extract 9 of the present book. The idea of gravitational collapse, however, is one which could already have been considered by Newton. Since Newton's law of gravitation states that the forces between any two masses are always attractive, unlike the situation in the electrostatic case, it follows that a large mass of matter acted on by no other forces than the gravitational will always collapse towards the centre of mass.

It is not difficult to work out some order of magnitude results about such a collapse. Suppose that we have initially a uniform sphere of gravitating matter acted on only by gravitational force. The matter will therefore collapse towards the centre of the sphere. We can start by making the assumption that there is no overtaking, so that matter which is originally at a greater distance from the centre always remains at a greater distance. When we have derived our solutions it will then be possible to verify that this assumption is satisfied.

The matter which is originally at a distance R from the centre forms a shell which is under the attraction of all the matter which is inside it. That is to say, the amount of matter attracting this shell is

$$\tfrac{4}{3}\pi R^3 \varrho,$$

where ϱ is the original uniform density. Our assumption of no overtaking means that this is the matter which is always attracting the shell and, as is well known in Newtonian gravitation, this matter acts as if it were all concentrated at the centre. Accordingly, when the shell is at a distance r from the centre, the attraction on it will be per unit mass

$$\tfrac{4}{3} G\pi R^3 \varrho / r^2.$$

By a well-known calculation the potential energy of this attraction is

$$-\tfrac{4}{3} G\pi R^3 \varrho / r,$$

and therefore the total energy of the matter which was originally at a distance R from the centre will be

$$\frac{1}{2}v^2 - \frac{4}{3}\,\frac{G\pi R^3 \varrho}{r}$$

per unit mass.

This total energy is, of course, constant, so that if we suppose that, by some inexplicable means, the sphere starts from rest, we get for the speed of the matter at any time

$$v^2 = \frac{8}{3}G\pi R^3\varrho\left(\frac{1}{r}-\frac{1}{R}\right).$$

However, the inward velocity is obviously given by

$$v = -\frac{dr}{dt}$$

so that the time taken for the shell to reach the centre is

$$t = -\int_R^0 \left(\frac{8}{3}G\pi R^3\varrho\frac{R-r}{Rr}\right)^{-\frac{1}{2}} dr.$$

This integral is quite easy to evaluate by making the substitution

$$r = R\cos^2\theta,$$

and the corresponding time is

$$t = \left(\frac{3\pi}{32G\varrho}\right)^{\frac{1}{2}}.$$

It is important to notice that this time does not depend upon the value of R, so that all the matter reaches the centre at the same time. Accordingly, our assumption that there is no overtaking has been satisfied. It is also of interest to note the size of this time. If we start with a sphere which is initially of the density of water and is moving under its gravitational attraction alone, all the matter in the sphere will apparently have reached the centre in less than half an hour.

Such results were not regarded as very important in Newtonian gravitation because of the artificial assumption made in deriving them that only gravitational forces were acting. It is obvious that for ordinary matter other forces must always be present. The principal of equivalence makes some difference to this, however. The more the mass of a body is increased by the addition of further matter, the more the gravitational forces increase. Although these forces are never the only ones acting, the other forces which act, which are chemical and nuclear forces, all saturate to a certain extent. Accordingly, for sufficiently large collections of matter we can be sure that the gravitational forces will be all important. This is the situation considered by Oppenheimer and Snyder and the extract quoted goes quite a long way to working out the dynamics of the situation.

At the same time it is interesting to try to put this picture, derived by theoretical calculations within general relativity, into a larger framework provided by physics as a whole. Matter in fact consists of elementary particles of various kinds, and the mass of the matter mainly resides in the baryons. It is convenient to consider a collection of baryons and in the calculations to disregard the appropriate number of electrons which have to be added in order to keep the matter neutral. It is then possible to do certain calculations about the equilibrium states of numbers of baryons (Wheeler, 1962; Harrison *et al.*, 1965). It has, for example, been calculated that a collection of 560 baryons has an equilibrium state, that is to say a state of lowest energy, which is of the form of 10 atoms of iron of atomic weight 56. These atoms are arranged in a certain crystal lattice. The reason that the matter is iron lies in the nuclear forces. The crystal lattice in which the atoms are found is determined by the chemical forces.

Now imagine the calculation done for a larger number of baryons; even when there are as many as 56×10^{41} baryons the state of equilibrium is not very different. It consists of a sphere, of iron again, of radius about 8 km and if we continue in this way, constantly carrying out the calculations for larger and larger numbers of baryons, the gravitational forces eventually overcome the weaker of the other two forces acting, that is the chemical ones. It turns out that there is an instability first at the point where the chemical forces are overcome. The central density of the sphere is then about 5×10^{8} g cm^{-3}.

The electrons at the centre of the sphere have now been squeezed to such a small volume that they combine with the protons to make neutrons. If we add any more matter, the body, which is essentially a star, collapses with the central pressure rising still further, more electrons are crushed and the matter sinks further inwards. It might be thought that the result of this catastrophic interaction is that the system would tend inwards to a stable situation in which all the protons had become neutrons and we should have a neutron star. This is not true, however; the mass at which the chemical forces are overcome is about 1.2 times that of the sun, but for very high densities the star has a critical mass of only 0.7 of the sun's mass. At these high densities the gravitational forces then overcome even the nuclear ones. If one starts with matter of about nuclear density and adds even more mass to it, a second collapse point arises; the central density now becomes 10^{18} g cm^{-3}.

If still further matter is added after this, the nuclear and chemical forces both become entirely negligible compared with the gravitational ones and we are back at the simple problem discussed earlier in this section. There are now no forces left which could prevent the catastrophic collapse described.

However, the situation in general relativity is considerably more complicated than that in Newtonian mechanics because, as we have seen before, the field round a spherically symmetrical body has, at least in an ordinary coordinate system, a singularity at a finite distance from the origin—the so-called Schwarzschild singularity. If a body is to collapse indefinitely then it seems as if it will eventually pass inside its own Schwarzschild singularity. Of course the critical radius in the Schwarzschild solution was noticed as early as 1916, but it was not regarded as a serious defect in the theory since for any ordinary body such as the sun the Schwarzschild sphere was so far inside the matter as to be quite unreachable. The actual solution for the sun as a whole consisted of the Schwarzschild solution *outside* the matter and some other solution corresponding to the matter present inside the sun. The investigations of collapse initiated by Oppenheimer and Snyder have changed all this. When a body reaches the second crushing point and engages in its final catastrophic collapse, it seems as if there is nothing to stop all the matter eventually being inside the critical radius.

If, indeed, this is possible, then we should have extremely massive bodies and these bodies would be surrounded by the Schwarzschild sphere which further investigation shows to have rather unusual properties. Signals, such as light signals, can be traced into the sphere from outside, although there are certain difficulties about the length of time which they take, but from inside the sphere no signals can come outside. Accordingly, such a body, even if it produced light and heat like a star, would be completely invisible to us and we could only detect it by its strong gravitational field. Such bodies have come to be known as black holes, and current astronomy is on the look out for them.

It should be remarked, however, that not all workers in the subject agree with this analysis of the collapse situation. We can see why this is, even within the framework of the Newtonian theory, if we use some of the ideas of special relativity (McCrea, 1964). If a sphere of mass M and radius R has a mass m brought up from infinity to its surface, the potential energy lost by the particle is GMm/R. If we import from special relativity the energy mass relation $E = mc^2$, or $E = m$, when $c = 1$, this loss of potential energy will show up as a loss of mass of amount $(GMm)/R$. Accordingly, only the amount of mass $m[1-(GM)/R]$ is actually added to the sphere. In this mixed theory, a critical situation arises when $R = GM$; no more mass can then be added. In general relativity a factor 2 is introduced; the critical radius is now the Schwarzschild radius, and accordingly those workers who accept this analysis refuse to believe in the prediction of black holes by the theory.

To conclude this chapter it is appropriate to try to look ahead at the direction in which future developments in general relativity may take place. An outstanding problem is the relationship between general relativity and quantum mechanics. These two subjects have developed in quite different ways employing different techniques; it is very hard at this moment to see how these can be related to each other. Attempts have been made of various kinds, but it is too early to give any reasoned assessment of them; accordingly, we can do no more than note that this is an outstanding problem which will require to be tackled at some time. Leaving that on one side, the most striking development of the last 10 years has been an entirely new method of setting out the working in general relativity. This may seem a comparatively trivial matter, but when one comes down to the calculation of even such a simple problem as the spherically symmetrical gravitational field which was given in an earlier chapter, the reader will be convinced that any easing of the problem of calculation is to be welcomed. This is all the more so in theoretical investigations in the subject. It appears to me that the most important step forward has been that of Penrose (1960) in the paper reproduced as Extract 10 of the present book, in which he introduced the spinor technique to general relativity.

In understanding Penrose's paper it is important to notice a certain difference of emphasis in the treatment of spinors from that in special relativity. In the corresponding book to this on special relativity, spinors were dealt with in the following way. The rotation group in three dimensions, which is a sub-group of the Lorentz group, was studied in some detail on page 79 of that book. It was found that, by using the transformation of a symmetric tensor of rank 2, it was possible to deduce that there would be new quantities which were not tensor quantities and which transformed under a two-valued representation of the rotation group. These quantities, with two components, are called spinors. It is very easy to extend the working given in the earlier book to the whole Lorentz group; sometimes it is convenient to use the proper Lorentz group, i.e. excluding time reflection and space reversal, in which case two-component spinors are appropriate; sometimes it is useful to use the full Lorentz group, including improper transformations, and then we need four-component spinors of the type studied by Dirac.

Whichever we use, however, the treatment in the previous book was firmly grounded on the two-valuedness of the representation under which the spinors transform and, although it was not stated there, this was based on a particular property of the orthogonal group in relation to what is known as its universal covering group.

A number of attempts were made from 1930 onwards (Infeld and van der Waerden, 1933; Bade and Jehle, 1953; Corson, 1953) to extend the spinor technique to general relativity, but all of these attempts were founded on the two-valuedness of the spinor representation. Since they regarded the spinor technique as essentially to do with the structure of the Lorentz group, it was necessary to have the Lorentz group in the theory. But general relativity is invariant under a quite different group and accordingly these earlier techniques operated by introducing, as well as the ordinary coordinate system of the theory, a local orthogonal coordinate system at every point. The transformations of one of these local orthogonal coordinate systems into another one at the same point then formed a group isomorphic to the Lorentz group, and these earlier authors considered that it was here that the spinor representations might enter.

Although this was all perfectly correct and straightforward to carry out, nothing of great value arose from it. It was Penrose's genius to see that the importance of the spinor technique in general relativity was not particularly related to the two-valuedness of the representation under which it transformed. Its real value was in the way in which it simplified the very heavy algebra connected with the Riemann tensor. It is easy to see why it is so valuable by using again the electromagnetic analogy. In the treatment in the book on special relativity we explained how a four-vector corresponds to a pair of one-component spinors. There is however, another connected correspondence, i.e. between the real skew-symmetric tensors of rank two and the symmetric spinors of rank 2. In fact, as Penrose shows in equation (1.3) of his paper, we may write instead of the electromagnetic bivector F_{ab} a corresponding four-index spinor

$$F_{AB'CD'} = \tfrac{1}{2}\{\phi_{AC}\varepsilon_{B'D'} + \varepsilon_{AC}\bar{\phi}_{B'D'}\}.$$

It is to be noticed here that the spinor ϕ_{AC} has only three independent components, since it is symmetric. These components are, of course, complex, but none the less this way of writing the electric field has produced a considerable simplification. Of course, this grouping of the electric field into three complex quantities is essentially the same as the one employed above in discussing the classification of electric fields when the quantity Z was introduced.

An important quantity in electromagnetic theory is the Maxwell stress tensor which arises as a rather complicated kind of product of the electric field bivector with itself. The corresponding quantity in spinor notation is easily written down and turns out to be a particularly simple kind of product of the spinor ϕ_{AC} with its complex conjugate. This alone shows the value of the spinor algebra in dealing with electromagnetic theory.

Now the curvature tensor is skew-symmetric in both pairs of indices and can therefore be regarded as a linear mapping of the set of all bivectors at a point into this same set by an equation of the form

$$F_{ab} \rightarrow G_{ab} = R_{abcd}F^{cd}.$$

Such a linear mapping will, of course, be a simple function of the spinor ϕ_{AC}, although, because both ϕ_{AC} and its complex conjugate are involved in specifying the electric field, it cannot be simply a linear function of ϕ. In fact, to get the general idea at this point the reader need only turn back to the earlier treatment in terms of **Z**. The situation here is exactly the same, but the spinor technique makes it a little more straightforward. For example, such a linear function would have the form

$$\phi_{AB} \rightarrow \psi_{AB} = \lambda_{ABCD}\phi^{CD},$$

and it is easy to see that this is not sufficiently general, since the operator must be symmetric in both pairs of suffixes. That means each pair of suffixes can have only three independent values, but it is also known that the Riemann tensor, and therefore the operator, is symmetric under interchange of the pairs of suffixes so that the total number of independent components available here would be six. These are, of course, complex components and so would correspond to twelve components of the Riemann tensor at most. Since the curvature tensor has twenty independent components, this cannot be enough to describe the general field. In fact, just as in the calculations with Z two such quantities are required, as Penrose shows in equation (2.2) of his paper.

However, it is not the general gravitational field with which one is concerned in free space in general relativity but the gravitational field which satisfies Einstein's field equation, and these equations quite simply can be shown to imply, according to Penrose in equation (3.3), $\phi_{ABCD} = 0$ exactly as, in the **Z** calculation, they implied $B = 0$, together with the condition χ_{ABCD} completely symmetric.

As Penrose remarks, it is very striking that the gravitational field can be described by a single totally symmetric spinor with four indices.

On page 179 of his paper, another striking result appears. By taking the product of this spinor with itself, another quantity occurs which when translated into tensor notation corresponds to a rather complicated type of product of the curvature tensor with itself, a type of product which has been interpreted as a kind of energy. The simplification here is of the same type as that with the Maxwell stress tensor in the electromagnetic theory, but even more considerable. Similarly, in section 4 of the paper the classification of the Riemann tensor by Petrov, which we described above with the help of the electromagnetic analogy, is carried out in an extremely simple manner by means of spinors.

Of course, all these results are really only finding new ways of deriving relations about the curvature tensor which have been found already with much greater labour. But the fact that the labour can be cut down in this way is extremely significant in a subject in which so very much heavy computation is required, as can be seen in the last part of the paper where Penrose even applies his new ideas to the construction of analytic solutions. Moreover, the techniques employed here are essentially free of any coordinate system and they may therefore be applied to a problem of considerably greater subtlety and complexity than any mentioned up to now. All orthodox treatments of general relativity tend to begin with the assumption that the events to be described can be plotted in a four-dimensional differentiable manifold. Without going into technical details, this means that the space–time framework for the events is always *assumed* and, moreover, it is assumed to have comfortable smooth properties at every point, although, of course, with the proviso that if there is anything peculiar, such as a point particle, some kind of singularities can arise. It is only a matter of elementary prudence to remember, when beginning the subject in this way, that the attempts so far to quantize gravitation, i.e. to render it consistent with quantum mechanics, have all failed, and since quantization is a process for making a continuous theory consistent with the discreteness involved in quantum mechanics, one reason for this failure could be the overstrong assumption of the existence of the differentiable manifold. If this is so, then we need to construct the differentiable manifold, or rather something which approximates to it, explicitly as we go along instead of taking it for granted from the beginning (Penrose, 1967, 1968, 1971). The coordinate-free technique, with its freedom from the differentiable manifold assumption and its algebraic power, provides a possibility of doing this. It is too early yet to say more than that the possibility exists but if, indeed, it is successfully carried out in future years, it is likely to rest heavily on this paper of Penrose.

References

ADAMS, M. G. (1952) *Mon. Not. R. Astr. Soc.* **112,** 149.

ADAMS, M. G. (1955) *Mon. Not. R. Astr. Soc.* **115,** 405.

ADAMS, M. G. (1958) *Mon. Not. R. Astr. Soc.* **118,** 106.

ADAMS, M. G. (1959) *Non. Not. R. Astr. Soc.* **119,** 460.

BADE, W. L. and JEHLE, H. (1953) *Rev. Mod. Phys.* **25,** 714.

BEISBROEK, G. (1950) *Astr. J.* **55,** 49 and **55,** 247 (dealing with the 1947 eclipse).

BEISBROEK, G. (1953) *Astr. J.* **58,** 87 (dealing with the 1952 eclipse).

BONDI, H., VAN DER BERG, M. G. J. and METZNER, A. W. K. (1962) *Proc. Roy. Soc.* A, **269,** 21 (Extract 8 of the present book.)

BONNOR, W. B. (1954) *Z. Astrophys* **35,** 10.

BONNOR, W. B. (1959) *Phil. Trans.* A, **251,** 233.

CARMELI, M., FICKLER, S. I., WITTEN, L. (eds.) (1970) *Relativity*, Plenum Press, New York, London. A recent account of Weber's work is to be found on p. 133.

CHRISTOFFEL, E. B. (1869) *Z. reine angew. Math.* **70,** 46.

CLEMENCE, G. M. (1947) *Rev. Mod. Phys.* **19,** 361 gives the results for Mercury. It is also of interest to consult those for Venus, DUNCOMBE, R. L. (1956) *Astr. J.* **61,** 174; and for the Earth, MORGAN, H. P. (1945), *Astr. J.* **51,** 127.

CLIFFORD, W. K. (1873) *Nature (London)* **183,** 14 (Extract 1 of the present book).

CLIFFORD, W. K. (1876) *Proc. Camb. Phil. Soc.* **2,** 157 (Extract 2 of the present book).

CORSON, E. M. (1953) *Introduction to Tensors, Spinors and Relativistic Wave Equations*, Blackie, London.

DEBEVER, R. (1958) *Bull. Soc. math. Belg.* **10,** 112.

EDDINGTON, SIR A. S. (1924) *Mathematical Theory of Relativity*, Cambridge. (A suggestion for a relativistic use of a biquadratic metric is on p. 224 of the second edition.)

EINSTEIN, A. (1905) *Annln. Phys.* **17,** 891.

EINSTEIN, A. (1911) *Annln. Phys.* **35,** 898 (Extract 3 of the present book).

EINSTEIN, A. (1916) *Annln. Phys.* **49,** 769 (Extract 4 of the present book).

EINSTEIN, A. (1917) *Sber. preuss. Akad.*, p. 142.

EINSTEIN, A. (1927) *Sber. preuss. Akad.*, p. 235.

EINSTEIN, A. and GROMMER, J. (1927) *Sber. preuss. Akad.*, p. 2.

EINSTEIN, A. and INFELD, L. (1949) *Can. J. Math.* **1,** 209 (Extract 5 of the present book).

EINSTEIN, A., INFELD, L. and HOFFMANN, B. (1938) *Ann. Math.* **39,** 66.

FARADAY, M. (1852) *Royal Inst. Proc.* 11 June.

FOCK, V. (1957) *Rev. Mod. Phys.* **29**, 325 (Extract 6 of the present book).
GALILEO (1638) *Dialogue Concerning Two New Sciences*, Amsterdam (completed 1636). The relevant passage occurs as chapter 16 of *Galileo Galilei* by R. J. SEEGER, in this series, Pergamon, Oxford, 1966.
GOLDBERG, J. N. and SACHS, R. (1962) *Acta phys. pol.* **22**, 13.
HARRISON, K., THORNE, K., WAKANO, M. and WHEELER, J. A. (1965) *Gravitational Theory and Gravitational Collapse*, Chicago.
HOLTON, G. (1965) *Proceedings of the International Conference on Relativistic Theories of Gravitation*, Vol. 1, London. Gives a sketch of the complicated relationship between Einstein, Mach himself, and Mach's writings.
HUBBLE, K. (1929) *Proc. Natn. Acad. Sci. USA* **15**, 168.
INFELD, L. and VAN DER WAERDEN, B. L. (1933) *Sber. preuss. Akad.* **9**, 380.
LEVI-CIVITA, T. (1926) *Math. Annln.* **97**, 291.
LOVELOCK, D. (1970) *Archs. Ration Mech. Analysis* **36**, 293 (contains the essential theory).
LOVELOCK, D. (1971) *J. Math. Phys.* **12**, 498. (See also H. A. Buchdahl, *Proc. Comb. Phil. Soc.* **68**, 179 (1970)).
MACH, E. (1960) *Science of Mechanics*, translated T. J. MCCORMACK, Open Court.
MAXWELL, J. C. (1892) *Treatise on Electricity and Magnetism*, 3rd edn., Oxford.
MCCREA, W. H. (1964) *Astroph. norv.* **9**, 89.
MÖSSBAUER, R. L. (1958) *Z. Phys.* **151**, 124.
NEWMAN, E. and PENROSE, R. (1962) *J. Math. Phys.* **3**, 566.
NOETHER, E. (1918) *Göttingen Nach.*, p. 235.
OPPENHEIMER, J. R., and SNYDER, H. (1939) *Phys. Rev.* **56**, 455.
PENROSE, R. (1960) *Ann. Phys.* **10**, 171 (Extract 10 of the present book).
PENROSE, R. (1967) *J. Math. Phys.* **8**, 345.
PENROSE, R. (1968) *Int. J. Theor. Phys.* **1**, 61.
PENROSE, R. (1971) Angular momentum: an approach to combinatorial space–time in *Quantum Theory and Beyond*, (ed. Ted Bastin), Cambridge.
PETROV, A. Z. (1954) *Sci. Notices Kazan Univ.* **114**, 55.
PIRANI, F. A. E. (1956) *Acta phys. pol.* **22**, 13.
PIRANI, F. A. E. (1957) *Phys. Rev.* **105**, 1089 (Extract 7 of the present book).
POUND, R. V. and REBKA, G. A. (1959) *Phys. Rev. Letters* **3**, 439 (Extract 11 of the present book).
ROBINSON, I. and TRAUTMANN, A. (1962) *Proc. Roy. Soc* A, **265**, 463.
SACHS, R. (1961) *Proc. Roy. Soc* A, **264**, 309.
SCHWARZSCHILD, K. (1916) *Sber. preuss. Akad.*, p. 189.
SITTER, W. DE (1917) *Proc. Acad. Amst.* **19**, 1217 (and see also *Mon. Not. R. Astr. Soc.* **78**, 3).
TULCZYJEW, W. (1965) *Proceedings of the International Conference on Relativistic Theories of Gravitation*, Vol. 2, London.
WEBER, J. (1961) *General Relativity and Gravitational Waves*, New York.
WHEELER, J. A. (1962) *Geometrodynamics*, New York.
WIRTZ, C. (1922) *Astr. Nachr.* **216**, 451.

PART II

NOTES ON EXTRACT 1

EXTRACT 1 is a translation by W. K. Clifford of the introduction to Riemann's famous discussion on the foundations of differential geometry. Both Clifford and Riemann were completely familiar with certain striking properties of the differential geometry of surfaces, mostly due to Gauss, which are now less well known. These results, though hardly mentioned in Extract 1, are implicit in the whole discussion. These notes aim to re-write the results, on which Riemann was drawing, in a modern notation.

A surface S is a set of points with two degrees of freedom (just as a curve has one degree of freedom) and so any point $\mathbf{r}$ of S can be expressed in terms of two parameters. It is convenient to write $\mathbf{r} = \mathbf{r}(u^1, u^2)$, putting the suffixes on the parameters at the top. If we displace ourselves to a neighbouring point of the surface by making small changes in the parameters, the displacement is

$$d\mathbf{r} = \frac{\partial \mathbf{r}}{\partial u^1} du^1 + \frac{\partial \mathbf{r}}{\partial u^2} du^2 = \mathbf{r}_i \, du^i$$

with the summation convention over $i = 1, 2$, defining

$$\mathbf{r}_i = \frac{\partial \mathbf{r}}{\partial u^i}.$$

The length of this displacement is then

$$ds^2 = d\mathbf{r}^2 = \mathbf{r}_i \cdot \mathbf{r}_j \, du^i \, du^j = g_{ij} \, du^i \, du^j$$

(say), and it is important to notice that g_{ij} is an *intrinsic* property (that is to say, it does not depend on how S is embedded in the surrounding three-dimensional space). One can think of intrinsic properties conveniently in terms of what can be measured by imaginary beings living on S. In this picture, the parameters u^i must be thought of as a coordinate system on S and then g_{ij} can be measured in terms of the lengths of displacements between (u^1, u^2) and nearby points.

The two vectors $\mathbf{r}_i$ serve as two vectors in the tangent plane to S, and so define the tangent plane. Any vector $\mathbf{A}$ in the tangent plane can then be written

$$\mathbf{A} = \mathbf{r}_i A^i$$

for suitable components A^i, and its derivatives are

$$\mathbf{A}_{,j} = \frac{\partial \mathbf{A}}{\partial u^j} = \mathbf{r}_{ij} A^i + \mathbf{r}_i A^i_{,j},$$

where

$$\mathbf{r}_{ij} = \frac{\partial \mathbf{r}_i}{\partial u^j}.$$

We define

$$\mathbf{r}_{ij} = \Gamma^k_{ij}\mathbf{r}_k + l_{ij}\mathbf{n},$$

where $\mathbf{n}$ is written for the unit normal to the surface, so $\mathbf{r}_i \cdot \mathbf{n} = 0$. We would not expect either of the symmetric arrays of coefficients l_{ij} or Γ^k_{ij} to be intrinsic; in fact, l_{ij} is not, but it is surprising that Γ^k_{ij} is. This can be seen as follows:

$$\mathbf{r}_i \cdot \mathbf{r}_j = g_{ij}.$$

Hence, differentiating with respect to u^k,

$$(\Gamma^p_{ik}\mathbf{r}_p + l_{ik}\mathbf{n}) \cdot \mathbf{r}_j + \mathbf{r}_i \cdot (\Gamma^p_{jk}\mathbf{r}_p + l_{jk}\mathbf{n}) = g_{ij,\,k},$$

i.e.

$$g_{pj}\Gamma^p_{ik} + g_{ip}\Gamma^p_{jk} = g_{ij,\,k}.$$

Denoting for a moment, $g_{pj}\Gamma^p_{ik}$ by $[j]$ this has the form

$$[j]+[i] = g_{ij,\,k},$$

whence

$$[k]+[j] = g_{jk,\,i},$$

$$[i]+[k] = g_{ki,\,j},$$

and so

$$[k] = \tfrac{1}{2}(g_{ki,\,j} - g_{ij,\,k} + g_{jk,\,i}) = g_{pk}\Gamma^p_{ij}.$$

Assuming (as usual) that the determinant

$$g = |g_{pk}| \neq 0,$$

these equations can be solved and so Γ^p_{ij} is defined in terms of g_{ij} and its first derivatives, which we have already seen to be intrinsic.

The geometer is always interested in some measure of curvature of the surface (say, in terms of the radius of a suitable defined closest fitting sphere). A suitable measure can be defined by drawing curves on the surface and discussing their properties. Let $\mathbf{t}$ be the unit tangent, $d\mathbf{r}/ds$, of any curve on the surface, so that

$$\mathbf{t} = \mathbf{r}_i \frac{du^i}{ds}.$$

Let $\mathbf{n}$ be, as before, the unit normal to the surface and let $\mathbf{t}\wedge\mathbf{n} = \mathbf{b}$. Then $(\mathbf{t}, \mathbf{n}, \mathbf{b})$ forms a triad of unit vectors at right angles and so has an "angular velocity" (see Chapter 1) except that the independent variable here is s instead of t. Call

$$\Omega = \Omega_1\mathbf{t} + \Omega_2\mathbf{n} + \Omega_3\mathbf{b},$$

this "angular velocity", so that, multiplying out,

$$\mathbf{t}' = \Omega_3\mathbf{n} - \Omega_2\mathbf{b},$$

$$\mathbf{n}' = -\Omega_3\mathbf{t} + \Omega_1\mathbf{b},$$

$$\mathbf{b}' = \Omega_2\mathbf{t} - \Omega_1\mathbf{n},$$

where primes are used to denote differentiation with respect to arc length. From the second of these equations (which simply gives an expression for the rate of change of the normal to S), Ω_1 and Ω_3 must be properties of S related to the particular direction $\mathbf{t}$ of displacement, but not depending on the curve further than that. The properties of the particular curve (curvature and so on) are determined by Ω_2. Curves characterized by $\Omega_2 = 0$ at all points are called *geodesics*. Since $\mathbf{t}' = \mathbf{r}''$ is then normal to the surface, i.e. has no component in the tangent plane, it follows that the displacement of the projection of a point of the curve on to the tangent plane differs from that along the tangent by small quantities of the *third* order in the arc length, i.e. that the curvature of the projection (which depends on the second-order terms) vanishes at the point. In a well-defined sense, then, the geodesics are the "straightest" curves that can be drawn on the surface.

This characterization of geodesics, however, does not show that the property in question is intrinsic; i.e. it does not make clear that the "straightest" character can be judged by the beings dwelling on S. This will be clear, however, from the fact that, if $\Omega_2 = 0$, then $\mathbf{t}' = \mathbf{r}''$ is parallel to $\mathbf{n}$. Now

$$\mathbf{r}'' = \left(\mathbf{r}_i \frac{du^i}{ds}\right)' = \mathbf{r}_{ij} \frac{du^i}{ds}\frac{du^j}{ds} + \mathbf{r}_i \frac{d^2u^i}{ds^2}.$$

Using the expression above for $\mathbf{r}_{ij}$,

$$\mathbf{r}'' = r_p \left(\frac{d^2u^p}{ds^2} + \Gamma^p_{ij}\frac{du^i}{ds}\frac{du^j}{ds}\right) + l_{ij}\frac{du^i}{ds}\frac{du^j}{ds}\,\mathbf{n}.$$

From this we see at once that, if $\Omega_2 = 0$, then

$$\frac{d^2u^p}{ds^2} + \Gamma^p_{ij}\frac{du^i}{ds}\frac{du^j}{ds} = 0,$$

and since Γ^p_{ij} is an intrinsic array, the property of being a geodesic is intrinsic.

Further, the non-intrinsic part of the same equation, as it were, gives

$$\mathbf{r}''\cdot\mathbf{n} = \mathbf{t}'\cdot\mathbf{n} = \Omega_3 = l_{ij}\frac{du^i}{ds}\frac{du^j}{ds},$$

from the same equation. This may be used to investigate how Ω_3 depends on direction. Near any point P of S let us adopt such parameters u^i that the coordinate lines ($u^i =$ const) are at right angles and also displacement along each line is measured by the change of parameter along that line. That is,

$$\left(\frac{du^1}{ds}\right)^2 + \left(\frac{du^2}{ds}\right)^2 = 1,$$

so we can put

$$\frac{du^1}{ds} = \cos\theta, \quad \frac{du^2}{ds} = \sin\theta,$$

where θ is the angle between the direction in which Ω_3 is being measured and the axis of u^1. Then $\Omega_3 = l_{11}\cos^2\theta + 2l_{12}\cos\theta\sin\theta + l_{22}\sin^2\theta$.

This has a turning-value when $d\Omega_3/d\theta = 0$, i.e.

$$2(l_{22}-l_{11}) \sin\theta \cos\theta + 2l_{12} \cos 2\theta = 0,$$

or

$$\tan 2\theta = \frac{2l_{12}}{l_{11}-l_{22}}.$$

Notice that, except in the special case when $l_{11} = l_{22}$ and $l_{12} = 0$ (in which case Ω_3 is independent of θ), there are always two directions at right angles that satisfy this equation. Now let these two directions (the principal directions of curvature) be chosen as the two coordinate directions at P. Then, in the new coordinate system

$$\Omega_3 = \varkappa_1 \cos^2\theta + \varkappa_2 \sin^2\theta,$$

where $\varkappa_1$, $\varkappa_2$ are the values of l_{11}, l_{22} in the new expression, and obviously the new $l_{12} = 0$. Here $\varkappa_1$, $\varkappa_2$ are called the principal curvatures, and this expression for the curvature in any direction in terms of them is Euler's theorem. But, as noted at the beginning of the paragraph, $\varkappa_1$, $\varkappa_2$ are not intrinsic properties.

The problem to which Gauss directed his attention in this field was to find an intrinsic quantity which was a measure of the curvature. Such a quantity must be a function of $\varkappa_1$, $\varkappa_2$, from Euler's theorem. Gauss's investigation is on the following lines.

Returning to the equation, in general coordinates,

$$\Omega_3 = l_{ij} \frac{du^i}{ds} \frac{du^j}{ds},$$

it is convenient to use the fact that $ds^2 = g_{ij}\, du^i\, du^j$, and so write

$$\Omega_3 = \frac{l_{pq}\, du^p\, du^q}{g_{ij}\, du^i\, du^j}.$$

The corresponding turning values can now be derived by remarking that, for any arbitrarily chosen Ω_3, the direction (du^1, du^2) is determined by

$$(l_{pq} - \Omega_3 g_{pq})\, du^p\, du^q = 0$$

a quadratic in du^1/du^2. The turning values arise when this has equal roots, and the condition for this is the vanishing of the determinant

$$|l_{pq} - \Omega_3\, g_{pq}| = 0$$

which gives a quadratic equation

$$g\Omega_3^2 - 2(l, g)\Omega_3 + l = 0,$$

if we write g, l for the determinants $|g_{pq}|$, $|l_{pq}|$, and (l, g) for the remaining coefficient. The roots of this equation are $\varkappa_1$, $\varkappa_2$ and we have

$$H = \frac{1}{2}(\varkappa_1 + \varkappa_2) = \frac{(l, g)}{g}, \quad G = \varkappa_1 \varkappa_2 = \frac{l}{g}.$$

H is called the mean curvature, G the Gaussian curvature.

Now, in fact, H is not intrinsic, but Gauss was able to show that G was, and so G provides the sought-for intrinsic measure of curvature. This can be seen as follows:
From the expression

$$\mathbf{r}_{ij} = \Gamma^{p}_{ij}\mathbf{r}_p + l_{ij}\mathbf{n}$$

by differentiating with respect to u^k,

$$\mathbf{r}_{ij,k} = \Gamma^{p}_{ij,k}\mathbf{r}_p + \Gamma^{p}_{ij}\Gamma^{q}_{pk}\mathbf{r}_q + \Gamma^{p}_{ij}l_{pk}\mathbf{n} + \mathbf{n}l_{ij,k} + l_{ij}\mathbf{n}_{,k}.$$

Interchanging j, k and subtracting,

$$0 = \mathbf{r}_q R^{q}_{ikj} + (\Gamma^{p}_{ij}l_{pk} - \Gamma^{p}_{ik}l_{pj})\mathbf{n} + l_{ij}\mathbf{n}_{,k} - l_{ik}\mathbf{n}_{,j},$$

where we have written

$$R^{q}_{ikj} = \Gamma^{q}_{ij,k} - \Gamma^{q}_{ik,j} + \Gamma^{p}_{ij}\Gamma^{q}_{pk} - \Gamma^{p}_{ik}\Gamma^{q}_{pj},$$

an expression which is evidently intrinsic, since Γ_{ij} is intrinsic. Considering the components of this expression in the tangent plane, it follows that, in particular

$$\mathbf{r}_q R^{q}_{212} + [l_{22}\mathbf{n}_{,1} - l_{12}\mathbf{n}_{,2}] = 0,$$

where the square brackets denote that only the component in the tangent plane is taken.

Now $$\mathbf{n} = \phi\mathbf{r}_1 \wedge \mathbf{r}_2,$$

where $$\phi^{-2} = (\mathbf{r}_1 \wedge \mathbf{r}_2)^2 = \mathbf{r}_1^2\mathbf{r}_2^2 - (\mathbf{r}_1 \cdot \mathbf{r})_2^2$$
$$= g_{11}g_{22} - g_{12}^2 = g$$

(again writing g for $|g_{ij}|$), so that

$$\mathbf{n}\sqrt{(g)} = \mathbf{r}_1 \wedge \mathbf{r}_2.$$

Differentiating,

$$\sqrt{(g)}[\mathbf{n}\sqrt{(g)}]_{,k} = l_{1k}(\mathbf{r}_1 \wedge \mathbf{r}_2) \wedge \mathbf{r}_2 + l_{2k}\mathbf{r}_1 \wedge (\mathbf{r}_1 \wedge \mathbf{r}_2)$$
$$= (l_{2k}g_{12} - l_{1k}g_{22})\mathbf{r}_1 + (l_{1k}g_{12} - l_{2k}g_{11})\mathbf{r}_2.$$

Hence, in all,

$$g[l_{22}\mathbf{n}_{,1} - l_{12}\mathbf{n}_{,2}] = g_{12}(l_{11}l_{22} - l_{12}^2)\mathbf{r}_2 - g_{22}(l_{11}l_{22} - l_{12}^2)\mathbf{r}_1.$$

Taking scalar products of the original equation with $\mathbf{r}_1$, $\mathbf{r}_2$ respectively gives

$$g_{1q}R^{q}_{212} + \frac{1}{g}(g_{12}^2 - g_{11}g_{22})(l_{11}l_{22} - l_{12}^2) = 0,$$

i.e. $$l = g_{1q}R^{q}_{212} \quad \text{and} \quad g_{2q}R^{2}_{212} = 0.$$

Hence $$G = \frac{l}{g} = \frac{g_{1q}R^{q}_{212}}{g},$$

which is evidently an intrinsic expression. Gauss, who was not noted for overrating his own work, described this as *theorema egregium*, a theorem out of the common herd.

EXTRACT I†

On the Hypotheses which Lie at the Bases of Geometry

W. K. CLIFFORD

[Translation of a paper by Riemann]

Plan of the investigation

IT is known that geometry assumes, as things given, both the notion of space and the first principles of constructions in space. She gives definitions of them which are merely nominal, while the true determinations appear in the form of axioms. The relation of these assumptions remains consequently in darkness; we neither perceive whether and how far their connection is necessary, nor, *a priori*, whether it is possible.

From Euclid to Legendre (to name the most famous of modern reforming geometers) this darkness was cleared up neither by mathematicians nor by such philosophers as concerned themselves with it. The reason of this is doubtless that the general notion of multiply extended magnitudes (in which space-magnitudes are included) remained entirely unworked. I have in the first place, therefore, set myself the task of constructing the notion of a multiply extended magnitude out of general notions of magnitude. It will follow from this that a multiply extended magnitude is capable of different measure-relations, and consequently that space is only a particular case of a triply extended magnitude. But hence flows as a necessary conse-

† *Nature* (*London*) **183,** 14 (1873).

quence that the propositions of geometry cannot be derived from general notions of magnitude, but that the properties which distinguish space from other conceivable triply extended magnitudes are only to be deduced from experience. Thus arises the problem, to discover the simplest matters of fact from which the measure-relations of space may be determined; a problem which from the nature of the case is not completely determinate, since there may be several systems of matters of fact which suffice to determine the measure-relations of space—the most important system for our present purpose being that which Euclid has laid down as a foundation. These matters of fact are—like all matters of fact—not necessary, but only of empirical certainty; they are hypotheses. We may therefore investigate their probability, which within the limits of observation is of course very great, and inquire about the justice of their extension beyond the limits of observation, on the side both of the infinitely great and of the infinitely small.

I. Notion of an n-ply extended magnitude

In proceeding to attempt the solution of the first of these problems, the development of the notion of a multiply extended magnitude, I think I may the more claim indulgent criticism in that I am not practised in such undertakings of a philosophical nature where the difficulty lies more in the notions themselves than in the construction; and that besides some very short hints on the matter given by Privy Councillor Gauss in his second memoir on Biquadratic Residues, in the *Göttingen Gelehrte Anzeige*, and in his Jubilee-book, and some philosophical researches of Herbart, I could make use of no previous labours.

§ 1. Magnitude-notions are only possible where there is an antecedent general notion which admits of different specialisations. According as there exists among these specialisations a continuous path from one to another or not, they form a *continuous* or *discrete* manifoldness: the individual specialisations are called in the first case points, in the second case elements, of the manifoldness. Notions whose specialisations form a *discrete* manifoldness are so common that at least in the cultivated languages any things being given it is always possible to find a notion in which they are included. (Hence mathematicians might unhesitatingly found the theory of discrete magnitudes upon the postulate that certain given things are to be regarded as equivalent.) On the other hand, so few and far between are the occasions for forming notions whose specialisations make up a *continuous* manifoldness, that the only simple notions whose specialisations form a multiply extended manifoldness are the positions of perceived objects and colours. More frequent occasions for the creation and development of these notions occur first in the higher mathematic.

Definite portions of a manifoldness, distinguished by a mark or by a boundary, are called Quanta. Their comparison with regard to quantity is accomplished in the case of discrete magnitudes by counting, in the case of continuous magnitudes by measuring. Measure consists in the superposition of the magnitudes to be compared; it therefore requires a means of using one magnitude as the standard for another. In the absence of this, two magnitudes can only be compared when one is a part of the other; in which case also we can only determine the more or less and not the how much. The researches which can in this case be instituted about them form a general division of the science of magnitude in which magnitudes are regarded not as existing independently of position and not as expressible in terms of a unit, but as regions in a manifoldness. Such researches have become a necessity for many parts of mathematics, e.g., for the treatment of many-valued analytical functions; and the want of them is no doubt a chief cause why the celebrated theorem of Abel and the achievements of Lagrange, Pfaff, Jacobi for the general theory of differential equations, have so long remained unfruitful. Out of this general part of the science of extended magnitude in which nothing is assumed but what is contained in the notion of it, it will suffice for the present purpose to bring into prominence two points; the first of which relates to the construction of the notion of a multiply extended manifoldness, the second relates to the reduction of determinations of place in a given manifoldness to determinations of quantity, and will make clear the true character of an n-fold extent.

§ 2. If in the case of a notion whose specialisations form a continuous manifoldness, one passes from a certain specialisation in a definite way to another, the specialisations passed over form a simply extended manifoldness, whose true character is that in it a continuous progress from a point is possible only on two sides, forwards or backwards. If one now supposes that this manifoldness in its turn passes over into another entirely different, and again in a definite way, namely so that each point passes over into a definite point of the other, then all the specialisations so obtained form a doubly extended manifoldness. In a similar manner one obtains a triply extended manifoldness, if one imagines a doubly extended one passing over in a definite way to another entirely different; and it is easy to see how this construction may be continued. If one regards the variable object instead of the determinable notion of it, this construction may be described as a composition of a variability of $n+1$ dimensions out of a variability of n dimensions and a variability of one dimension.

§ 3. I shall now show how conversely one may resolve a variability whose region is given into a variability of one dimension and a variability of fewer dimensions. To this end let us suppose a variable piece of a manifoldness of one dimension—reckoned from a fixed origin, that the values of it may be comparable with one another—which has for every point of the given manifoldness a definite value, varying continuously with the point; or, in other words, let us take a continuous function of position within the given manifoldness, which, moreover, is not constant throughout any part of that manifoldness. Every system of points where the function has a constant value, forms then a continuous manifoldness of fewer dimensions than the given one. These manifoldnesses pass over continuously into one another as the function changes; we may therefore assume that out of one of them the others proceed, and speaking generally this may occur in such a way that each point passes over into a definite point of the other; the cases of exception (the study of which is important) may here be left unconsidered. Hereby the determination of position in the given manifoldness is reduced to a determination of quantity and to a determination of position in a manifoldness of less dimensions. It is now easy to show that this manifoldness has $n-1$ dimensions when the given manifoldness is n-ply extended. By repeating then this operation n times, the determination of position in an n-ply extended manifoldness is reduced to n determinations of quantity, and therefore the determination of position in a given manifoldness is reduced to a finite number of determinations of quantity *when this is possible.* There are manifoldnesses in which the determination of position requires not a finite number, but either an endless series or a continuous manifoldness of determinations of quantity. Such manifoldnesses are, for example, the possible determinations of a function for a given region, the possible shapes of a solid figure, etc.

II. Measure-relations of which a manifoldness of *n* dimensions is capable on the assumption that lines have a length independent of position, and consequently that every line may be measured by every other

Having constructed the notion of a manifoldness of n dimensions, and found that its true character consists in the property that the determination of position in it may be reduced to n determinations of magnitude, we come to the second of the problems proposed above, viz. the study of the measure-relations of which such a manifoldness is capable, and of the conditions which suffice to determine them. These measure-relations can only be studied in abstract notions of quantity, and their dependence on one another can only be represented by formulae. On certain assumptions, however, they are decomposable into relations which, taken separately, are capable of geometric representation; and thus it becomes possible to express geometrically the calculated results. In this way, to come to solid ground, we cannot, it is true, avoid abstract considerations in our formulae, but at least the results of calculation may subsequently be presented in a geometric form. The foundations of these two parts of the question are established in the celebrated memoir of Gauss, *Disquisitiones generales circa superficies curvas.*

§ 1. Measure-determinations require that quantity should be independent of position, which may happen in various ways. The hypothesis which first presents itself, and which I shall here develop, is that according to which the length of lines is independent of their position, and consequently every line is measurable by means of every other. Position-fixing being reduced to quantity-fixings, and the position of a point in the n-dimensioned manifoldness being consequently expressed by means of n variables $x_1, x_2, x_3, \ldots x_n$, the determination of a line comes to the giving of these quantities as functions of one variable. The problem consists then in establishing a mathematical expression for the length of a line, and to this end we must consider the quantities x as expressible in terms of certain units. I shall treat this problem only under certain restrictions, and I shall confine myself in the first place to lines in which the ratios of the increments dx of the respective variables vary continuously. We may then conceive these lines broken up into elements, within which the ratios of the quantities dx may be regarded as constant; and the problem is then reduced to establishing for each point a general expression for the linear element ds starting from that point, an expression which will thus contain the quantities x and the quantities dx. I shall suppose, secondly, that the length of the linear element, to the first order, is unaltered when all the points of this element undergo the same infinitesimal displacement, which implies at the same time that if all the quantities dx are increased in the same ratio, the linear element will vary also in the same ratio. On these suppositions, the linear element may be any homogeneous function of the first degree of the quantities dx, which is unchanged when we change the signs of all the dx, and in which the arbitrary constants are continuous functions of the quantities x. To find the simplest cases, I shall seek first an expression for manifoldnesses of $n-1$ dimensions which are everywhere equidistant from the origin of the linear element; that is, I shall seek a continuous function of position whose values distinguish them from one another. In going outwards from the origin, this must either increase in all directions or decrease in all directions; I assume that it increases in all directions, and therefore has a minimum at that point. If, then, the first and second differential coefficients of this function are finite, its first differential must vanish, and the second differential cannot become negative; I assume that it is always positive. This differential expression, then, of the second order remains constant when *ds* remains constant, and increases in the duplicate ratio when the *dx*, and therefore also *ds*, increase in the same ratio; it must therefore be ds^2 multiplied by a constant, and consequently *ds* is the square root of an always positive integral homogeneous function of the second order of the quantities *dx*, in which the coefficients are continuous functions of the quantities *x*. For Space, when the position of points is expressed by rectilinear co-ordinates, $ds = \sqrt{\Sigma(dx)^2}$; Space is therefore included in this simplest case. The next case in simplicity includes those manifoldnesses in which the line-element may be expressed as the fourth root of a quartic differential expression. The investigation of this more general kind would require no really different principles, but would take considerable time and throw little new light on the theory of space, especially as the results cannot be geometrically expressed; I restrict myself, therefore, to those manifoldnesses in which the line-element is expressed as the square root of a quadric differential expression. Such an expression we can transform into another similar one if we substitute for the *n* independent variables functions of *n* new independent variables. In this way, however, we cannot transform any expression into any other; since the expression contains $\frac{1}{2}n(n+1)$ coefficients which are arbitrary functions of the independent variables; now by the introduction of new variables we can only satisfy *n* conditions, and therefore make no more than *n* of the coefficients equal to given quantities. The remaining $\frac{1}{2}n(n-1)$ are then entirely determined by the nature of the continuum to be represented, and consequently $\frac{1}{2}n(n-1)$ functions of positions are required for the determination of its measure-relations. Manifoldnesses in which, as in the Plane and in Space, the line-element may be reduced to the form $\sqrt{(\Sigma\, dx^2)}$, are therefore only a particular case of the manifoldnesses to be here investigated; they require a special name, and therefore these manifoldnesses in which the square of the line-element may be expressed as the sum of the squares of complete differentials

I will call *flat*. In order now to review the true varieties of all the continua which may be represented in the assumed form, it is necessary to get rid of difficulties arising from the mode of representation, which is accomplished by choosing the variables in accordance with a certain principle.

§ 2. For this purpose let us imagine that from any given point the system of shortest lines going out from it is constructed; the position of an arbitrary point may then be determined by the initial direction of the geodesic in which it lies, and by its distance measured along that line from the origin. It can therefore be expressed in terms of the ratios dx_0 of the quantities dx in this geodesic, and of the length s of this line. Let us introduce now instead of the dx_0 linear functions dx of them, such that the initial value of the square of the line-element shall equal the sum of the squares of these expressions, so that the independent variables are now the length s and the ratios of the quantities dx. Lastly, take instead of the dx quantities $x_1, x_2, x_3, \ldots x_n$ proportional to them, but such that the sum of their squares $= s^2$. When we introduce these quantities, the square of the line-element is $\Sigma\, dx^2$ for infinitesimal values of the x, but the term of next order in it is equal to a homogeneous function of the second order of the $\frac{1}{2}n(n-1)$ quantities $(x_1dx_2-x_2dx_1), (x_1dx_3-x_3dx_1)\ldots$ an infinitesimal, therefore, of the fourth order; so that we obtain a finite quantity on dividing this by the square of the infinitesimal triangle, whose vertices are $(0, 0, 0, \ldots), (x_1, x_2, x_3, \ldots), (dx_1, dx_2, dx_3, \ldots)$. This quantity retains the same value so long as the x and the dx are included in the same binary linear form, or so long as the two geodesics from 0 to x and from 0 to dx remain in the same surface-element; it depends therefore only on place and direction. It is obviously zero when the manifold represented is flat, i.e., when the squared line-element is reducible to Σdx^2, and may therefore be regarded as the measure of the deviation of the manifoldness from flatness at the given point in the given surface-direction. Multiplied by $-\frac{3}{4}$ it becomes equal to the quantity which Privy Councillor Gauss has called the total curvature of a surface. For the determination of the measure-relations of a manifoldness capable of representation in the assumed form we found that $\frac{1}{2}n(n-1)$ place-functions were necessary; if, therefore, the curvature at each point in $\frac{1}{2}n(n-1)$ surface-directions is given, the measure-relations of the continuum may be determined from them—provided there be no identical relations among these values, which in fact, to speak generally, is not the case. In this way the measure-relations of a manifoldness in which the line-element is the square root of a quadric differential may be expressed in a manner wholly independent of the choice of independent variables. A method entirely similar may for this purpose be applied also to the manifoldness in which the line-element has a less simple expression, e.g., the fourth root of a quartic differential. In this case the line-element, generally speaking, is no longer reducible to the form of the square root of a sum of squares, and therefore the deviation from flatness in the squared line-element is an infinitesimal of the second order, while in those manifoldnesses it was of the fourth order. This property of the last-named continua may thus be called flatness of the smallest parts. The most important property of these continua for our present purpose, for whose sake alone they are here investigated, is that the relations of the twofold ones may be geometrically represented by surfaces, and of the morefold ones may be reduced to those of the surfaces included in them; which now requires a short further discussion.

§ 3. In the idea of surfaces, together with the intrinsic measure-relations in which only the length of lines on the surfaces is considered, there is always mixed up the position of points lying out of the surface. We may, however, abstract from external relations if we consider such deformations as leave unaltered the length of lines—i.e., if we regard the surface as bent in any way without stretching, and treat all surfaces so related to each other as equivalent. Thus, for example, any cylindrical or conical surface counts as equivalent to a plane, since it may be made out of one by mere bending, in which the intrinsic measure-relations remain, and all theorems about a plane—therefore the whole of planimetry—retain their validity. On the other hand they count as essentially different from the sphere, which cannot be changed into a plane without stretching. According to our previous investigation the intrinsic measure-relations of a twofold extent in which the line-element may be expressed as the square root of a quadric differential, which is the case with surfaces, are characterised by the total curvature. Now this quantity in the case of surfaces is capable of a visible interpretation, viz., it is the product of the two curvatures of the surface, or multiplied by the area of a small geodesic triangle, it is equal to the spherical excess of the same. The first definition assumes the proposition that the product of the two radii of curvature is unaltered by mere bending; the second, that in the same place the area of a small triangle is proportional to its spherical excess. To give an intelligible meaning to the curvature of an n-fold extent at a given point and in a given surface-direction through it, we must start from the fact that a geodesic proceeding from a point is entirely determined when its initial direction is given. According to this we obtain a determinate surface if we prolong all the geodesics proceeding from the given point and lying initially in the given surface-direction; this surface has at the given point a definite curvature, which is also the curvature of the n-fold continuum at the given point in the given surface-direction.

§ 4. Before we make the application to space, some considerations about flat manifoldnesses in general are necessary; i.e., about those in which the square of the line-element is expressible as a sum of squares of complete differentials.

In a flat n-fold extent the total curvature is zero at all points in every direction; it is sufficient, however (according to the preceding investigation), for the determination of measure-relations, to know that at each point the curvature is zero in $\frac{1}{2}n(n-1)$ independent surface-directions. Manifoldnesses whose curvature is constantly zero may be treated as a special case of those whose curvature is constant. The common character of these continua whose curvature is constant may be also expressed thus, that figures may be moved in them without stretching. For clearly figures could not be arbitrarily shifted and turned round in them if the curvature at each point were not the same in all directions. On the other hand, however, the measure-relations of the manifoldness are entirely determined by the curvature; they are therefore exactly the same in all directions at one point as at another, and consequently the same constructions can be made from it: whence it follows that in aggregates with constant curvature figures may have any arbitrary position given them. The measure-relations of these manifoldnesses depend only on the value of the curvature, and in relation to the analytic expression it may be remarked that if this value is denoted by α, the expression for the line-element may be written

$$\frac{1}{1+\frac{1}{4}\alpha\Sigma x^2}\sqrt{(\Sigma\, dx^2)}.$$

§ 5. The theory of *surfaces* of constant curvature will serve for a geometric illustration. It is easy to see that surfaces whose curvature is positive may always be rolled on a sphere whose radius is unity divided by the square root of the curvature; but to review the entire manifoldness of these surfaces, let one of them have the form of a sphere and the rest the form of surfaces of revolution touching it at the equator. The surfaces with greater curvature than this sphere will then touch the sphere internally, and take a form like the outer portion (from the axis) of the surface of a ring; they may be rolled upon zones of spheres having less radii, but will go round more than once. The surfaces with less positive curvature are obtained from spheres of larger radii, by cutting out the lune bounded by two great half-circles and bringing the section-lines together. The surface with curvature zero will be a cylinder standing on the equator; the surfaces with negative curvature will touch the cylinder externally and be formed like the inner portion (towards the axis) of the surface of a ring. If we regard these surfaces as *locus in quo* for surface-regions moving in them, as Space is *locus in quo* for bodies, the surface-regions can be moved in all these surfaces without stretching. The surfaces with positive curvature can always be so formed that surface-regions may also be moved arbitrarily about upon them without *bending*, namely (they may be formed) into sphere-surfaces; but not those with negative curvature. Besides this independence of surface-regions from position there is in surfaces of zero curvature also an independence of *direction* from position, which in the former surfaces does not exist.

III. Application to Space

§ 1. By means of these inquiries into the determination of the measure-relations of an n-fold extent the conditions may be declared which are necessary and sufficient to determine the metric properties of space, if we assume the independence of line-length from position and expressibility of the line-element as the square root of a quadric differential, that is to say, flatness in the smallest parts.

First, they may be expressed thus: that the curvature at each point is zero in three surface-directions; and thence the metric properties of space are determined if the sum of the angles of a triangle is always equal to two right angles.

Secondly, if we assume with Euclid not merely an existence of lines independent of position, but of bodies also, it follows that the curvature is everywhere constant; and then the sum of the angles is determined in all triangles when it is known in one.

Thirdly, one might, instead of taking the length of lines to be independent of position and direction, assume also an independence of their length and direction from position. According to this conception changes or differences of position are complex magnitudes expressible in three independent units.

§ 2. In the course of our previous inquiries, we first distinguished between the relations of extension or partition and the relations of measure, and found that with the same extensive properties, different measure-relations were conceivable; we then investigated the system of simple size-fixings by which the measure-relations of space are completely determined, and of which all propositions about them are a necessary consequence; it remains to discuss the question how, in what degree, and to what extent these assumptions are borne out by experience. In this respect there is a real distinction between mere extensive relations, and measure-relations; in so far as in the former, where the possible cases form a discrete manifoldness, the declarations of experience are indeed not quite certain, but still not inaccurate; while in the latter, where the possible cases form a continuous manifoldness, every determination from experience remains always inaccurate: be the probability ever so great that it is nearly exact. This consideration becomes important in the extensions of these empirical determinations beyond the limits of observation to the infinitely great and infinitely small; since the latter may clearly become more inaccurate beyond the limits of observation, but not the former.

In the extension of space-construction to the infinitely great, we must distinguish between *unboundedness* and *infinite extent*, the former belongs to the extent relations, the latter to the measure-relations. That space is an unbounded three-fold manifoldness, is an assumption which is developed by every conception of the outer world; according to which every instant the region of real perception is completed and the possible positions of a sought object are constructed, and which by these applications is for ever confirming itself. The unboundedness of space possesses in this way a greater empirical certainty than any external experience. But its infinite extent by no means follows from this; on the other hand if we assume independence of bodies from position, and therefore ascribe to space constant curvature, it must necessarily be finite provided this curvature has ever so small a positive value. If we prolong all the geodesics starting in a given surface-element, we should obtain an unbounded surface of constant curvature, i.e., a surface which in a *flat*-manifoldness of three dimensions would take the form of a sphere, and consequently be finite.

§ 3. The questions about the infinitely great are for the interpretation of nature useless questions. But this is not the case with the questions about the infinitely small. It is upon the exactness with which we follow phenomena into the infinitely small that our knowledge of their causal relations essentially depends. The progress of recent centuries in the knowledge of mechanics depends almost entirely on the exactness of the construction which has become possible through the invention of the infinitesimal calculus, and through the simple principles discovered by Archimedes, Galileo, and Newton, and used by modern physic. But in the natural sciences which are still in want of simple principles for such constructions, we seek to discover the causal relations by following the phenomena into great minuteness, so far as the microscope permits. Questions about the measure-relations of space in the infinitely small are not therefore superfluous questions.

If we suppose that bodies exist independently of position, the curvature is everywhere constant, and it then results from astronomical measurements that it cannot be different from zero; or at any rate its reciprocal must be an area in comparison with which the range of our telescopes may be neglected. But if this independence of bodies from position does not exist, we cannot draw conclusions from metric relations of the great, to those of the infinitely small; in that case the curvature at each point may have an arbitrary value in three directions, provided that the total curvature of every measurable portion of space does not differ sensibly from zero. Still more complicated relations may exist if we no longer suppose the linear element expressible as the square root of a quadric differential. Now it seems that the empirical notions on which the metrical determinations of space are founded, the notion of a solid body and of a ray of light, cease to be valid for the infinitely small. We are therefore quite at liberty to suppose that the metric relations of space in the infinitely small do not conform to the hypotheses of geometry; and we ought in fact to suppose it, if we can thereby obtain a simpler explanation of phenomena.

The question of the validity of the hypotheses of geometry in the infinitely small is bound up with the question of the ground of the metric relations of space. In this last question, which we may still regard as belonging to doctrine of space, is found the application of the remark made above; that in a discrete manifoldness, the ground of its metric relations is given in the notion of it, while in a continuous manifoldness, this ground must come from outside. Either therefore the reality which underlies space must form a discrete manifoldness, or we must seek the ground of its metric relations outside it, in binding forces which act upon it.

The answer to these questions can only be got by starting from the conception of phenomena which has hitherto been justified by experience, and which Newton assumed as a foundation, and by making in this conception the successive changes required by facts which it cannot explain. Researches starting from general notions, like the investigation we have just made, can only be useful in preventing this work from being hampered by too narrow views, and progress in knowledge of the interdependence of things from being checked by traditional prejudices.

This leads us into the domain of another science, of physic, into which the object of this work does not allow us to go to-day.

Synopsis

PLAN of the Inquiry:

I. Notion of an *n*-ply extended magnitude.

§ 1. Continuous and discrete manifoldnesses. Defined parts of a manifoldness are called Quanta. Division of the theory of continuous magnitude into the theories,

(1) Of mere region-relations, in which an independence of magnitudes from position is not assumed;

(2) Of size-relations, in which such an independence must be assumed.

§ 2. Construction of the notion of a one-fold, two-fold, *n*-fold extended magnitude.

§ 3. Reduction of place-fixing in a given manifoldness to quantity-fixings. True character of an *n*-fold extended magnitude.

II. Measure-relations of which a manifoldness of *n*-dimensions is capable on the assumption that lines have a length independent of position, and consequently that every line may be measured by every other.

§ 1. Expression for the line-element. Manifoldnesses to be called Flat in which the line-element is expressible as the square root of a sum of squares of complete differentials.

§ 2. Investigation of the manifoldness of *n*-dimensions in which the line-element may be represented as the square root of a quadric differential. Measure of its deviation from flatness (curvature) at a given point in a given surface-direction. For the determination of its measure-relations it is allowable and sufficient that the curvature be arbitrarily given at every point in $\frac{1}{2}n(n-1)$ surface directions.

§ 3. Geometric illustration.

§ 4. Flat manifoldnesses (in which the curvature is everywhere = 0) may be treated as a special case of manifoldnesses with constant curvature. These can also be defined as admitting an independence of n-fold extents in them from position (possibility of motion without stretching).

§ 5. Surfaces with constant curvature.

III. Application to Space.

§ 1. System of facts which suffice to determine the measure-relations of space assumed in geometry.

§ 2. How far is the validity of these empirical determinations probable beyond the limits of observation towards the infinitely great?

§ 3. How far towards the infinitely small? Connection of this question with the interpretation of nature.

NOTES ON EXTRACT 2

CLIFFORD here goes some distance, in a purely speculative manner, building on his study of Riemann's paper (Extract 1 above), towards a physical theory like general relativity. His idea is that curvature represents matter, and its change represents the motion of matter.

EXTRACT 2[†]

On the Space-theory of Matter

W. K. CLIFFORD

(Abstract)

Riemann has shewn that as there are different kinds of lines and surfaces, so there are different kinds of space of three dimensions; and that we can only find out by experience to which of these kinds the space in which we live belongs. In particular, the axioms of plane geometry are true within the limits of experiment on the surface of a sheet of paper, and yet we know that the sheet is really covered with a number of small ridges and furrows, upon which (the total curvature not being zero) these axioms are not true. Similarly, he says although the axioms of solid geometry are true within the limits of experiment for finite portions of our space, yet we have no reason to conclude that they are true for very small portions; and if any help can be got thereby for the explanation of physical phenomena, we may have reason to conclude that they are not true for very small portions of space.

I wish here to indicate a manner in which these speculations may be applied to the investigation of physical phenomena. I hold in fact

(1) That small portions of space *are* in fact of a nature analogous to little hills on a surface which is on the average flat; namely, that the ordinary laws of geometry are not valid in them.

(2) That this property of being curved or distorted is continually being passed on from one portion of space to another after the manner of a wave.

† *Proc. Camb. Phil. Soc.* **2,** 157 (1876).

(3) That this variation of the curvature of space is what really happens in that phenomenon which we call the *motion of matter*, whether ponderable or etherial.

(4) That in the physical world nothing else takes place but this variation, subject (possibly) to the law of continuity.

I am endeavouring in a general way to explain the laws of double refraction on this hypothesis, but have not yet arrived at any results sufficiently decisive to be communicated.

NOTES ON EXTRACT 3

In this 1911 paper Einstein has progressed a great way towards understanding the physical basis of general relativity, but he has presumably not worked out the mathematics to his satisfaction since he proposes to submit "only a few quite elementary reflections". The technical details were to take four more years to work out.

EXTRACT 3†

On the Effect of Gravitation on the Propagation of Light

A. EINSTEIN

IN A memoir published 4 years ago[1] I tried to answer the question whether the propagation of light is influenced by gravitation. I return to this theme because my previous presentation of the subject does not satisfy me, and, moreover, because I now see that one of the most important consequences of my former treatment can be tested experimentally. It follows from the present theory that rays of light passing close to the sun are deflected by its gravitational field, so that the angular distance between the sun and a fixed star appearing near to it is apparently increased by nearly a second of arc.

In the course of these considerations, further results arise which relate to gravitation. But as the exposition of the whole matter would be rather difficult to follow, only a few quite elementary reflections will be given in the following pages from which the reader will readily be able to see the suppositions of the theory and its line of thought. The relations here deduced, even if the theoretical foundation is sound, are valid only in the first approximation.

§ 1. A Hypothesis on the Physical Nature of the Gravitational Field

In a homogeneous gravitational field (acceleration of gravity γ) let there be a stationary system of co-ordinates K, orientated so that the

† *Annln. Phys.* **35,** 898 (1911).

[1] A. Einstein, *Jahrbuch für Radioakt. und Elektronik*, 4, 1907.

lines of force of the gravitational field run in the negative direction of the z-axis. In a space free of gravitational fields let there be a second system of co-ordinates K', moving with uniform acceleration γ in the positive direction of its z-axis. To avoid unnecessary complications, let us disregard the theory of relativity and regard both systems from the customary point of view of kinematics and the movements occurring in them from that of ordinary mechanics.

Relative to K, as well as relative to K', material points which are not subjected to the action of other material points move in keeping with the equations

$$\frac{d^2x}{dt^2} = 0, \quad \frac{d^2y}{dt^2} = 0, \quad \frac{d^2z}{dt^2} = -\gamma.$$

For the accelerated system K' this follows directly from Galileo's principle, but for the system K, at rest in a homogeneous gravitational field, from the experience that all bodies in such a field are equally and uniformly accelerated. This experience, of the equal fall of all bodies in the gravitational field, is one of the most universal which the observation of nature has yielded; but in spite of that the law has not found any place in the foundations of our physical picture of the world.

But we arrive at a very satisfactory interpretation of this law of experience if we assume that the systems K and K' are physically exactly equivalent, that is, if we assume that we may just as well regard the system K as being in a space free from gravitational fields, if we then regard K as uniformly accelerated. This assumption of exact physical equivalence makes it impossible for us to speak of the absolute acceleration of the system of reference, just as the usual theory of relativity prevents our speaking of the absolute velocity of a system;[2] and it makes the equal falling of all bodies in a gravitational field seem obvious.

[2] Of course we cannot replace any arbitrary gravitational field by a state of motion of the system without a gravitational field, any more than, by a transformation of relativity, we can transform all points of a medium in any kind of motion to rest.

So long as we restrict ourselves to purely mechanical processes where Newton's mechanics holds, we are certain of the equivalence of the systems K and K'. But this view of ours will only have a deeper significance if the systems K and K' are equivalent with respect to all physical processes, that is, if the laws of nature with respect to K are in entire agreement with those with respect to K'. By assuming this to be so, we arrive at a principle which, if it is really true, has great heuristic importance. For by theoretical consideration of processes which take place relatively to a system of reference with uniform acceleration, we obtain information as to how the processes take place in a homogeneous gravitational field. We shall now show, first of all, from the standpoint of the ordinary theory of relativity, what degree of probability is inherent in our hypothesis.

§ 2. On the Weight of Energy

One result yielded by the theory of relativity is that the inertial mass of a body increases with the energy it contains; if the increase of energy amounts to E, the increase in inertial mass is equal to E/c^2, when c denotes the velocity of light. Now is there an increase of gravitating mass corresponding to this increase of inertial mass? If not, then a body would fall in the same gravitational field with varying acceleration according to the energy it contained. That highly satisfactory result of the theory of relativity by which the law of the conservation of mass is merged in the law of conservation of energy could not be maintained, because it would compel us to abandon the law of the conservation of mass in its old form for inertial mass, but maintain it for gravitating mass.

This must be regarded as very improbable. On the other hand, the usual theory of relativity does not provide us with any argument from which to infer that the weight of a body depends on the energy contained in it. But we shall show that our hypothesis of the equivalence of the systems K and K' gives us the weight of energy as a necessary consequence.

Let two material systems S_1 and S_2, provided with instruments of measurement, be situated on the z-axis of K at the distance h from each other,[3] so that the gravitation potential in S_2 is greater than that in S_1 by γh. Let a definite quantity of energy E be emitted from S_2 towards S_1. Let the quantities of energy in S_1 and S_2 be measured by devices which—brought to one place in the system z and there compared—are perfectly alike. As to the process of this conveyance of energy by radiation we can make no *a priori* assertion, because we do not know the influence of the gravitational field on the radiation and the measuring instruments in S_1 and S_2.

But by our postulate of the equivalence of K and K' we are able, in place of the system K in a homogeneous gravitational field, to set the gravitation-free system K', which moves with uniform acceleration in the direction of positive z, and with the z-axis of which the material systems S_1 and S_2 are rigidly connected.

We judge of the process of the transference of energy by radiation from S_2 to S_1 from a system K_0, which is to be free from acceleration.

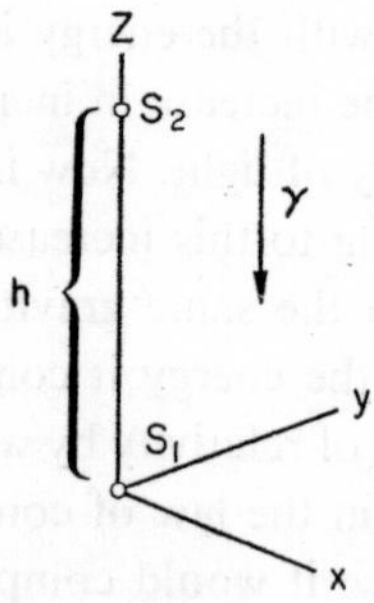

FIG. 1.

At the moment when the radiation energy E_2 is emitted from S_2 toward S_1, let the velocity of K' relatively to K_0 be zero. The radiation will arrive at S_1 when the time h/c has elapsed (to a first approximation). But at this moment the velocity of S_1 relatively to K_0 is $\gamma h/c = v$. Therefore by the ordinary theory of relativity the radiation arriving

[3] The dimensions of S_1 and S_2 are regarded as infinitely small in comparison with h.

at S_1 does not possess the energy E_2, but a greater energy E_1, which is related to E_2 to a first approximation by the equation

$$E_1 = E_2\left(1+\frac{v}{c}\right) = E_2\left(1+\gamma\frac{h}{c^2}\right). \tag{1}$$

By our assumption exactly the same relation holds if the same process takes place in the system K, which is not accelerated, but is provided with a gravitational field. In this case we may replace γh by the potential Φ of the gravitation vector in S_2, if the arbitrary constant of Φ in S_1 is equated to zero. We then have the equation

$$E_1 = E_2 + \frac{E_2}{c^2}\Phi. \tag{1a}$$

This equation expresses the law of energy for the process under observation. The energy E_1 arriving at S_1 is greater than the energy E_2, measured by the same means, which was emitted in S_2, the excess being the potential energy of the mass E_2/c^2 in the gravitational field. Accordingly, for the fulfilment of the principle of energy we have to ascribe to the energy E, before its emission in S_2, a potential energy due to gravity, which corresponds to the gravitational mass E/c^2. Our assumption of the equivalence of K and K' thus removes the difficulty mentioned at the beginning of this paragraph which is left unsolved by the ordinary theory of relativity.

The meaning of this result is shown particularly clearly if we consider the following cycle of operations:

1. The energy E, as measured in S_2, is emitted in the form of radiation in S_2 towards S_1, where, by the result just obtained, the energy $E(1+\gamma h/c^2)$, as measured in S_1, is absorbed.

2. A body W of mass M is lowered from S_2 to S_1, work $M\gamma h$ being done in the process.

3. The energy E is transferred from S_1 to the body W while W is in S_1. Let the gravitational mass M be thereby changed so that it acquires the value M'.

4. Let W be again raised to S_2, work $M'\gamma h$ being done in the process.

5. Let E be transferred from W back to S_2.

The effect of this cycle is simply that S_1 has undergone the increase of energy $E\gamma h/c^2$, and that the quantity of energy $M'\gamma h - M\gamma h$ has been conveyed to the system in the form of mechanical work. By the principle of energy, we must therefore have

$$E\gamma \frac{h}{c^2} = M'\gamma h - M\gamma h$$

or

$$M' - M = E/c^2. \tag{1b}$$

The increase in gravitational mass is thus equal to E/c^2, and therefore equal to the increase in inertial mass as given by the theory of relativity.

The result emerges still more directly from the equivalence of the systems K and K', according to which the gravitational mass with respect to K is exactly equal to the inertial mass with respect to K'; energy must therefore possess a gravitational mass which is equal to its inertial mass. If a mass M_0 be suspended on a spring balance in the system K', the balance will indicate the apparent weight $M_0\gamma$ on account of the inertia of M_0. If the quantity of energy E be transferred to M_0, the spring balance, by the law of the inertia of energy, will indicate $(M_0 + E/c^2)\gamma$. By reason of our fundamental assumption exactly the same thing must occur when the experiment is repeated in the system K, that is, in the gravitational field.

§ 3. Time and the Velocity of Light in the Gravitational Field

If the radiation emitted in the uniformly accelerated system K' in S_2 toward S_1 had the frequency ν_2 relatively to the clock in S_2, then, relatively to S_1, at its arrival in S_1 it no longer has the frequency ν_2 relatively to an identical clock in S_1, but a greater frequency ν_1, such that to a first approximation

$$\nu_1 = \nu_2 \left(1 + \gamma \frac{h}{c^2}\right). \tag{2}$$

For if we again introduce the unaccelerated system of reference K_0, relatively to which, at the time of the emission of light, K' has no velocity, then S_1, at the time of arrival of the radiation at S_1, has, relatively to K_0, the velocity $\gamma h/c$, from which, by Doppler's principle, the relation as given results immediately.

In agreement with our assumption of the equivalence of the systems K' and K, this equation also holds for the stationary system of co-ordinates K, provided with a uniform gravitational field, if the transference by radiation takes place as described. It follows, then, that a ray of light emitted in S_2 with a definite gravitational potential, and possessing at its emission the frequency ν_2—compared with a clock in S_2—will, at its arrival in S_1, possess a different frequency ν_1—measured by an identical clock in S_1. For γh we substitute the gravitational potential Φ of S_2—that of S_1 being taken as zero—and assume that the relation which we have deduced for the homogeneous gravitational field also holds for other forms of field. Then

$$\nu_1 = \nu_2\left(1+\frac{\Phi}{c^2}\right). \tag{2a}$$

This result (which by our deduction is valid to a first approximation) permits, in the first place, of the following application. Let ν_0 be the vibration-number of an elementary light-generator, measured by a delicate clock at the same place. Let us imagine them both at a place on the surface of the sun (where our S_2 is located). Of the light there emitted a portion reaches the Earth (S_1), where we measure the frequency of the arriving light with a clock U in all respects resembling the one just mentioned. Then by (2a),

$$\nu = \nu_0\left(1+\frac{\Phi}{c^2}\right),$$

where Φ is the (negative) difference of gravitational potential between the surface of the sun and the earth. Thus according to our view the spectral lines of sunlight, as compared with the corresponding spectral lines of terrestrial sources of light, must be somewhat displaced toward the red, in fact by the relative amount

$$\frac{\nu_0-\nu}{\nu_0} = -\frac{\Phi}{c^2} = 2.10^{-6}.$$

If the conditions under which the solar bands arise were exactly known, this shifting would be susceptible of measurement. But as other influences (pressure, temperature) affect the position of the centres of the spectral lines, it is difficult to discover whether the inferred influence of the gravitational potential really exists.[4]

On a superficial consideration equation (2), or (2a), respectively, seems to assert an absurdity. If there is constant transmission of light from S_2 to S_1, how can any other number of periods per second arrive in S_1 than is emitted in S_2? But the answer is simple. We cannot simply regard ν_2 or ν_1 as frequencies (as the number of periods per second) since we have not yet determined the time in system K. What ν_2 denotes is the number of periods with reference to the time-unit of the clock U in S_2, while ν_1 denotes the number of periods per second with reference to the identical clock in S_1. Nothing compels us to assume that the clocks U in different gravitation potentials must be regarded as going at the same rate. On the contrary, we must certainly define the time in K in such a way that the number of wave crests and troughs between S_2 and S_1 is independent of the absolute value of time; for the process under observation is by nature a stationary one. If we did not satisfy this condition, we should arrive at a definition of time by the application of which time would enter explicitly into the laws of nature, and this would certainly be unnatural and unpractical. Therefore the two clocks in S_1 and S_2 do not both give the "time" correctly. If we measure time in S_1 with the clock U, then we must measure time in S_2 with a clock which goes $1+\Phi/c^2$ times more slowly than the clock U when it is compared with U at one and the

[4] L. F. Jewell (*Journ. de Phys.*, 6, 1897, p. 84) and particularly Ch. Fabry and H. Boisson (*Comptes rendus*, 148, 1909, pp. 688–690) have actually found such displacements of fine spectral lines toward the red end of the spectrum, of the order of magnitude here calculated, but have ascribed them to an effect of pressure in the absorbing layer.

same place. For when measured by such a clock the frequency of the ray of light which is considered above is at its emission in S_2

$$\nu_2\left(1+\frac{\Phi}{c^2}\right)$$

and is therefore, by (2a), equal to the frequency ν_1 of the same ray of light on its arrival in S_1.

This has a consequence which is of fundamental importance for our theory. For if we measure the velocity of light at different places in the accelerated, gravitation-free system K', employing clocks U of identical constitution, we obtain the same magnitude at all these places. The same holds good, by our fundamental assumption, for the system K as well. But from what has just been said we must use clocks of different constitution for measuring time at places with differing gravitation potential. For measuring time at a place which, relatively to the origin of the co-ordinates, has the gravitation potential Φ, we must employ a clock which—when removed to the origin of co-ordinates—goes $(1+\Phi/c^2)$ times more slowly than the clock used for measuring time at the origin of co-ordinates. If we call the velocity of light at the origin of co-ordinates c_0, then the velocity of light c at a place with the gravitation potential Φ will be given by the relation

$$c = c_0\left(1+\frac{\Phi}{c^2}\right). \tag{3}$$

The principle of the constancy of the velocity of light holds good according to this theory in a different form from that which usually underlies the ordinary theory of relativity.

§ 4. Bending of Light-rays in the Gravitational Field

From the proposition which has just been proved, that the velocity of light in the gravitational field is a function of position, we may easily infer, by means of Huyghens's principle, that light-rays propagated across a gravitational field undergo deflexion. For let E be a wave front of a plane light-wave at the time t, and let P_1 and P_2 be two points in that plane at unit distance from each other. P_1 and P_2 lie in the plane of the paper, which is chosen so that the differential coefficient of Φ, taken in the direction of the normal to the plane, vanishes,

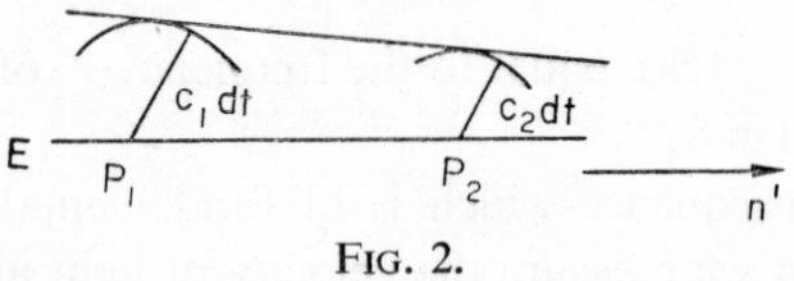

FIG. 2.

and therefore also that of c. We obtain the corresponding wave front at time $t+dt$, or, rather, its line of section with the plane of the paper, by describing circles round the points P_1 and P_2 with radii c_1dt and c_2dt respectively, where c_1 and c_2 denote the velocity of light at the points P_1 and P_2 respectively, and by drawing the tangent to these circles. The angle through which the light-ray is deflected in the path cdt is therefore

$$(c_1-c_2)\,dt = -\frac{\partial c}{\partial n'}\,dt,$$

if we calculate the angle positively when the ray is bent toward the side of increasing n'. The angle of deflexion per unit of path of the light-ray is thus

$$-\frac{1}{c}\frac{\partial c}{\partial n'}, \quad \text{or by (3)} \quad -\frac{1}{c^2}\frac{\partial \Phi}{\partial n'}.$$

Finally, we obtain for the deflexion of a light-ray toward the side n' on any path (s) the expression

(4) $$a = -\frac{1}{c^2}\int \frac{\partial \Phi}{\partial n'}\,ds.$$

We might have obtained the same result by directly considering the propagation of a ray of light in the uniformly accelerated system K', and transferring the result to the system K, and thence to the case of a gravitational field of any form.

By equation (4) a ray of light passing along by a heavenly body suffers a deflexion to the side of the diminishing gravitational potential, that is, on the side directed toward the heavenly body, of the magnitude

$$a = \frac{1}{c^2}\int_{\theta=-\frac{1}{2}\pi}^{\theta=\frac{1}{2}\pi} \frac{kM}{r^2}\cos\theta\, ds = 2\frac{kM}{c^2\Delta},$$

where k denotes the constant of gravitation, M the mass of the heavenly body, Δ the distance of the ray from the centre of the body.

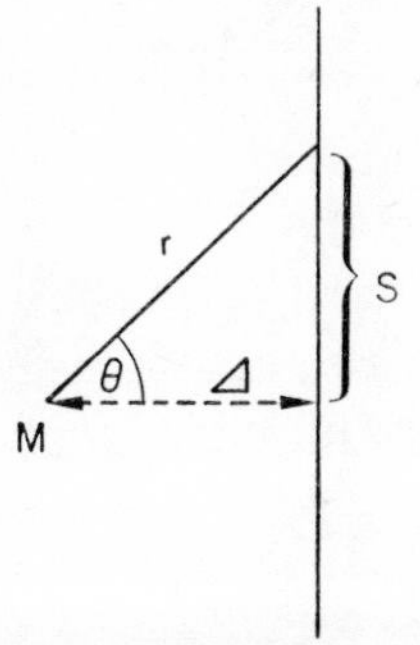

FIG. 3.

A ray of light going past the sun would accordingly undergo deflexion to the amount of 4.10^{-6} = 83 seconds of arc. The angular distance of the star from the centre of the Sun appears to be increased by this amount. As the fixed stars in the parts of the sky near the Sun are visible during total eclipses of the Sun, this consequence of the theory may be compared with experience. With the planet Jupiter the displacement to be expected reaches to about $\frac{1}{100}$ of the amount given. It would be a most desirable thing if astronomers would take up the question here raised. For apart from any theory there is the question whether it is possible with the equipment at present available to detect an influence of gravitational fields on the propagation of light.

NOTES ON EXTRACT 4

THIS paper is the definitive presentation of Einstein's theory. (In the present extract some 23 pages are omitted; they contain merely technical development.) So complete is it, that it might be read as a standard textbook on the subject.

EXTRACT 4[†]

The Foundations of General Relativity Theory

A. EINSTEIN

THE theory set out in the following is perhaps the most extensive generalisation amongst the theories known to-day as relativity theories. These last I take as the heirs to the first "special relativity" theory, which I assume is familiar to the reader. The generalisation of relativity theory is carried out very easily in the form given to special relativity by Minkowski. This mathematician was the first to recognise clearly the formal equivalence of space and time coordinates and use it for a convenient exposition of the theory. The necessary mathematical apparatus for the general relativity theory is already completed in the absolute differential calculus. This rests on the researches of *Gauss*, *Riemann* and *Christoffel* on non-Euclidean manifolds, and was systematised by *Ricci* and *Levi-Civita*, who have already applied it to problems of theoretical physics. In part B of the present paper I have explained as much of this apparatus as is needed, and is not well known to physicists, as simply and transparently as I can. In this way no study of the mathematical literature is necessary for the understanding of the present paper. Finally my thanks are due here to my friend, the mathematician *Grossman*, who not only helped me in the study of the extensive mathematical literature, but led me by his researches to the discovery of the field equations.

[†] *Annaln. Phys.* **49**, 769 (1916)

A. Fundamental Considerations on the Postulate of Relativity

§ 1. *Remarks on Special Relativity Theory*

The special theory of relativity is based on the following postulate, which is also satisfied by the mechanics of Galileo and Newton:

If a system of co-ordinates K is chosen so that, in relation to it, physical laws hold good in their simplest form, the *same* laws also hold good in relation to any other system of co-ordinates K' moving in uniform translation relatively to K. This postulate we call the "special principle of relativity". The word "special" is meant to intimate that the principle is restricted to the case when K' has a motion of uniform translation relatively to K, and that the equivalence of K' and K does not extend to the case of non-uniform motion of K' relatively to K.

Thus special relativity does not depart from classical mechanics through the postulate of relativity, but through the postulate of the constancy of the velocity of light *in vacuo*, from which, in combination with the special principle of relativity, there follow, in the well-known way, the relativity of simultaneity, the Lorentz transformation, and the related laws for the behaviour of moving bodies and clocks.

The modification to which special relativity has subjected the theory of space and time is indeed far reaching, but one important point has remained unaffected. The laws of geometry, even according to special relativity, are to be interpreted directly as laws relating to the possible relative positions of solid bodies at rest; and more generally, the laws of kinematics are to be interpreted as laws which describe the relations of measuring bodies and clocks. To two selected material points of a stationary rigid body there always corresponds a distance of quite definite length, which is independent of the locality and orientation of the body, and is also independent of the time. To two selected positions of the hands of a clock at rest relatively to the privileged system of reference there always corresponds an interval of time of a definite length, which is independent of position and time. We shall soon see that the general theory of relativity cannot adhere to this simple physical interpretation of space and time.

§ 2. *The Need for an Extension of the Postulate of Relativity*

In classical mechanics, and no less in the special theory of relativity, there is an inherent epistemological defect which was, perhaps for the first time, clearly pointed out by Ernst Mach. We will elucidate it by the following example:—Two fluid bodies of the same size and nature hover freely in space at so great a distance from each other and from all other masses that only those gravitational forces need be taken into account which arise from the interaction of different parts of the same body. Let the distance between the two bodies be invariable, and in neither of the bodies let there be any relative movements of the parts with respect to one another. But let either mass, as judged by an observer at rest relatively to the other mass, rotate with constant angular velocity about the line joining the masses. This is a verifiable relative motion of the two bodies. Now let us imagine that each of the bodies has been surveyed by means of measuring instruments at rest relatively to itself, and let the surface of S_1 prove to be a sphere, and that of S_2 an ellipsoid of revolution. Thereupon we put the question—What is the reason for this difference in the two bodies? No answer can be admitted as epistemologically satisfactory,[1] unless the reason given is an *observable fact of experience*. The law of causality has not the significance of a statement as to the world of experience, except when *observable facts* ultimately appear as causes and effects.

Newtonian mechanics does not give a satisfactory answer to this question. It pronounces as follows:—The laws of mechanics apply to the space R_1, in respect to which the body S_1 is at rest, but not to the space R_2, in respect to which the body S_2 is at rest. But the privileged space R_1 of Galileo, thus introduced, is a merely *factitious*

[1] Of course an answer may be satisfactory from the point of view of epistemology, and yet be unsound physically, if it is in conflict with other experiences.

cause, and not a thing that can be observed. It is therefore clear that Newton's mechanics does not really satisfy the requirement of causality in the case under consideration, but only apparently does so, since it makes the factitious cause R_1 responsible for the observable difference in the bodies S_1 and S_2.

The satisfactory answer can only be that the physical system consisting of S_1 and S_2 reveals within itself no imaginable cause to which the differing behaviour of S_1 and S_2 can be referred. The cause must therefore lie *outside* this system. We have to take it that the general laws of motion, which in particular determine the shapes of S_1 and S_2, must be such that the mechanical behaviour of S_1 and S_2 is partly conditioned, in quite essential respects, by distant masses which we have not included in the system under consideration. These distant masses and their motions relative to S_1 and S_2 must then be regarded as the seat of the causes (which must be susceptible to observation) of the different behaviour of our two bodies S_1 and S_2. They take over the rôle of the factitious cause R_1. Of all imaginable spaces R_1, R_2, etc., in any kind of motion relatively to one another, there is none which we may look upon as privileged *a priori* without reviving the above-mentioned epistemological objection. *The laws of physics must be of such a nature that they apply to systems of reference in any kind of motion.* In this way we arrive at an extension of the postulate of relativity.

In addition to this weighty argument from the theory of knowledge, there is a well-known physical fact which favours an extension of the theory of relativity. Let K be a Galilean system of reference, i.e. a system relative to which (at least in the four-dimensional region under consideration) a mass, sufficiently distant from other masses, moves with uniform motion in a straight line. Let K' be a second system of reference which is moving relatively to K in *uniformly accelerated* translation. Then, relatively to K', a mass sufficiently distant from other masses would have an accelerated motion such that its acceleration and direction of acceleration are independent of the material composition and physical state of the mass.

Does this permit an observer at rest relatively to K' to infer that he is on a "really" accelerated system of reference? The answer is in the negative; for the above-mentioned relation of freely movable masses to K' may be interpreted equally well in the following way. The system of reference K' is unaccelerated, but the space–time region in question is in a gravitational field, which generates the accelerated motion of the bodies relatively to K'.

This view is made possible for us by our experience that there exists a field of force, namely, the gravitational field, which possesses the remarkable property of imparting the same acceleration to all bodies.[2] The mechanical behaviour of bodies relatively to K' is the same as is observed in the case of systems which we are used to regarding as "stationary" or as "privileged." Therefore, from the physical standpoint, the assumption that the systems K and K' may both with equal right be looked upon as "stationary", is a very natural one. That is to say, they have an equal title as systems of reference for the physical description of phenomena.

It will be seen from these reflections that in pursuing the general theory of relativity we shall be led to a theory of gravitation, since we are able to "produce" a gravitational field merely by changing the system of co-ordinates. It will also be obvious that the principle of the constancy of the velocity of light *in vacuo* must be modified, since we easily recognize that the path of a ray of light with respect to K' must in general be curvilinear, if with respect to K light is propagated in a straight line with a definite constant velocity.

§ 3. *The Space–Time Continuum. Requirement of General Co-variance for the Equations Expressing General Laws of Nature*

In classical mechanics, as well as in the special theory of relativity, the co-ordinates of space and time have a direct physical meaning. To say that a point-event has the X_1 coordinate x_1 means that the projection of the point-event on the axis of X_1, determined by rigid rods and in accordance with the rules of Euclidean geometry, is obtain-

[2] Eötvös has proved experimentally that the gravitational field has this property with great accuracy.

ed by measuring off a given rod (the unit of length) x_1 times from the origin of coordinates along the axis of X_1. To say that a point-event has the X_4 co-ordinate $x_4 = t$, means that a standard clock, made to measure time in a definite unit period, and which is stationary relatively to the system of co-ordinates and practically coincident in space with the point-event,[3] will have measured off $x_4 = t$ periods at the occurrence of the event.

This view of space and time has always been in the minds of physicists, even if, as a rule, unconsciously. This is clear from the part which these concepts play in physical measurements; it must also have underlaid the reader's consideration of the preceding paragraph (§ 2) for him to connect any meaning with what he read. But we shall now show that we must put it aside and replace it by a more general view, in order to be able to carry through the postulate of general relativity, if the special theory of relativity applies to the special case of the absence of a gravitational field.

In a space which is free of gravitational fields we introduce a Galilean system of reference K (x, y, z, t), and also a system of co-ordinates K' (x', y', z', t') in uniform rotation relatively to K. Let the origins of both systems, as well as their Z-axes permanently coincide. We shall show that for a space–time measurement in the system K' the above definition of the physical meaning of lengths and times cannot be maintained. For reasons of symmetry it is clear that a circle around the origin in the X, Y plane of K may at the same time be regarded as a circle in the X', Y' plane of K'. We suppose that the circumference and diameter of this circle have been measured with a unit measure infinitely small compared with the radius, and that we have the quotient of the two results. If this experiment were performed with a measuring-rod at rest relatively to the Galilean system K, the quotient would be π. With a measuring-rod at rest relatively to K', the quotient would be greater than π. This is readily understood if we envisage the whole process of measuring from the "stationary" system K, and

[3] We assume the possibility of verifying "simultaneity" for events immediately proximate in space, or—to speak more precisely—for immediate proximity or coincidence in space–time, without giving a definition of this fundamental concept.

take into consideration that the measuring-rod applied to the periphery undergoes a Lorentzian contraction, while the one applied along the radius does not. Hence Euclidean geometry does not apply to K'. The notion of co-ordinates defined above, which presupposes the validity of Euclidean geometry, therefore breaks down in relation to the system K'. So, too, we are unable to introduce a time corresponding to physical requirements in K', indicated by clocks at rest relatively to K'. To convince ourselves of this impossibility, let us imagine two clocks of identical constitution placed, one at the origin of co-ordinates, and the other at the circumference of the circle, and both observed from the "stationary" system K. By a familiar result of the special theory of relativity, the clock at the circumference—judged from K—goes more slowly than the other, because the former is in motion and the latter at rest. An observer at the common origin of co-ordinates, capable of observing the clock at the circumference by means of light, would therefore see it lagging behind the clock beside him. As he will not decide to let the velocity of light along the path in question depend explicitly on the time, he will interpret his observations as showing that the clock at the circumference "really" goes more slowly than the clock at the origin. So he will be obliged to define time in such a way that the rate of a clock depends upon where the clock may be.

We therefore reach this result:—In the general theory of relativity, space and time cannot be defined in such a way that differences of the spatial co-ordinates can be directly measured by the unit measuring-rod, or differences in the time co-ordinate by a standard clock.

The method hitherto employed for specifying co-ordinates in the space–time continuum in a definite manner thus breaks down, and there seems to be no other way which would allow us to adapt systems of co-ordinates to the four-dimensional universe so that we might expect from their application a particularly simple formulation of the laws of nature. So there is nothing for it but to regard all imaginable systems of co-ordinates, on principle, as equally suitable for the description of nature. This comes to requiring that:

The general laws of nature are to be expressed by equations which hold good for all systems of co-ordinates, that is, are covariant with respect to any substitutions whatever (generally covariant).

It is clear that a physics which satisfies this postulate will also be suitable for the general postulate of relativity. For amongst *all* substitutions are those which correspond to all relative motions of three-dimensional systems of co-ordinates. That this requirement of general covariance, which takes away from space and time the last remnant of physical objectivity, is a natural one, will be seen from the following reflexion. All our space–time verifications invariably amount to a determination of space–time coincidences. If, for example, events consisted merely in the motion of material points, then ultimately nothing would be observable but the meetings of two or more of these points. Moreover, the results of our measurings are nothing but verifications of such meetings of the material points of our measuring instruments with other material points, coincidences between the hands of a clock and points on the clock dial, and observed point-events happening at the same place at the same time.

The introduction of a system of reference serves no other purpose than to facilitate the description of the totality of such coincidences. We allot to the world four space–time variables x_1, x_2, x_3, x_4 in such a way that for every point-event there is a corresponding set of values of the variables $x_1 \ldots x_4$. To two coincident point-events there corresponds the same set of values of the variables $x_1 \ldots x_4$, i.e. coincidence is characterized by the identity of the co-ordinates. If, in place of the variables $x_1 \ldots x_4$, we introduce functions of them, x'_1, x'_2, x'_3, x'_4, as a new system of co-ordinates, so that the sets of values are made to correspond to one another in a one-valued way, the equality of all four co-ordinates in the new system will also serve as an expression for the space–time coincidence of the two point-events. As all our physical experience can be ultimately reduced to such coincidences, there is no immediate reason for preferring certain systems of co-ordinates to others, that is to say, we arrive at the requirement of general covariance.

§ 4. *The Relation of the Four Co-ordinates to Measurement in Space and Time*

It is not my purpose in this discussion to represent the general theory of relativity as a system that is as simple and logical as possible, and with the minimum number of axioms. Rather my main object is to develop this theory in such a way that the reader will feel that the path we have entered upon is psychologically the natural one, and that the underlying assumptions will seem to have the highest possible degree of security. With this aim in view let it now be granted that:

For infinitely small four-dimensional regions the theory of relativity in the restricted sense is appropriate, if the co-ordinates are suitably chosen.

For this purpose we must choose the acceleration of the infinitely small ("local") system of co-ordinates so that no gravitational field occurs; this is possible for an infinitely small region. Let X_1, X_2, X_3, be the space co-ordinates and X_4 the corresponding time co-ordinate measured in the appropriate unit.[4] If a rigid rod is imagined to be given as the unit measure, the co-ordinates, with a given orientation of the system of co-ordinates, have a direct physical meaning in the sense of the special theory of relativity. By the special theory of relativity the expression

$$ds^2 = -dX_1^2 - dX_2^2 - dX_3^2 + dX_4^2 \tag{1}$$

then has a value which is independent of the orientation of the local system of co-ordinates, and is ascertainable by measurements of space and time. The magnitude of the linear element pertaining to infinitely near points of the four-dimensional continuum we call *ds*. If the *ds* belonging to the element $dX_1 \ldots dX_4$ is positive, we follow Minkowski in calling it time-like; if it is negative, we call it space-like.

To the "linear element" in question, or to the two infinitely near point-events, there will also correspond definite differentials $dx_1 \ldots dx_4$

[4] The unit of time is to be chosen so that the velocity of light *in vacuo* as measured in the "local" system of co-ordinates is to be equal to unity.

of the four-dimensional co-ordinates of any chosen system of reference. If this system, as well as the "local" system, is given for the region under consideration, the dX_ν will allow themselves to be represented here by definite linear homogeneous expressions of the dx_σ:

$$dX_\nu = \sum_\sigma a_{\nu\sigma}\, dx_\sigma. \tag{2}$$

Inserting these expressions in (1), we obtain

$$ds^2 = \sum_{\tau\sigma} g_{\sigma\tau}\, dx_\sigma\, dx_\tau, \tag{3}$$

where the $g_{\sigma\tau}$ will be functions of the x_σ. These can no longer be dependent on the orientation and the state of motion of the "local" system of co-ordinates, for ds^2 is a quantity ascertainable by rod-clock measurement of infinitely near point-events in space–time, and defined independently of any particular choice of co-ordinates. The $g_{\sigma\tau}$ are to be chosen here so that $g_{\sigma\tau} = g_{\tau\sigma}$; the summation is to extend over all values of σ and τ, so that the sum consists of 4×4 terms, of which twelve are equal in pairs.

The case of the usual relativity theory arises out of the case here considered, if it is possible, by reason of the particular relations between the $g_{\sigma\tau}$ in a finite region, to choose the system of reference in the finite region in such a way that the $g_{\sigma\tau}$ assume the constant values

$$\left.\begin{matrix} -1 & 0 & 0 & 0 \\ 0 & -1 & 0 & 0 \\ 0 & 0 & -1 & 0 \\ 0 & 0 & 0 & +1 \end{matrix}\right\}. \tag{4}$$

We shall find below that the choice of such co-ordinates is, in general, not possible for a finite region.

From the considerations of § 2 and § 3 it follows that the quantities $g_{\tau\sigma}$ are to be regarded from a physical standpoint as the quantities which describe the gravitational field relative to the chosen system of reference. For, if we now assume the special theory of relativity to apply to a certain four-dimensional region with the co-ordinates properly chosen, then the $g_{\sigma\tau}$ have the values given in (4). A free material point then moves, relative to this system, with uniform motion in a straight line. Then if we introduce new space–time co-ordinates x_1, x_2, x_3, x_4, by means of an arbitrary substitution, the $g_{\mu\nu}$ in this new system will no longer be constants, but functions of space and time. At the same time the motion of the free material point will be represented in the new co-ordinates as a curvilinear non-uniform motion, and the law of this motion will be independent of the nature of the moving particle. We shall therefore interpret this motion as a motion under the influence of a gravitational field. We thus find the occurrence of a gravitational field connected with a space–time variability of the $g_{\sigma\tau}$. So, too, in the general case, when we are no longer able by a suitable choice of co-ordinates to apply the special theory of relativity to a finite region, we shall hold fast to the view that the $g_{\sigma\tau}$ describe the gravitational field.

Thus, according to general relativity theory, gravitation occupies an exceptional position with regard to other forces, particularly the electromagnetic forces, since the ten functions representing the gravitational field at the same time define the metrical properties of the space measured.

[Here there follows an exposition of the mathematical techniques on much the same lines as is given in the main text of the present book. In what follows, the equation

$$A_{\mu\nu;\sigma} = A_{\mu\nu,\sigma} - \Gamma^{\tau}_{\sigma\mu} A_{\tau\nu} - \Gamma^{\tau}_{\sigma\nu} A_{\mu\tau}$$

which, in Einstein's notation is written

$$A_{\mu\nu\sigma} = \frac{\partial A_{\mu\nu}}{\partial x^{\sigma}} - \begin{Bmatrix} \sigma\mu \\ \tau \end{Bmatrix} A_{\tau\nu} - \begin{Bmatrix} \sigma\nu \\ \tau \end{Bmatrix} A_{\mu\tau} \qquad (27)$$

is referred to.]

§ 12. *The Riemann–Christoffel Tensor*

We now seek the tensor which can be obtained from the fundamental tensor *alone*, by differentiation. At first sight the solution seems obvious. We place the fundamental tensor of the $g_{\mu\nu}$ in (27) instead of any given tensor $A_{\mu\nu}$, and thus have a new tensor, namely, the extension of the fundamental tensor. But we may easily convince ourselves that this extension vanishes identically. We reach our goal, however, in the following way. In (27) place

$$A_{\mu\nu} = \frac{\partial A_\mu}{\partial x_\nu} - \{\mu\nu, \varrho\} A_\varrho,$$

i.e. the extension of any four-vector A_μ. Then (with a somewhat different naming of the indices) we get the tensor of the third rank

$$A_{\mu\sigma\tau} = \frac{\partial^2 A_\mu}{\partial x_\sigma \partial x_\tau} - \{\mu\sigma, \varrho\} \frac{\partial A_\varrho}{\partial x_\tau} - \{\mu\tau, \varrho\} \frac{\partial A_\varrho}{\partial x_\sigma} - \{\sigma\tau, \varrho\} \frac{\partial A_\mu}{\partial x_\varrho}$$
$$+ \left[-\frac{\partial}{\partial x_\tau} \{\mu\sigma, \varrho\} + \{\mu\tau, \alpha\} \{\alpha\sigma, \varrho\} + \{\sigma\tau, \alpha\} \{\alpha\mu, \varrho\} \right] A.$$

This expression suggests forming the tensor $A_{\mu\sigma\tau} - A_{\mu\tau\sigma}$. For, if we do so, the following terms of the expression for $A_{\mu\sigma\tau}$ cancel those of $A_{\mu\tau\sigma}$: the first, the fourth, and the member corresponding to the last term in square brackets; because all these are symmetrical in σ and τ. The same holds good for the sum of the second and third terms. Thus we obtain

$$A_{\mu\sigma\tau} - A_{\mu\tau\sigma} = B^\varrho_{\mu\sigma\tau} A_\varrho \tag{42}$$

where

$$B^\varrho_{\mu\sigma\tau} = -\frac{\partial}{\partial x_\tau} \{\mu\sigma, \varrho\} + \frac{\partial}{\partial x_\sigma} \{\mu\tau, \varrho\} - \{\mu\sigma, \alpha\} \{\alpha\tau, \varrho\}$$
$$+ \{\mu\tau, \alpha\} \{\alpha\sigma, \varrho\}. \tag{43}$$

The essential feature of the result is that on the right side of (42) the A_ϱ occur alone, without their derivatives. From the tensor character of $A_{\mu\sigma\tau} - A_{\mu\tau\sigma}$ in conjunction with the fact that A_ϱ is an arbitrary vector, it follows that $B^\varrho_{\mu\sigma\tau}$ is a tensor (the Riemann–Christoffel tensor).

The mathematical significance of this tensor is as follows: If the continuum is of such a nature that there is a co-ordinate system with reference to which the $g_{\mu\nu}$ are constants, then all the $B^\varrho_{\mu\sigma\tau}$ vanish. If we choose any new system of co-ordinates in place of the original ones, the $g_{\mu\nu}$ referred thereto will not be constants, but in consequence of its tensor nature, the transformed components of $B^{\varrho}_{\mu\sigma\tau}$ will still vanish in the new system. Thus the vanishing of the Riemann tensor is a necessary condition that, by an appropriate choice of the system of reference, the $g_{\mu\nu}$ may be constants. In our problem this corresponds to the case in which[5] with a suitable choice of the system of reference, the special theory of relativity holds good for a *finite* region of the continuum.

Contracting (43) with respect to the indices τ and ϱ we obtain the covariant tensor of second rank

$$G_{\mu\nu} = B^{\varrho}_{\mu\nu\varrho} = R_{\mu\nu} + S_{\mu\nu}$$

where

$$\left.\begin{aligned} R_{\mu\nu} &= -\frac{\partial}{\partial x_\alpha}\{\mu\nu, \alpha\} + \{\mu\alpha, \beta\}\{\nu\beta, \alpha\} \\ S_{\mu\nu} &= \frac{\partial^2 \log\sqrt{-g}}{\partial x_\mu \, \partial x_\nu} - \{\mu\nu, \alpha\}\frac{\partial \log\sqrt{-g}}{\partial x_\alpha} \end{aligned}\right\} \tag{44}$$

Note on the Choice of Co-ordinates.—It has already been observed [in the part omitted in this extract] that the choice of co-ordinates may with advantage be made so that $\sqrt{-g} = 1$. A glance at the equations obtained in the last two sections shows that by such a choice the laws of formation of tensors undergo an important simplification. This applies particularly to $G_{\mu\nu}$, the tensor just developed, which plays a fundamental part in the theory to be set forth. For this specialization of the choice of co-ordinates brings about the vanishing of $S_{\mu\nu}$, so that the tensor $G_{\mu\nu}$ reduces to $R_{\mu\nu}$.

On this account I shall hereafter give all relations in the simplified form which this specialization of the choice of co-ordinates brings with it. It will then be an easy matter to revert to the *generally* covariant equations, if this seems desirable in a special case.

[5] The mathematicians have proved that this is also a *sufficient* condition.

C. Theory of the Gravitational Field

§ 13. *Equations of Motion of a Mass-point in the Gravitational Field. Expression for the Field-Components of Gravitation*

A freely movable body not subjected to external forces moves, according to special relativity, in a straight line and uniformly. This is also the case, according to general relativity, for a part of four-dimensional space in which the system of co-ordinates K_0, may be, and is, so chosen that they have the special constant values given in (4).

If we consider precisely this motion from any chosen system of co-ordinates K_1, the body, observed from K_1, moves, according to the considerations in § 2, in a gravitational field. The law of motion with respect to K_1 results without difficulty from the following consideration. With respect to K_0 the law of motion corresponds to a four-dimensional straight line, i.e. to a geodetic line. Now since the geodetic line is defined independently of the system of reference, its equations will also be the equation of motion of the mass-point with respect to K_1. If we set

$$\Gamma^{\tau}_{\mu\nu} = -\{\mu\nu, \tau\} \tag{45}$$

the equation of the motion of the point with respect to K_1, becomes

$$\frac{d^2x_\tau}{ds^2} = \Gamma^{\tau}_{\mu\nu}\frac{dx_\mu}{ds}\frac{dx_\nu}{ds}. \tag{46}$$

We now make the assumption, which readily suggests itself, that this covariant system of equations also defines the motion of the point in the gravitational field in the case when there is no system of reference K_0, with respect to which the special theory of relativity holds good in a finite region. We have all the more justification for this assumption as (46) contains only *first* derivatives of the $g_{\mu\nu}$, between which even in the special case of the existence of K_0, no relations subsist.[6]

If the $\Gamma^{\tau}_{\mu\nu}$ vanish, then the point moves uniformly in a straight line. These quantities therefore condition the deviation of the motion from uniformity. They are the components of the gravitational field.

§ 14. *The Field Equations of Gravitation in the Absence of Matter*

We make a distinction in what follows between "gravitational field" and "matter" in this way: we denote everything but the gravitational field as "matter". Our use of the word therefore includes not only matter in the ordinary sense, but the electromagnetic field as well.

Our next task is to find the field equations of gravitation in the absence of matter. Here we again apply the method employed in the preceding paragraph in formulating the equations of motion of the material point. A special case in which the required equations must in any case be satisfied is that of the special theory of relativity, in which the $g_{\mu\nu}$ have certain constant values. Let this be the case in a certain finite region relative to a definite system of co-ordinates K_0. Relative to this system all the components of the Riemann tensor $B^{\varrho}_{\mu\sigma\tau}$, defined in (43), vanish. For the space under consideration they then vanish in any other system of co-ordinates.

Thus the required equations of the matter-free gravitational field must in any case be satisfied if all $B^{\varrho}_{\mu\sigma\tau}$ vanish. But this condition goes too far. For it is clear that, e.g., the gravitational field generated by a point in its environment certainly cannot be "transformed away" by any choice of the system of co-ordinates, i.e. it cannot be transformed to the case of constant $g_{\mu\nu}$.

This prompts us to require for the matter-free gravitational field that the symmetrical tensor $G_{\mu\nu}$, derived from the tensor $B^{\varrho}_{\mu\nu\tau}$, shall vanish. Thus we obtain ten equations for the ten quantities $g_{\mu\nu}$, which are satisfied in the special case of the vanishing of all $B^{\varrho}_{\mu\nu\tau}$. With the

[6] It is only between the second (and first) derivatives that, by § 12, the relations $B^{\varrho}_{\mu\sigma\tau} = 0$ subsist.

choice which we have made of a system of co-ordinates, and taking (44) into consideration, the equations for the matter-free field are

$$\left.\begin{array}{c}\dfrac{\partial \Gamma^{\alpha}_{\mu\nu}}{\partial x_{\alpha}}+\Gamma^{\alpha}_{\mu\beta}\Gamma^{\beta}_{\nu\alpha}=0\\ \sqrt{-g}=1\end{array}\right\}. \tag{47}$$

It must be pointed out that there is only a minimum of arbitrariness in the choice of these equations. For besides $G_{\mu\nu}$ there is no tensor of second rank which is formed from the $g_{\mu\nu}$ and its derivatives, contains no derivations higher than second, and is linear in these second derivatives.[7]

These equations, which proceed, by the method of pure mathematics, from the requirement of the general theory of relativity, give us, in combination with the equations of motion (46), to a first approximation Newton's law of attraction, and to a second approximation the explanation of the motion of the perihelion of the planet Mercury discovered by Leverrier (as it remains after corrections for perturbation have been made). These facts must, in my opinion, be taken as convincing proof of the correctness of the theory.

§ 15. *The Hamiltonian Function for the Gravitational Field. Laws of Momentum and Energy*

To show that the field equations correspond to the laws of momentum and energy, it is most convenient to write them in the following Hamiltonian form:

$$\left.\begin{array}{c}\delta\displaystyle\int H\,d\tau=0\\ H=g^{\mu\nu}\Gamma^{\alpha}_{\mu\beta}\Gamma^{\beta}_{\nu\alpha}\\ \sqrt{-g}=1\end{array}\right\}, \tag{47a}$$

[7] Properly speaking, this can be affirmed only of the tensor

$$G_{\mu\nu}+\lambda g_{\mu\nu}g^{\alpha\beta}G_{\alpha\beta},$$

where λ is a constant. If, however, we set this tensor $= 0$, we come back again to the equations $G_{\mu\nu}=0$.

where, on the boundary of the finite four-dimensional region of integration which we have in view, the variations vanish.

We first have to show that the form (47a) is equivalent to the equations (47). For this purpose we regard H as a function of the $g^{\mu\nu}$ and the $g^{\mu\nu}_{\sigma}(=\partial g^{\mu\nu}/\partial x_{\sigma})$.

Then in the first place

$$\delta H = \Gamma^{\alpha}_{\mu\beta}\Gamma^{\beta}_{\nu\alpha}\delta g^{\mu\nu}+2g^{\mu\nu}\Gamma^{\alpha}_{\mu\beta}\delta\Gamma^{\beta}_{\nu\alpha}$$
$$= -\Gamma^{\alpha}_{\mu\beta}\Gamma^{\beta}_{\nu\alpha}\delta g^{\mu\nu}+2\Gamma^{\alpha}_{\mu\beta}\delta(g^{\mu\nu}\Gamma^{\beta}_{\nu\alpha}).$$

But

$$\delta(g^{\mu\nu}\Gamma^{\beta}_{\nu\alpha}) = -\frac{1}{2}\,\delta\left[g^{\mu\nu}g^{\beta\lambda}\left(\frac{\partial g_{\nu\lambda}}{\partial x_{\alpha}}+\frac{\partial g_{\alpha\lambda}}{\partial x_{\nu}}-\frac{\partial g_{\alpha\nu}}{\partial x_{\lambda}}\right)\right].$$

The terms arising from the last two terms in round brackets are of different sign, and result from each other (since the denomination of the summation indices is immaterial) through interchange of the indices μ and β. They cancel each other in the expression for δH, because they are multiplied by the quantity $\Gamma^{\alpha}_{\mu\beta}$, which is symmetrical with respect to the indices μ and β. Thus there remains only the first term in round brackets to be considered, so that we obtain

$$\delta H = -\Gamma^{\alpha}_{\mu\beta}\Gamma^{\beta}_{\nu\alpha}\delta g^{\mu\nu}+\Gamma^{\alpha}_{\mu\beta}\delta g^{\mu\beta}_{\alpha}.$$

Thus

$$\left.\begin{aligned}\frac{\partial H}{\partial g^{\mu\nu}} &= -\Gamma^{\alpha}_{\mu\beta}\Gamma^{\beta}_{\nu\alpha}\\ \frac{\partial H}{\partial g^{\mu\nu}_{\sigma}} &= \Gamma^{\sigma}_{\mu\nu}\end{aligned}\right\}. \tag{48}$$

Carrying out the variation in (47a), we get in the first place

$$\frac{\partial}{\partial x_{\alpha}}\left(\frac{\partial H}{\partial g^{\mu\nu}_{\alpha}}\right)-\frac{\partial H}{\partial g^{\mu\nu}} = 0, \tag{47b}$$

which, on account of (48), agrees with (47), as was to be proved.

If we multiply (47b) by $g^{\mu\nu}_{\sigma}$, then because

$$\frac{\partial g^{\mu\nu}_{\sigma}}{\partial x_{\alpha}} = \frac{\partial g^{\mu\nu}_{\alpha}}{\partial x_{\sigma}}$$

and, consequently,

$$g_\sigma^{\mu\nu} \frac{\partial}{\partial x_\alpha}\left(\frac{\partial H}{\partial g_\alpha^{\mu\nu}}\right) = \frac{\partial}{\partial x_\alpha}\left(g_\sigma^{\mu\nu}\frac{\partial H}{\partial g_\alpha^{\mu\nu}}\right) - \frac{\partial H}{\partial g_\alpha^{\mu\nu}}\frac{\partial g_\alpha^{\mu\nu}}{\partial x_\sigma},$$

we obtain the equation

$$\frac{\partial}{\partial x_\alpha}\left(g_\sigma^{\mu\nu}\frac{\partial H}{\partial g_\alpha^{\mu\nu}}\right) - \frac{\partial H}{\partial x_\sigma} = 0$$

or[8]

$$\left.\begin{aligned} \frac{\partial t_\sigma^\alpha}{\partial x_\alpha} &= 0 \\ -2\varkappa t_\sigma^\alpha &= g_\sigma^{\mu\nu}\frac{\partial H}{\partial g_\alpha^{\mu\nu}} - \delta_\sigma^\alpha H \end{aligned}\right\}, \tag{49}$$

where, on account of (48), and the second equation of (47),

$$\varkappa t_\sigma^\alpha = \tfrac{1}{2}\delta_\sigma^\alpha g^{\mu\nu}\Gamma_{\mu\beta}^\lambda \Gamma_{\nu\lambda}^\beta - g^{\mu\nu}\Gamma_{\mu\beta}^\alpha \Gamma_{\nu\alpha}^\beta. \tag{50}$$

It is to be noticed that t_σ^α is not a tensor; on the other hand (49) applies to all systems of co-ordinates for which $\sqrt{-g} = 1$. This equation expresses the law of conservation of momentum and of energy for the gravitational field. Actually the integration of this equation over a three-dimensional volume V yields the four equations

$$\frac{d}{dx_4}\int t_\sigma^4\, dV = \int (lt_\sigma^1 + mt_\sigma^2 + nt_\sigma^3)\, dS, \tag{49a}$$

where l, m, n denote the direction-cosines of the inward drawn normal at the element dS of the bounding surface (in the sense of Euclidean geometry). We recognize in this the expression of the laws of conservation in their usual form. The quantities t_σ^α we call the "energy components" of the gravitational field.

I will now give equations (47) in a third form, which is particularly useful for a vivid grasp of our subject. By multiplication of the field equations (47) by $g^{\nu\sigma}$ these are obtained in the "mixed" form. Note

[8] The reason for the introduction of the factor $-2\varkappa$ will be apparent later.

that

$$g^{\nu\sigma}\frac{\partial \Gamma^{\alpha}_{\mu\nu}}{\partial x_\alpha} = \frac{\partial}{\partial x_\alpha}(g^{\nu\sigma}\Gamma^{\alpha}_{\mu\nu}) - \frac{\partial g^{\nu\sigma}}{\partial x_\alpha}\Gamma^{\alpha}_{\mu\nu},$$

which quantity is equal to

$$\frac{\partial}{\partial x_\alpha}(g^{\nu\sigma}\Gamma^{\alpha}_{\mu\nu}) - g^{\nu\beta}\Gamma^{\sigma}_{\alpha\beta}\Gamma^{\alpha}_{\mu\nu} - g^{\sigma\beta}\Gamma^{\nu}_{\beta\alpha}\Gamma^{\alpha}_{\mu\nu},$$

or (with different symbols for the summation indices)

$$\frac{\partial}{\partial x_\alpha}(g^{\sigma\beta}\Gamma^{\alpha}_{\mu\beta}) - g^{\gamma\varrho}\Gamma^{\sigma}_{\gamma\beta}\Gamma^{\beta}_{\varrho\mu} - g^{\nu\sigma}\Gamma^{\alpha}_{\mu\beta}\Gamma^{\beta}_{\nu\alpha}.$$

The third term of this expression cancels with the one arising from the second term of the field equations (47); using relation (50), the second term may be written

$$\varkappa(t^{\sigma}_{\mu} - \tfrac{1}{2}\delta^{\sigma}_{\mu}t),$$

where $t = t^{\alpha}_{\alpha}$. Thus instead of equations (47) we obtain

$$(51) \qquad \left.\begin{aligned} \frac{\partial}{\partial x_\alpha}(g^{\sigma\beta}\Gamma^{\alpha}_{\mu\beta}) &= -\varkappa\left(t^{\sigma}_{\mu} - \frac{1}{2}\delta^{\sigma}_{\mu}t\right) \\ \sqrt{-g} &= 1 \end{aligned}\right\}.$$

§ 16. *The General Form of the Field Equations of Gravitation*

The field equations for matter-free space formulated in § 15 are to be compared with the field equation

$$\nabla^2\phi = 0$$

of Newton's theory. We require the equation corresponding to Poisson's equation

$$\nabla^2\phi = 4\pi\varkappa\varrho,$$

where ϱ denotes the density of matter.

The special theory of relativity has led to the conclusion that inertial mass is nothing more or less than energy, which finds its complete mathematical expression in a symmetrical tensor of second rank, the energy-tensor. Thus in the general theory of relativity we must introduce a corresponding energy-tensor of matter T_σ^α, which, like the energy-components t_σ^α [equations (49) and (50)] of the gravitational field, will have mixed character, but will pertain to a symmetrical covariant tensor.[9]

The system of equation (51) shows how this energy-tensor (corresponding to the density ϱ in Poisson's equation) is to be introduced into the field equations of gravitation. For if we consider a complete system (e.g. the solar system), the total mass of the system, and therefore its total gravitating action as well, will depend on the total energy of the system, and therefore on the ponderable energy together with the gravitational energy. This will be expressed by introducing into (51), in place of the energy-components of the gravitational field alone, the sums $t_\mu^\sigma + T_\mu^\sigma$ of the energy-components of matter and of gravitational field. Thus instead of (51) we obtain the tensor equation

$$\left.\begin{aligned} &\frac{\partial}{\partial x_\alpha}(g^{\sigma\beta}\Gamma_{\mu\beta}^\alpha) = -\varkappa\left[(t_\mu^\sigma + T_\mu^\sigma) - \frac{1}{2}\,\delta_\mu^\sigma(t+T)\right], \\ &\sqrt{-g} = 1 \end{aligned}\right\}, \tag{52}$$

where we have set $T = T_\mu^\mu$ (Laue's scalar). These are the required general field equations of gravitation in mixed form. Working back from these, we have in place of (47)

$$\left.\begin{aligned} &\frac{\partial}{\partial x_\alpha}\Gamma_{\mu\nu}^\alpha + \Gamma_{\mu\beta}^\alpha \Gamma_{\nu\alpha}^\beta = -\varkappa\left(T_{\mu\nu} - g_{\mu\nu}T\right), \\ &\sqrt{-g} = 1 \end{aligned}\right\}. \tag{53}$$

It must be admitted that this introduction of the energy-tensor of matter is not justified by the relativity postulate alone. For this reason we have here deduced it from the requirement that the energy of the gravitational field shall act gravitatively in the same way as any other kind of energy. But the strongest reason for the choice of these equa-

[9] $g_{\alpha\tau}T_\sigma^\alpha = T_{\sigma\tau}$ and $g^{\sigma\beta}T_\sigma^\alpha = T^{\alpha\beta}$ are to be symmetrical tensors.

tions lies in their consequence, that the equations of conservation of momentum and energy, corresponding exactly to equations (49) and (49a), hold good for the components of the total energy. This will be shown in § 17.

§ 17. *The Laws of Conservation in the General Case*

Equation (52) may readily be transformed so that the second term on the right-hand side vanishes. Contract (52) with respect to the indices μ and σ, and after multiplying the resulting equation by $\frac{1}{2}\delta^{\sigma}_{\mu}$, subtract it from equation (52). This gives

$$\text{(52a)} \qquad \frac{\partial}{\partial x_\alpha}\left(g^{\sigma\beta}\Gamma^{\alpha}_{\mu\beta}-\frac{1}{2}\,\delta^{\sigma}_{\mu}g^{\lambda\beta}\Gamma^{\alpha}_{\lambda\beta}\right) = -\varkappa(t^{\sigma}_{\mu}+T^{\sigma}_{\mu}).$$

On this equation we perform the operation $\partial/\partial x_\sigma$. We have

$$\frac{\partial^2}{\partial x_\alpha\,\partial x_\sigma}(g^{\sigma}\Gamma^{\alpha}_{\beta\mu}) = -\frac{1}{2}\,\frac{\partial^2}{\partial x_\alpha\,\partial x_\sigma}\left[x^{\sigma\beta}g^{\alpha\lambda}\left(\frac{\partial g_{\mu\lambda}}{\partial x_\beta}+\frac{\partial g_{\beta\lambda}}{\partial x_\mu}-\frac{\partial g_{\mu\beta}}{\partial x_\lambda}\right)\right].$$

The first and third terms of the round brackets yield contributions which cancel one another, as may be seen by interchanging, in the contribution of the third term, the summation indices α and ϱ on the one hand, and β and λ on the other. The second term may be re-modelled so that we have

$$\text{(54)} \qquad \frac{\partial^2}{\partial x_\alpha\,\partial x_\sigma}(g^{\sigma\beta}\Gamma^{\alpha}_{\mu\beta}) = \frac{1}{2}\,\frac{\partial^3 g^{\alpha\beta}}{\partial x_\alpha\,\partial x_\beta\,\partial x_\mu}.$$

The second term on the left-hand side of (52a) yields in the first place

$$-\frac{1}{2}\,\frac{\partial^2}{\partial x_\alpha\,\partial x_\mu}(g^{\lambda\beta}\Gamma^{\alpha}_{\lambda\beta})$$

or

$$\frac{1}{4}\,\frac{\partial^2}{\partial x_\alpha\,\partial x_\mu}\left[g^{\lambda\beta}g^{\alpha\delta}\left(\frac{\partial g_{\delta\lambda}}{\partial x_\beta}+\frac{\partial g_{\delta\beta}}{\partial x_\lambda}-\frac{\partial g_{\lambda\beta}}{\partial x_\delta}\right)\right].$$

With the choice of co-ordinates which we have made, the term deriving from the last term in round brackets disappears. The other two may be combined, and together they give

$$-\frac{1}{2}\,\frac{\partial^3 g^{\alpha\beta}}{\partial x_\alpha\,\partial x_\beta\,\partial x_\mu},$$

so that using (54), we have the identity

$$\frac{\partial^2}{\partial x_\alpha\,\partial x_\sigma}\left(g^{\sigma\beta}\Gamma^\alpha_{\mu\beta}-\frac{1}{2}\,\delta^\sigma_\mu g^{\lambda\beta}\Gamma^\alpha_{\lambda\beta}\right)\equiv 0. \tag{55}$$

From (55) and (52a), it follows that

$$\frac{\partial(t^\sigma_\mu+T^\sigma_\mu)}{\partial x_\sigma}=0. \tag{56}$$

Thus it follows from our field equations of gravitation that the laws of conservation of momentum and energy are satisfied. This may be seen most easily from the consideration which leads to equation (49a); except that here, instead of the energy components t^σ_μ of the gravitational field, we have to introduce the totality of the energy components of matter and gravitational field.

§ 18. *The Laws of Momentum and Energy for Matter, as a Consequence of the Field Equations*

Multiplying (53) by $\partial g^{\mu\nu}/\partial x_\sigma$, we obtain, by the method adopted in § 15, in view of the vanishing of

$$g_{\mu\nu}\frac{\partial g^{\mu\nu}}{\partial x_\sigma},$$

the equation

$$\frac{\partial t^\alpha_\sigma}{\partial x_\alpha}+\frac{1}{2}\,\frac{\partial g^{\mu\nu}}{\partial x_\sigma}T_{\mu\nu}=0,$$

or, in view of (56),

$$\frac{\partial T^\alpha_\sigma}{\partial x_\alpha}+\frac{1}{2}\,\frac{\partial g^{\mu\nu}}{\partial x_\sigma}T_{\mu\nu}=0. \tag{57}$$

With the choice of system of co-ordinates which we have made, this equation predicates nothing more nor less than the vanishing of the divergence of the material energy-tensor. Physically, the occurrence of the second term on the left-hand side shows that laws of conservation of momentum and energy do not apply in the strict sense for matter alone, or else that they apply only when the $g^{\mu\nu}$ are constant, i.e. when the field intensities of gravitation vanish. This second term is an expression for momentum, and for energy, as transferred per unit of volume and time from the gravitational field to matter. This is brought out still more clearly by re-writing (57) as

$$\frac{\partial T^{\alpha}_{\sigma}}{\partial x_{\alpha}} = -\Gamma^{\beta}_{\alpha\sigma} T^{\alpha}_{\beta}. \tag{57a}$$

The right side expresses the energetic effect of the gravitational field on matter.

Thus the field equations of gravitation contain four conditions which govern the course of material phenomena. They give the equations of material phenomena completely, if the latter is capable of being characterized by four differential equations independent of one another.[10]

D. Material Phenomena

The mathematical aids developed in part B [omitted in this extract] enable us forthwith to generalize the physical laws of matter (hydrodynamics, Maxwell's electrodynamics), as they are formulated in special relativity, so that they will fit in general relativity. When this is done, the general principle of relativity does not indeed afford us a further limitation of possibilities; but it makes us acquainted with the influence of the gravitational field on all processes, without our having to introduce any new hypothesis whatever.

Hence it comes about that it is not necessary to introduce definite assumptions as to the physical nature of matter (in the narrower

[10] On this question cf. H. Hilbert, *Nachr. d. K. Gesellsch. d. Wiss. zu Göttingen, Math.-phys. Klasse*, 1915, p. 3.

sense). In particular it must remain an open question whether the theory of the electromagnetic field in conjunction with that of the gravitational field furnishes a sufficient basis for the theory of matter or not. The general postulate of relativity is unable in principle to tell us anything about this. It must remain to be seen, during the working out of the theory, whether electromagnetics and the doctrine of gravitation are able in collaboration to perform what the former by itself is unable to do.

§ 19. *Euler's Equations for a Frictionless Adiabatic Fluid*

Let p and ϱ be two scalars, the former of which we call the "pressure", the latter the "density" of a fluid; and let an equation subsist between them. Let the contravariant symmetrical tensor

$$T^{\alpha\beta} = -g^{\alpha\beta}p + \varrho \frac{dx_\alpha}{ds}\frac{dx_\beta}{ds} \tag{58}$$

be the contravariant energy-tensor of the fluid. To it belongs the covariant tensor

$$T_{\mu\nu} = -g_{\mu\nu}p + g_{\mu\alpha}g_{\mu\beta}\frac{dx_\alpha}{ds}\frac{dx_\beta}{ds}\varrho, \tag{58a}$$

as well as the mixed tensor[11]

$$T_\sigma^\alpha = -\delta_\sigma^\alpha + g_{\sigma\beta}\frac{dx_\beta}{ds}\frac{dx_\alpha}{ds}\varrho. \tag{58b}$$

Inserting the right-hand side of (58b) in (57a), we obtain the Eulerian hydrodynamical equations of the general theory of relativity. They give, in theory, a complete solution of the problem of motion, since the four equations (57a), together with the given equation between p and ϱ, and the equation

$$g_{\alpha\beta}\frac{dx_\alpha}{ds}\frac{dx_\beta}{ds} = 1,$$

[11] For an observer using a system of reference in the sense of the special theory of relativity for an infinitely small region, and moving with it, the density of energy T_4^4 equals $\varrho - p$. This gives the definition of ϱ. Thus ϱ is not constant for an incompressible fluid.

are sufficient, $g_{\alpha\beta}$ being given, to define the six unknowns

$$p, \varrho, \frac{dx_1}{ds}, \frac{dx_2}{ds}, \frac{dx_3}{ds}, \frac{dx_4}{ds}.$$

If the $g_{\mu\nu}$ are also unknown, the equations (53) are brought in. These are eleven equations for defining the ten functions $g_{\mu\nu}$ so that these functions appear over-defined. We must remember, however, that the equations (57a) are already contained in the equations (53), so that the latter represent only seven independent equations. There is good reason for this lack of definition, in that the wide freedom of the choice of co-ordinates causes the problem to remain mathematically undefined to such a degree that three of the functions of space may be chosen at will. [There follows an investigation of Maxwell's equations.]

§ 21. *Newton's Theory as a First Approximation*

As has already been mentioned more than once, special relativity as a special case of the general theory is characterized by the $g_{\mu\nu}$ having the constant values (4). From what has already been said, this means complete neglect of the effects of gravitation. We arrive at a closer approximation to reality by considering the case where the $g_{\mu\nu}$ differ from the values of (4) by quantities which are small compared with 1, and neglecting small quantities of second and higher order. (First stage of approximation.)

It is further to be assumed that in the space–time region under consideration the $g_{\mu\nu}$ at spatial infinity, with a suitable choice of co-ordinates, tend toward the values (4); i.e. we are considering gravitational fields which may be regarded as generated exclusively by matter in the finite region.

It might be thought that these approximations must lead us to Newton's theory. But to that end we still need to approximate the fundamental equations. We consider the motion of a mass-point. In the case of special relativity the components

$$\frac{dx_1}{ds}, \frac{dx_2}{ds}, \frac{dx_3}{ds}$$

may take on any values. This signifies that any velocity

$$v = \sqrt{\left[\left(\frac{dx_1}{dx_4}\right)^2 + \left(\frac{dx_2}{dx_4}\right)^2 + \left(\frac{dx_3}{dx_4}\right)^2\right]}$$

may occur, which is less than the velocity of light *in vacuo*. If we restrict ourselves to the case which is almost the only one in our experience, of v being small as compared with the velocity of light; this denotes that the components

$$\frac{dx_1}{ds}, \frac{dx_2}{ds}, \frac{dx_3}{ds}$$

are to be treated as small quantities, while dx_4/ds, to the second order of small quantities, is equal to one. (Second stage of approximation.)

Now we remark that in the first stage of approximation the magnitudes $\Gamma^{\tau}_{\mu\nu}$ are all small magnitudes of at least the first order. A glance at (46) thus shows that in this equation, in the second stage of approximation, we have to consider only terms for which $\mu = \nu = 4$. Restricting ourselves to terms of lowest order we first obtain in place of (46) the equations

$$\frac{d^2x_\tau}{dt^2} = \Gamma^{\tau}_{44},$$

where we have set $ds = dx_4 = dt$; or with restriction to terms which in the first stage of approximation are of first order:

$$\frac{d^2x_\tau}{dt^2} = [44, \tau] \quad (\tau = 1, 2, 3),$$

$$\frac{d^2x_4}{dt^2} = -[44, 4].$$

If in addition we suppose the gravitational field to be a quasistatic field, by confining ourselves to the case where the motion of the matter generating the gravitational field is slow (in comparison with the velocity of the propagation of light), we may neglect on the right-hand side differentiations with respect to the time in comparison with those with respect to the space co-ordinates, so that we have

$$\frac{d^2x_\tau}{dt^2} = -\frac{1}{2}\frac{\partial g_{44}}{\partial x_\tau} \quad (\tau = 1, 2, 3). \tag{67}$$

This is the equation of motion of the material point according to Newton's theory, in which $\frac{1}{2}g_{44}$ plays the part of the gravitational potential. What is remarkable in this result is that the component g_{44} of the fundamental tensor alone defines, to a first approximation, the motion of the mass-point.

We now turn to the field equations (53). Here we have to take into consideration that the energy-tensor of "matter" is almost exclusively defined by the density of matter in the narrower sense, i.e. by the second term of the right-hand side of (58) [or, respectively, (58a) or (58b)]. If we form the approximation in question, all the components vanish with the one exception of $T_{44} = \varrho = T$. On the left-hand side of (53) the second term is a small quantity of second order; the first yields, to the approximation in question,

$$\frac{\partial}{\partial x_1}[\mu\nu, 1] + \frac{\partial}{\partial x_2}[\mu\nu, 2] + \frac{\partial}{\partial x_3}[\partial\nu, 3] - \frac{\partial}{\partial x_4}[\mu\nu, 4].$$

For $\mu = \nu = 4$, this gives, with the omission of terms differentiated with respect to time,

$$-\frac{1}{2}\left(\frac{\partial^2 g_{44}}{\partial x_1^2} + \frac{\partial^2 g_{44}}{\partial x_2^2} + \frac{\partial^2 g_{44}}{\partial x_3^2}\right) = -\frac{1}{2}\nabla^2 g_{44}.$$

The last of equations (53) thus yields

$$\nabla^2 g_{44} = \varkappa\varrho. \tag{68}$$

The equations (67) and (68) together are equivalent to Newton's law of gravitation.

By (67) and (68) the expression for the gravitational potential becomes

$$-\frac{\varkappa}{8\pi}\int\frac{\varrho d\tau}{r}, \tag{68a}$$

while Newton's theory, with the unit of time which we have chosen, gives

$$-\frac{K}{c^2}\int\frac{\varrho d\tau}{r}$$

in which K denotes the constant $6{\cdot}7\times 10^{-8}$, usually called the constant of gravitation. By comparison we obtain

$$\varkappa = \frac{8\pi K}{c^2} = 1{\cdot}87\times 10^{-27} \tag{69}$$

§ 22. *Behaviour of Rods and Clocks in the Static Gravitational Field. Bending of Light-rays. Motion of the Perihelion of a Planetary Orbit*

To arrive at Newton's theory as a first approximation we had to calculate only one component, g_{44}, of the ten $g_{\mu\nu}$ of the gravitational field, since this component alone enters into the first approximation, (67), of the equation for the motion of the material point in the gravitational field. From this, however, it is already apparent that other components of the $g_{\mu\nu}$ must differ from the values given in (4) by small quantities of the first order. This is required by the condition $g = -1$.

For a field-producing point mass at the origin of co-ordinates, we obtain, to the first approximation, the radially symmetrical solution

$$\left.\begin{aligned} g_{\varrho\sigma} &= -\delta_{\varrho\sigma} - \alpha\frac{x_\varrho x_\sigma}{r^3} \qquad (\varrho, \sigma = 1, 2, 3) \\ g_{\varrho 4} &= g_{4\varrho} = 0 \qquad (\varrho = 1, 2, 3) \\ g_{44} &= 1 - \frac{\alpha}{r} \end{aligned}\right\}, \tag{70}$$

where $\delta_{\varrho\sigma}$ is 1 or 0, respectively, accordingly as $\varrho = \sigma$ or $\varrho \neq \sigma$, and r is the quantity $+\sqrt{(x_1^2+x_2^2+x_3^2)}$. On account of (68a)

$$\alpha = \frac{\varkappa M}{4\pi}, \tag{70a}$$

if M denotes the field-producing mass. It is easy to verify that the field equations (outside the mass) are satisfied to the first order of small quantities.

We now examine the influence exerted by the field of the mass M upon the metrical properties of space. The relation

$$ds^2 = g_{\mu\nu}\, dx_\mu\, dx_\nu,$$

always holds between the "locally" (§ 4) measured lengths and times ds on the one hand, and the differences of co-ordinates dx_ν on the other hand.

For a unit-measure of length laid "parallel" to the axis of x, for example, we should have to set $ds^2 = -1$; $dx_2 = dx_3 = dx_4 = 0$. Therefore $-1 = g_{11}\, dx_1^2$. If, in addition, the unit-measure lies on the axis of x, the first of equations (70) gives

$$g_{11} = -\left(1+\frac{\alpha}{r}\right).$$

From these two relations it follows that, correct to a first order of small quantities,

$$dx = 1-\frac{\alpha}{2r}. \tag{71}$$

The unit measuring-rod thus appears a little shortened in relation to the system of co-ordinates by the presence of the gravitational field, if the rod is laid along a radius.

In an analogous manner we obtain the length of co-ordinates in tangential direction if, for example, we set

$$ds^2 = -1; \quad dx_1 = dx_3 = dx_4 = 0; \quad x_1 = r, \quad x_2 = x_3 = 0.$$

The result is

$$-1 = g_{22}\, dx_2^2 = -dx_2^2. \tag{71a}$$

With the tangential position, therefore, the gravitational field of the point of mass has no influence on the length of a rod.

Thus Euclidean geometry does not hold even to a first approximation in the gravitational field, if we wish to take one and the same rod, independently of its place and orientation, as a realization of the same interval; although, to be sure, a glance at (70a) and (69) shows that the deviations to be expected are much too slight to be noticeable in measurements of the earth's surface.

Further, let us examine the rate of a unit clock, which is arranged to be at rest in a static gravitational field. Here we have for a clock period $ds = 1$; $dx_1 = dx_2 = dx_3 = 0$.

Therefore

$$1 = g_{44}\, dx_4^2;$$

$$dx_4 = \frac{1}{\sqrt{g_{44}}} = \frac{1}{\sqrt{(1+(g_{44}-1))}} = 1 - \frac{1}{2}(g_{44}-1);$$

or

(72)
$$dx_4 = 1 + \frac{\varkappa}{8\pi}\int \varrho \frac{d\tau}{r}.$$

Thus the clock goes more slowly if set up in the neighbourhood of ponderable masses. From this it follows that the spectral lines of light reaching us from the surface of large stars must appear displaced towards the red end of the spectrum.[12]

We now examine the course of light-rays in the static gravitational field. By the special theory of relativity the velocity of light is given by the equation

$$-dx_1^2 - dx_2 - dx_3^2 + dx_4^2 = 0$$

and therefore by the general theory of relativity by the equation

(73)
$$ds^2 = g_{\mu\nu}\, dx_\mu\, dx_\nu = 0.$$

If the direction, i.e. the ratio $dx_1 : dx_2 : dx_3$ is given, equation (73) gives the quantities

$$\frac{dx_1}{dx_4}, \frac{dx_2}{dx_4}, \frac{dx_3}{dx_4}$$

[12] According to E. Freundlich, spectroscopical observations on fixed stars of certain types indicate the existence of an effect of this kind, but a crucial test of this consequence has not yet been made.

and accordingly the velocity

$$\sqrt{\left[\left(\frac{dx_1}{dx_4}\right)^2+\left(\frac{dx_2}{dx_4}\right)^2+\left(\frac{dx_3}{dx_4}\right)^2\right]} = \gamma$$

defined in the sense of Euclidean geometry. We easily recognize that the course of the light-rays must be bent with regard to the system of co-ordinates, if the $g_{\mu\nu}$ are not constant. If n is a direction perpendicular to the propagation of light, the Huyghens principle shows that the light-ray, envisaged in the plane (γ, n), has the curvature $-\partial\gamma/\partial n$.

We examine the curvature undergone by a ray of light passing by a mass M at the distance $\triangle$. If we choose the system of co-ordinates in agreement with the accompanying diagram, the total bending of the ray (calculated positively if concave towards the origin) is given to a sufficient approximation by

$$B = \int_{-\infty}^{+\infty} \frac{\partial\gamma}{\partial x_1}\, dx_2,$$

while (73) and (70) give

$$\gamma = \sqrt{\left(-\frac{g_{44}}{g_{22}}\right)} = 1 - \frac{\alpha}{2r}\left(1+\frac{x_2^2}{r^2}\right).$$

Carrying out the calculation, this gives

$$B = \frac{2\alpha}{\Delta} = \frac{\varkappa M}{2\pi\Delta}. \tag{74}$$

According to this, a ray of light going past the sun undergoes a deflexion of 1·7″ and a ray going past the planet Jupiter a deflexion of about ·02″.

If we calculate the gravitational field to a higher degree of approximation, and likewise (with corresponding accuracy) the orbital motion of a mass-point of relatively infinitely small mass, we find a deviation of the following kind from the Kepler–Newton laws of planetary motion. The orbital ellipse of a planet undergoes a slow

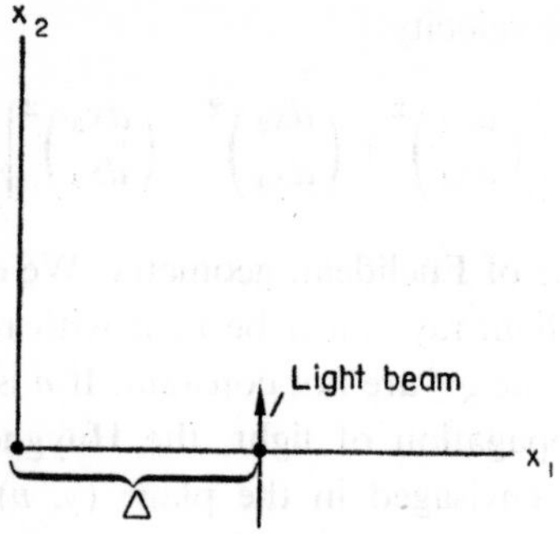

FIG. 1.

rotation, in the direction of motion, of amount

$$\varepsilon = 24\pi^3 \frac{\alpha^2}{T^2c^2(1-e^2)} \tag{75}$$

per revolution. In this formula *a* denotes the major semiaxis, *c* the velocity of light in the usual measurement, *e* the eccentricity, *T* the time of revolution in seconds.[13]

Calculation gives for the planet Mercury a rotation of the orbit of 43″ per century, corresponding exactly to astronomical observation (Leverrier); for the astronomers have discovered in the motion of the perihelion of this planet, after allowing for disturbances by other planets, an inexplicable remainder of this magnitude.

[13] For the calculation I refer to the original papers: A. Einstein, *Sitzungsber. d. Preuss. Akad. d. Wiss.*, 1915, p. 831; K. Schwarzschild, *ibid.*, 1916, p. 189.

NOTES ON EXTRACT 5

WHEN the general theory of relativity had been formulated, there were two distinct assumptions: (i) the geometry was Riemannian, and the physical meaning of this was that the path of a particle was a geodesic; (ii) the geometry satisfied the field equations. Now a small particle is represented by a singularity in the solution of the field equations (source), so Einstein felt that (ii) should determine (i) without further assumption. He had succeeded in proving this by 1938 but the present paper with Infeld gives a much simpler development.

EXTRACT 5[†]

On the Motion of Particles in General Relativity Theory

A. EINSTEIN and L. INFELD

[*Received February 12, 1949*]

1. Introduction

The gravitational field manifests itself in the motion of bodies. Therefore the problem of determining the motion of such bodies from the field equations alone is of fundamental importance. This problem was solved for the first time some ten years ago and then equations of motion for two particles were deduced [1]. A more general and simplified version of this problem was given shortly thereafter [2].

Mr. Lewison pointed out to us, that from our approximation procedure, it does not follow that the field equations can be solved up to an arbitrarily high approximation. This is indeed true. We believe that the present work not only removes this difficulty, but that it gives a new and deeper insight into the problem of motion. From the logical point of view the present theory is considerably simpler and clearer than the old one. But as always, we must pay for these logical simplifications by prolonging the chain of technical argument.

The subject matter is presented here from the beginning and the knowledge of previous work is not assumed. To facilitate the reading for those who have studied the previous papers we use here essentially the same notation as before.

† *Can. J. Math.* **1,** 209 (1949).

Let us start with some general remarks.

All attempts to represent matter by an energy-momentum tensor are unsatisfactory and we wish to free our theory from any particular choice of such a tensor. Therefore we shall deal here only with gravitational equations in empty space, and matter will be represented by singularities of the gravitational field.

In Newtonian mechanics, particles are represented as singularities of a scalar field φ, which satisfies Laplace's equation everywhere outside the singularities. Because the classical equation is linear, the field can be decomposed into partial fields, each part due to a single particle. Each particle *is* in a field due to all other particles. The theory is completed by the equation of motion, that is by putting the acceleration equal to the negative gradient of the field, the proportionality factor being a universal constant. Thus classical physics postulates the equations of motion independently of the field laws. The masses of the sources of the field are assumed to be independent of time. The laws of motion are supposed to be valid in an inertial system. Therefore space–time appears as an independent physical entity. The conceptual weakness of such a space–time background in the classical theory was already recognized by Newton.

If we compare this state of affairs with that in general relativity theory, in its original formulation, we see striking similarities and differences. Laplace's equation

$$\Delta\varphi = 0$$

is replaced by the gravitational equation

$$R_{hl} = 0,$$

which, however, unlike the classical equation, satisfies the general relativity principle. The classical principle of inertia becomes in relativity theory the principle of the geodesic line valid for a particle with infinitely small mass. True enough, the difficulty with the inertial system disappears in relativity theory, as does the independent physical reality of space–time. Yet the equations of motion still appear independently of the field equations.

Our aim is to investigate to what extent the field equations *alone* contain the equations of motion of particles; also to develop a method that will allow us to find these equations of motion up to an arbitrary approximation.

Let us start with a simple remark: a *linear* law always means that the motion of singularities is arbitrary. If to a world-line of a singularity with mass m_1 there belongs a field $F_{(1)}$ and if to a world-line of a singularity with mass m_2 there belongs a field $F_{(2)}$, then the superposition of these two fields, that is $F_{(1)}+F_{(2)}$ is also a solution of the linear field equations. In such a solution the same two world-lines would appear together that before appeared singly. Therefore the field with its linear laws cannot imply any interaction between the singularities. Thus only non-linear field equations can provide us with equations of motion because only non-linearity can express the interaction between singularities.

But the argument cannot be reversed. Non-linearity is necessary but not sufficient for the equations of motion to follow from the field equations.

The reason why the gravitational field equations do provide us with equations of motion lies not in their non-linear character alone, but also in the fact that these equations are not independent from each other. Indeed. among the ten components four are free, this being due to the freedom of choice in the co-ordinate system. The ten equations are valid, so to speak, only for six effective functions. They would be inconsistent were it not for the four (Bianchi) identities that they satisfy. This must be so for every relativistic system of equations derived from a variational principle. These identities are (besides the non-linearity) responsible for the *equations of motion being determined by the field equations.*

The ideas leading to the equations of motion are not easy and are mutually interwoven.

One of the essential ideas in this paper is the treatment of gravitational equations by a "new approximation method". In it we treat space and time differently. We regard the changes of the field in time as small compared with those in space. Only then do we arrive at a consistent, manageable set of equations that can be solved step by step. This idea is not new and was contained in the previous papers.

The other important idea is the deduction of the equations of motion, which are *ordinary* differential equations, from the field equations which are *partial* differential equations. This idea, treated here differently than in the previous papers, leads to the use of surface integrals taken around the singularities of the field. These surface integrals will depend only on the motion of the singularities and not on the shape of the surface.

These and other ideas will be treated in detail in this paper. To make them clear we have decided to delegate all the more tedious calculations to the Appendices. (If we refer, for example, to A.4, this means the Appendix belonging to Sec. 4.) But even so, many straightforward but long calculations had to be omitted. This is especially true for the calculations that lead beyond Newtonian motion. We included here a short section on this subject, just for the sake of completeness. But, as in [1], so here we have to refer those who would like to see the full calculations to the manuscript which is deposited at the Institute for Advanced Study.

Finally we should like to thank Mr. Lewison for his critical study of our previous papers, and Mr. Schild for a careful and critical reading of this manuscript.

2. Notations

Since in the greater part of our work we shall have to separate space and time, our notation will not be the usual four-dimensional one. We make the conventions: Latin indices take the values 1, 2, 3, and they refer to space co-ordinates only. Greek indices refer to both space and time, running over the values 0, 1, 2, 3. Repetition of indices implies summation.

The expression

$$g_{\mu\nu|\sigma} \text{ etc. stands for } \frac{\partial g_{\mu\nu}}{\partial x^\sigma} \text{ etc.} \tag{2.1}$$

At infinity the gravitational field takes the Galilean values $\eta_{\mu\nu}$, that is:

$$\eta_{mn} = -\delta_{mn}; \quad \eta_{0n} = 0; \quad \eta_{00} = 1. \tag{2.2}$$

We write:

$$g_{\mu\nu} = \eta_{\mu\nu} + h_{\mu\nu}; \quad g^{\mu\nu} = \eta^{\mu\nu} + h^{\mu\nu}, \tag{2.3}$$

where $h_{\mu\nu}$ represents the deviation of space–time from flat space and it is not assumed to be small.

The $h^{\mu\nu}$ can be calculated as functions of $h_{\mu\nu}$ by means of the relation

$$g_{\mu\sigma} g^{\mu\nu} = \delta^{\nu}_{\sigma}. \tag{2.4}$$

It turns out to be convenient to replace the h's by γ's which are their linear combinations:

$$\gamma_{\mu\nu} = h_{\mu\nu} - \tfrac{1}{2}\eta_{\mu\nu}\eta^{\sigma\varrho}h_{\sigma\varrho}, \tag{2.5}$$

or more explicitly:

$$\gamma_{00} = \tfrac{1}{2}h_{00} + \tfrac{1}{2}h_{ss}, \tag{2.6}$$

$$\gamma_{0n} = h_{0n}, \tag{2.7}$$

$$\gamma_{mn} = h_{mn} - \tfrac{1}{2}\delta_{mn}h_{ss} + \tfrac{1}{2}\delta_{mn}h_{00}. \tag{2.8}$$

This replacement is, of course, not very important but it does simplify the calculations.

Thus we can, throughout, replace the h's by the γ's. The equations of the gravitational field for empty space,

$$R_{\mu\nu} = 0, \tag{2.9}$$

can be written (see A.2) in the following way:

$$\Phi_{00} + 2\Lambda_{00} = 0, \tag{2.10}$$

$$\Phi_{0n} + 2\Lambda_{0n} = 0, \tag{2.11}$$

$$\Phi_{mn} + 2\Lambda_{mn} = 0, \tag{2.12}$$

where:

(2.13) $$\Phi_{00} = -\gamma_{00|ss},$$

(2.14) $$\Phi_{0m} = -\gamma_{0m|ss}+\gamma_{0s|sm},$$

(2.15) $$\Phi_{mn} = -\gamma_{mn|ss}+\gamma_{ms|ns}+\gamma_{ns|mn}-\delta_{mn}\gamma_{rs|rs},$$

and:

(2.16) $$2\Lambda_{00} = \gamma_{sr|rs}+2\Lambda'_{00},$$

(2.17) $$2\Lambda_{0m} = \gamma_{ms|s0}-\gamma_{00|m0}+2\Lambda'_{0m},$$

(2.18) $$2\Lambda_{mn} = -\gamma_{0m|0n}-\gamma_{0n|0m}+2\delta_{mn}\gamma_{0s|0s} +\gamma_{mn|00}-\delta_{mn}\gamma_{00|00}+2\Lambda'_{mn}.$$

In these formulae, all the linear terms are written out explicitly, while $\Lambda'_{\mu\nu}$ stands for all the non-linear terms in the γ's. The division of the linear expressions into those belonging to $\Phi_{\mu\nu}$ and those belonging to $\Lambda_{\mu\nu}$ may seem artificial at this moment. In anticipation of further development, we shall remark here, that, in the actual approximation procedure, by which we shall solve the gravitational equations, these linear terms collected in $\Lambda_{\mu\nu}$ will behave like the non-linear terms.

3. Lemma

We mentioned in the introduction that the differential equations of motion will be derived by forming surface integrals. The technique of calculating such surface integrals will reappear many times in this paper and it is based on a lemma to which we shall refer as *the* lemma. Here we shall give its formulation and its proof.

We have a set of functions:

(3.1) $$F_{(\alpha a\ldots)kl}.$$

It is immaterial whether these functions of x^μ have tensorial character, or not. The bracketed indices are Greek, or Latin, and they will not play any role in our argument. But we do assume that these functions are skew-symmetric in the indices k, l:

$$F_{(\ldots)kl} = -F_{(\ldots)lk}. \tag{3.2}$$

We now form an integral

$$\int_{(S_2)} F_{(\ldots)kl|l} n_k \, dS \tag{3.3}$$

over an *arbitrary two-dimensional closed* surface that does not pass through the singularities of the field. In (3.3)

$$n_k = \cos(x^k, \vec{n}) \tag{3.4}$$

are the components of the "normal unit" vector to the surface. The words "normal", and "unit" are used in the conventional sense to designate the corresponding functions of the co-ordinates, which are implied by these terms in Euclidean geometry. They have nothing to do with any particular metric.

Our lemma is:

$$\int_{(S_2)} F_{(\ldots)kl|l} n_k \, dS = 0. \tag{3.5}$$

We see that the integral (3.3) is certainly independent of the shape of the surface, because

$$F_{(\ldots)kl|lk} = 0, \tag{3.6}$$

and because of Green's theorem. We can also write the integral (3.3) in the form

$$\int_{(S_2)} \operatorname{curl}_n \vec{A} \, dS, \tag{3.7}$$

where

$$F_{(\ldots)23} = A_1; \quad F_{(\ldots)31} = A_2; \quad F_{(\ldots)12} = A_3.$$

But (3.7) and therefore (3.3) can be changed, by Stokes' theorem, into a line integral over the rim of the surface. If the surface is closed, the rim is of zero length. Therefore, our lemma as expressed by (3.5) is proved.

4. Surface Integrals

We treat particles of matter as singularities of the field. Let us assume p particles and the knowledge of their world lines. Thus we denote by

$$\overset{s}{\xi}{}^k(x^0); \qquad s = 1, 2, 3, \ldots, p, \tag{4.1}$$

the world-line of the sth singularity. Here and later, the index written on the top will always label the particular singularity.

The gravitational field, that is the γ's, will depend on the x^μ's but also on the ξ's and their time derivatives. The equations that the γ's fulfil are

$$\Phi_{\mu\nu}+2\Lambda_{\mu\nu} = 0. \tag{4.2}$$

At an arbitrary moment x^0, let us surround the sth singularity, and it alone, by a closed surface. Then:

$$\int^s (\Phi_{\mu k}+2\Lambda_{\mu k})n_k \, dS = 0, \tag{4.3}$$

where the s over the integral indicates here, and later too, that the integral is to be taken on a two-dimensional surface surrounding the sth singularity and it alone.

We shall show that

$$\int^s \Phi_{\mu k} n_k \, dS = 0. \tag{4.4}$$

Indeed it follows from the definition (2.14) and (2.15) of $\Phi_{\mu k}$ that it can be written in the following form:

$$\Phi_{\mu k} = F_{(\mu)kl|l}, \tag{4.5}$$

$$F_{(\mu)kl} = \gamma_{\mu l|k}-\gamma_{\mu k|l}-\delta_{\mu k}\gamma_{lr|r}+\delta_{\mu l}\gamma_{kr|r}. \tag{4.6}$$

But $F_{(\mu)kl}$ is skew-symmetric in k and l. Therefore (4.4) is fulfilled. From it and from (4.3) we deduce:

$$\int^s 2\Lambda_{\mu k} n_k \, dS = 0. \tag{4.7}$$

Also, because of the structure of $\Phi_{\mu k}$ we easily verify:

$$\Phi_{\mu n|n} = 0, \tag{4.8}$$

therefore also:

$$\Lambda_{\mu n|n} = 0. \tag{4.9}$$

Equation (4.9) tells us that no surface integral of the form (4.7) can depend on the shape of the surface. But equation (4.7) tells us more; namely that such an integral vanishes.

The $4p$ surface integrals in (4.7) can give us no relation between the space co-ordinates of the field, because the surface is entirely arbitrary. They can only give us relations between the co-ordinates of the singularities and their time derivatives. Thus we may have at most $4p$ differential equations. Anticipating the later development, we may remark here that these equations will determine $3p$ functions of time

$$\overset{s}{\xi}{}^k(x^0),$$

that is, the motion of singularities.

5. The Method of Approximation

The problem before us is to solve our field equations and to deduce the equations of motion. This we shall do by a new approximation procedure. Let us assume a function $\varphi(x^\mu, \lambda)$ developed into a power series in the parameter λ (for small λ):

$$\varphi(x^\mu, \lambda) = \lambda^0 \underset{0}{\varphi} + \lambda^1 \underset{1}{\varphi} + \lambda^2 \underset{2}{\varphi} + \ \dots \ = \sum_{l=0}^{\infty} \lambda^l \underset{l}{\varphi}. \tag{5.1}$$

The indices below indicate the *order* (l in λ^l is always the exponent, not the index).

If the function φ varies quickly in space, but slowly with x^0, then we are justified in not treating all its derivatives in the same fashion. The derivatives with respect to x^0 will be of a higher order than space derivatives. We can formalize the procedure by introducing an *auxiliary time* τ,

(5.2) $$\tau = x^0\lambda,$$

so that derivatives with respect to τ can be treated on the same footing as the space derivatives:

(5.3) $$\varphi_{|0} = \frac{\partial\varphi}{\partial x^0} = \frac{\partial\varphi}{\partial\tau}\lambda = \lambda\varphi_{,0}.$$

We conclude: the "stroke differentiation" of a quantity with respect to x^0, can be replaced by the "comma differentiation" with respect to τ if the power of λ with which this quantity is associated is simultaneously raised by one. To express this explicitly we use numbers under zeros, written after the comma, e.g.:

(5.4) $$\lambda^{2l}\underset{2l}{\gamma}_{mn|0} = \lambda^{2l+1}\underset{2l}{\gamma}_{mn,\underset{1}{0}} \quad \text{or:} \quad \lambda^{2l}\underset{2l}{\gamma}_{mn|00} = \lambda^{2l+2}\underset{2l}{\gamma}_{mn,\underset{2}{00}}.$$

From now on, all differentiations will be with respect to (τ, x^1, x^2, x^3) and they will be denoted by commas:

(5.5) $$\gamma_{\dots|s} = \gamma_{\dots,s}; \quad \gamma_{\dots|0} = \lambda\gamma_{\dots,\underset{1}{0}}.$$

Thus we shall develop all functions that appear in the field equations in power series in λ. We start with the γ's in the following way:

(5.6) $$\begin{cases} \gamma_{00} = \lambda^2\underset{2}{\gamma}_{00} + \lambda^4\underset{4}{\gamma}_{00} + \lambda^6\underset{6}{\gamma}_{00} + \dots, \\ \gamma_{0m} = \lambda^3\underset{3}{\gamma}_{0m} + \lambda^5\underset{5}{\gamma}_{0m} + \dots, \\ \gamma_{mn} = \lambda^4\underset{4}{\gamma}_{mn} + \lambda^6\underset{6}{\gamma}_{mn} + \dots. \end{cases}$$

Why do we start with different powers of λ? This is an assumption, but it can be justified heuristically. Assuming for a moment the usual energy momentum tensor for matter, we have, for a quasi-stationary field, approximately:

$$(5.7)\qquad \begin{cases} \Delta\gamma_{00} = -2\varrho, \\ \Delta\gamma_{0m} = -2\varrho \dfrac{dx^m}{d\tau}\lambda, \\ \Delta\gamma_{mn} = -2\varrho \dfrac{dx^m}{d\tau}\dfrac{dx^n}{d\tau}\lambda^2, \end{cases}$$

therefore

$$(5.8)\qquad \gamma_{mn} \sim \lambda\gamma_{0m} \sim \lambda^2\gamma_{00},$$

and it is pure convention that we start with λ^2 for γ_{00}.

The other question suggested by (5.6) is: why do we omit the odd powers of λ in the developments of γ_{00}, γ_{mn}, and the even powers in γ_{0m}? Indeed, we could have introduced all powers in (5.6). A more thorough investigation shows that our choice (5.6) means that what we are doing here is similar to the procedure in electromagnetic theory when we take not the retarded, but the half-retarded plus half-advanced potentials [3].

All the functions that will appear later are gained from the γ's by summation, multiplication, differentiation. Thus to every component, the following rule applies throughout: Any component having an $\begin{Bmatrix}\text{odd}\\ \text{even}\end{Bmatrix}$ number of zero suffixes will have only $\begin{Bmatrix}\text{odd}\\ \text{even}\end{Bmatrix}$ powers of λ in its expansion.

6. Field Equations and the Approximation Method

We go back to the field equations

$$(6.1)\qquad \Phi_{\mu\nu}+2\Lambda_{\mu\nu} = 0$$

into which we introduce the γ's in their power-series development. Thus the (00) equation in (6.1) can be written:

$$(6.2)\qquad \sum_l \lambda^{2l}\big(\underset{2l}{\Phi_{00}}+2\underset{2l}{\Lambda_{00}}\big) = 0.$$

Now we cut up (6.2), and the other field equations, into equations for each approximation step. We write them down in the following form:

(6.3a) $$\underset{2l-2}{\Phi_{00}} + 2\underset{2l-2}{\Lambda_{00}} = 0,$$

(6.3b) $$\underset{2l-1}{\Phi_{0m}} + 2\underset{2l-1}{\Lambda_{0m}} = 0,$$

(6.3c) $$\underset{2l}{\Phi_{mn}} + 2\underset{2l}{\Lambda_{mn}} = 0.$$

Let us analyse more closely the structure of (6.3). Remembering (2.13) to (2.15) we can write more explicitly:

(6.4a) $$\underset{2l-2}{\Phi_{00}} = -\underset{2l-2}{\gamma_{00,rr}},$$

(6.4b) $$\underset{2l-1}{\Phi_{0m}} = -\underset{2l-1}{\gamma_{0m,rr}} + \underset{2l-1}{\gamma_{0r,mr}},$$

(6.4c) $$\underset{2l}{\Phi_{mn}} = -\underset{2l}{\gamma_{mn,rr}} + \underset{2l}{\gamma_{mr,nr}} + \underset{2l}{\gamma_{nr,mr}} - \delta_{mn}\underset{2l}{\gamma_{rs,rs}},$$

and:

(6.5a) $$2\underset{2l-2}{\Lambda_{00}} = \underset{2l-2}{\gamma_{rs,rs}} + 2\underset{2l-2}{\Lambda'_{00}},$$

(6.5b) $$2\underset{2l-1}{\Lambda_{0m}} = -\underset{2l-2\;1}{\gamma_{00,0m}} + \underset{2l-2\;1}{\gamma_{mr,0r}} + 2\underset{2l-1}{\Lambda'_{0m}},$$

(6.5c) $$\begin{cases} 2\underset{2l}{\Lambda_{mn}} = -\underset{2l-1\;1}{\gamma_{0m,0n}} - \underset{2l-1\;1}{\gamma_{0n,0m}} + 2\delta_{mn}\underset{2l-1\;1}{\gamma_{0r,0r}} \\ \qquad + \underset{2l-2\;2}{\gamma_{mn,00}} - \delta_{mn}\underset{2l-2\;2}{\gamma_{00,00}} + 2\underset{2l}{\Lambda'_{mn}}. \end{cases}$$

Let us now assume that:

(6.6a) $$\underset{2}{\gamma_{00}} \cdots \underset{2l-4}{\gamma_{00}},$$

(6.6b) $$\underset{3}{\gamma_{0m}} \cdots \underset{2l-3}{\gamma_{0m}},$$

(6.6c) $$\underset{4}{\gamma_{mn}} \cdots \underset{2l-2}{\gamma_{mn}},$$

are all known. Then $\underset{2l-2}{\gamma_{00}}$ can be found from (6.3a). Indeed $\underset{2l-2}{\Lambda_{00}}$ contains only terms already known, since $\underset{2l-2}{\gamma_{mn}}$ is known and $\underset{2l-2}{\Lambda'_{00}}$ is non-linear and can therefore depend only on the known γ's. The same is true for (6.3b) and (6.3c). The unknown functions are contained in

Φ's; the known functions in the Λ's. The $\underset{2l-2}{\eta_{00}}$, already found from (6.3a), appears as a known function in $\underset{2l-1}{\Lambda_{0m}}$. Similarly $\underset{2l-1}{\gamma_{0m}}$ found from (6.3b) appears as known in $\underset{2l}{\Lambda_{mm}}$. Indeed we see now the reasons for our division of linear terms.

Thus our equations (6.3), if solved, will give us

$$\underset{2l-2}{\gamma_{00}},\ \underset{2l-1}{\gamma_{0m}},\ \underset{2l}{\gamma_{mn}}, \tag{6.7}$$

and if such a procedure converges, we can determine the field to any approximation we wish.

The important question to consider is: are the equations (6.3) always solvable?

7. The Divergence Condition

We go back to our equations (6.3). The first of them, that is

$$\underset{2l-2}{\Phi_{00}}+2\underset{2l-2}{\Lambda_{00}} = 0 \tag{7.1}$$

is, because of (6.4a) and (6.5a), a Poisson equation, where $\underset{2l-2}{\Lambda_{00}}$ is known. There is no difficulty in integrating this equation and finding $\underset{2l-2}{\gamma_{00}}$. Next we have (6.3b), and because of (6.4b), we see:

$$\underset{2l-1}{\Phi_{0m,m}} = 0. \tag{7.2}$$

Thus the next three equations can be integrated only if

$$\underset{2l-1}{\Lambda_{0m,m}} = 0. \tag{7.3}$$

But $\underset{2l-1}{\Lambda_{0m}}$ is already known. Therefore we must be sure that our procedure leads us to $\underset{2l-1}{\Lambda_{0m}}$ satisfying (7.3). Similarly the last six equations (6.3c) lead us because of

$$\underset{2l}{\Phi_{mn,n}} = 0 \tag{7.4}$$

to the integrability condition:

(7.5) $$\underset{2l}{\Lambda}_{mn,\,n} = 0.$$

We shall prove that (7.3) and (7.5) are satisfied, if the field equations are satisfied in *all the previous* approximations.

The tensor

(7.6) $$G_{\mu\nu} = R_{\mu\nu} - \tfrac{1}{2}g_{\mu\nu}R$$

satisfies the Bianchi identity

(7.7) $$G^{\mu}_{\nu|\mu} + \begin{Bmatrix} \alpha \\ \alpha\beta \end{Bmatrix} G^{\beta}_{\nu} - \begin{Bmatrix} \beta \\ \nu\alpha \end{Bmatrix} G^{\alpha}_{\beta} = 0.$$

We assume that all field equations up to the order $(2l-2)$ are satisfied, that is including

$$\underset{2l-2}{\Phi_{00}} + 2\underset{2l-2}{\Lambda_{00}} = 0.$$

We know, that putting $\Phi_{\mu\nu} + 2\Lambda_{\mu\nu} = 0$ is equivalent to putting $R_{\mu\nu} = 0$. From A.2 follows:

(7.8) $$\Phi_{\mu\nu} + 2\Lambda_{\mu\nu} = -2(R_{\mu\nu} - \tfrac{1}{2}\eta_{\mu\nu}\eta^{\alpha\beta}R_{\alpha\beta}),$$

which means that our $\Phi_{\mu\nu} + 2\Lambda_{\mu\nu}$ are a linear combination of the $R_{\mu\nu}$. Thus, if our field equations are satisfied, then we have:

(7.9) $$\begin{cases} \underset{2}{G_{00}} = \underset{4}{G_{00}} = \ldots = \underset{2l-2}{G_{00}} = 0, \\ \underset{3}{G_{0m}} = \underset{5}{G_{0m}} = \ldots = \underset{2l-3}{G_{0m}} = 0, \\ \underset{2}{G_{mn}} = \underset{4}{G_{mn}} = \ldots = \underset{2l-2}{G_{mn}} = 0. \end{cases}$$

Let us write down the *zero* Bianchi identity of the order $(2l-1)$. From the left-hand side of (7.7) we have, putting $\nu = 0$, the following linear terms:

(7.10) $$-\underset{2l-1}{G_{0m,\,m}} + \underset{2l-2\;1}{G_{00,\,0}}.$$

The non-linear part contains the products of the G's and the γ's. But because of (7.9), both the non-linear part of the Bianchi identity and the second expression in (7.10) vanish. Thus the *zero* Bianchi identity, together with the field equations give:

$$\underset{2l-1}{G_{0m,\,m}} = 0. \tag{7.11}$$

Because of (7.8), (7.6) and (7.2) this means:

$$\underset{2l-1}{\Lambda_{0m,\,m}} = 0. \tag{7.12}$$

Going on to the next approximation step, let us now assume that besides (7.9), we have also:

$$\underset{2l-1}{G_{0m}} = 0. \tag{7.13}$$

Putting into Bianchi identity (7.7) $\nu = m$, we have in the $2l$ order, because of (7.9) and (7.13):

$$\underset{2l}{G_{mn,\,n}} = 0 \tag{7.14}$$

and therefore because of (7.4), (7.8):

$$\underset{2l}{\Lambda_{mn,\,n}} = 0. \tag{7.15}$$

Thus the divergence conditions are satisfied in each approximation step, though not identically. They are satisfied because of the Bianchi identities and because of the previous field equations.

8. The Surface Condition and the Equations of Motion

We now approach the most essential part of our argument. We are faced with the task of solving the following system of equations:

$$\underset{2l-2}{\Phi_{00}} + 2\underset{2l-2}{\Lambda_{00}} = 0, \tag{8.1a}$$

$$\underset{2l-1}{\Phi_{0m}} + 2\underset{2l-1}{\Lambda_{0m}} = 0, \tag{8.1b}$$

$$\underset{2l}{\Phi_{mn}} + 2\underset{2l}{\Lambda_{mn}} = 0. \tag{8.1c}$$

We know that because of the Bianchi identities and because (as we assumed) similar equations had been solved in the previous approximations, we have

$$\underset{2l-1}{\Lambda_{0m,\,m}} = 0; \quad \underset{2l}{\Lambda_{mn,\,n}} = 0. \tag{8.2}$$

Let us also remember that there is no difficulty in solving (8.1a) which is a Poisson equation. But what about (8.1b) and (8.1c)?

Before we return to this fundamental question, we wish to discuss the *start* of our approximation procedure which determines the character of our calculations.

In (8.1) we put $l = 2$ and write the first two equations explicitly:

$$\underset{2}{\gamma_{00,\,ss}} = 0, \tag{8.3a}$$

$$-\underset{3}{\gamma_{0m,\,ss}} + \underset{3}{\gamma_{0s,\,ms}} = \underset{2}{\gamma_{00,}}\underset{1}{{}_{0m}}. \tag{8.3b}$$

The character of the entire solution will depend on the choice of the harmonic function we take as the solution of (8.3a). As we are interested in solutions representing particles, we shall write:

$$\begin{cases} \underset{2}{\gamma_{00}} = 2\varphi; \quad \varphi = \sum\limits_{s=1}^{p} \left\{ -2\underset{2}{\overset{s}{m}}\overset{s}{\psi} \right\}, \\ \overset{s}{\psi} = \left[\left(x^k - \overset{s}{\xi^k}\right)\left(x^k - \overset{s}{\xi^k}\right) \right]^{-\frac{1}{2}} = \left(\overset{s}{r}\right)^{-1}. \end{cases} \tag{8.4}$$

Here $\overset{s}{r}$ is the "distance" in space of a point from the sth singularity.

We leave it undecided, for the moment, whether $\underset{2}{\overset{s}{m}}$ is a function of time, or a constant. Now we introduce this $\underset{2}{\gamma_{00}}$ into (8.3b) and again obtain three equations for the three functions $\underset{3}{\gamma_{0m}}$. But is (8.3b) always solvable? True, the divergence of both sides vanishes. But this is not sufficient. The surface integral of the left-hand side of (8.3b) vanishes, as follows from the lemma. But then the surface integral of the right-hand side of (8.3b) must vanish too. If we calculate the surface integral around each singularity, we find (see A.4) that it vanishes only if

(8.5) $$\frac{d}{dr}\left(\overset{s}{\underset{2}{m}}\right) = \overset{s}{\underset{2}{m}}{}_{,\underset{1}{0}} = \overset{s}{\underset{3}{\dot{m}}} = 0,$$

that is if the $\underset{2}{m}$'s do not depend on time. This is so, because

(8.6) $$\overset{s}{\psi}{}_{,0} = -\overset{s}{\psi}{}_{,k}\overset{s}{\dot{\xi}}{}^{k}; \quad \left(\dot{\xi}^{k} = \frac{d\xi^{k}}{d\tau}\right)$$

and because only expressions proportional to r^{-2} can give a contribution to the surface integral. Thus, going back to (8.4), we have to assume that

(8.7) $$\overset{1}{\underset{2}{m}}, \overset{2}{\underset{2}{m}}, \overset{3}{\underset{2}{m}}, \ldots, \overset{p}{\underset{2}{m}}$$

are constant.

These constants (8.7) can be positive or negative. We shall assume that $\overset{s}{\underset{2}{m}}$ are positive. Indeed, by taking the first particle and removing all others, we see that $\overset{1}{\underset{2}{m}}$ is its *gravitational mass*, since for large r the field is that of a particle with gravitational mass $\overset{1}{\underset{2}{m}}$. This is the same constant of integration that appears in the Schwarzschild solution, since our field for one particle is that of a Schwarzschild singularity when r is large. *Thus we shall have to exclude from our solution negative gravitational masses. But then we must also exclude dipoles and poles of higher order.*

Yet if we try to solve (8.1) we see (the details will be presented later) that we cannot do so without adding certain poles and dipoles to $\underset{2l-2}{\gamma_{00}}$. This we shall have to do, in order to insure the integrability of (8.1) in each approximation. But then the solution of the total field will contain dipoles which are not allowed, since they represent physically meaningless solutions. We shall have to remove them *after* the total field has been calculated. This can be done by *restricting the motion of particles*. That is, the condition that the dipole field vanishes will give us $3p$ ordinary differential equations for the motion of p particles. Thus the motion is undetermined in the approximation procedure. It becomes determined after the approximation procedure is finished and the dipole fields are removed.

In practice, we find solutions both for the field and for the equations of motion only to a certain approximation, say $2n$. We obtain the equations of motion to the $2n$ approximation, by removing all the dipole fields to such an approximation.

Although we have developed our field equations with respect to an arbitrary parameter λ, this λ can be absorbed by the actual equations of motion through the change of scale in $\overset{s}{\underset{2}{m}}$ and τ, so that λ is absent from the final form of the equations.

We have given a general outline of our treatment. Turning to the details, let us see why (8.1) will not, generally, be integrable. We know, from the contents of Sec. 4, particularly from (4.4) that the surface integrals of the Φ functions vanish. Although this was proved for the total field it is equally true in each approximation step, since the proof made use only of the structure of the Φ's, which is the same for the total field, as for the field in each approximation. Thus we have:

(8.8) $$\int^{S} \underset{2l-1}{\Phi_{0r}} n_r \, dS = 0; \quad \int^{S} \underset{2l}{\Phi_{mr}} n_r \, dS = 0.$$

But then our equations (8.1) can be self-consistent, only if we have:

(8.9) $$\int^{S} 2\underset{2l-1}{\Lambda_{0r}} n_r \, dS = 0; \quad \int^{S} 2\underset{2l}{\Lambda_{mr}} n_r \, dS = 0.$$

But the Λ's in (8.1) are already known; they are functions of the *known* field calculated in the previous approximation steps. Therefore we can calculate the integrals (8.9) and find whether they vanish or not.

At this point it is convenient to introduce a new notation. Because of (8.2) the surface integrals (8.9) will not depend on the shape of the surface, but only on the singularities and their motion. Thus the surface integrals, even if they do not vanish, can be functions of τ only.

We write:

$$\frac{1}{4\pi}\int^{s} 2\underset{2l-1}{\Lambda_{0r}} n_r\, dS = \underset{2l-1}{\overset{s}{C}}{}_0(\tau) = \underset{2l-1}{\overset{s}{C}}{}_0, \tag{8.10}$$

$$\frac{1}{4\pi}\int^{s} 2\underset{2l}{\Lambda_{mr}} n_r\, dS = \underset{2l}{\overset{s}{C}}{}_m(\tau) = \underset{2l}{\overset{s}{C}}{}_m, \tag{8.11}$$

and assume that we have calculated the C's. If they vanish identically, and if they vanish always as we proceed with our approximation, then our equations are self-consistent.

Let us assume, however, that the C's in (8.10) and (8.11) are *not* zero. Then (8.1b, c) cannot be solved. There is no difficulty in solving (8.1a). This equation is of the form

$$\underset{2l-2}{\gamma_{00,rr}} = 2\underset{2l-1}{\Lambda_{00}}, \tag{8.12}$$

where the right-hand side is known. We see that the solution of this equation is determined only up to an additive harmonic function. Thus we can add to any solution either single "poles" or "poles" and "dipoles".

By adding single poles we can insure the integrability of (8.1b). Then by adding dipoles we can insure the integrability of (8.1c). We could have done all that in *one* step, adding poles and dipoles, but the division into two steps makes for a simpler presentation.

After finding $\underset{2l-2}{\gamma_{00}}$ from (8.12), we calculate $\underset{2l-1}{\overset{s}{C}}{}_0$ and, in general, find $\underset{2l-1}{\overset{s}{C}}{}_0 \neq 0$. We then replace in (8.1b):

$$\underset{2l-2}{\gamma_{00}} \text{ by } \underset{2l-2}{\gamma_{00}} - \sum_s 4 \underset{2l-2}{\overset{s}{m}}\overset{s}{\psi}, \tag{8.13}$$

where $\underset{2l-2}{\overset{s}{m}}$ are certain functions of time to be determined soon, and ψ's are the functions defined in (8.4). Of course this change in $\underset{2l-2}{\gamma_{00}}$

induces a change in $\underset{2l-1}{C_0}$. Indeed,

(8.14)
$$\begin{cases} \underset{2l-1}{2\Lambda_0} \quad \text{changes now to} \\ \underset{2l-1}{2\Lambda_{0m}} + \sum\limits_{s} \left(4 \underset{2l-2}{\overset{s}{m}} \overset{s}{\psi}\right)_{,\underset{1}{0m}}, \end{cases}$$

as follows from (6.5b) because $\underset{2l-2}{\gamma_{00}}$ appears in $\underset{2l-1}{\Lambda_{0m}}$ only as $-\underset{2l-2\ 1}{\gamma_{00,\,m0}}$. Now obviously the old surface integral

(8.15)
$$\frac{1}{4\pi} \int^{s} \underset{2l-1}{2\Lambda_{0r}} n_r \, dS = \underset{2l-1}{\overset{s}{C_0}}$$

changes into A.4

(8.16)
$$\underset{2l-1}{\overset{s}{C_0}} - 4 \underset{2l-1}{\overset{s}{\dot{m}}},$$

therefore it can be made zero by choosing

(8.17)
$$4 \underset{2l-1}{\overset{s}{\dot{m}}} = \underset{2l-1}{\overset{s}{C_0}}.$$

Thus by adding a pole we can insure the integrability of (8.1b). The next step is to insure the integrability of (8.1c). Thus we assume that $\underset{2l-2}{\gamma_{00}}, \underset{2l-1}{\gamma_{0m}}$ are known, that (8.1b) is integrable and we have once more to return to $\underset{2l-2}{\gamma_{00}}$ looking for a different solution of (8.1a) so as to insure the integrability of (8.1c) without destroying the integrability of (8.1b).

We replace now our $\underset{2l-2}{\gamma_{00}}$ (containing the additional poles) by

(8.18)
$$\underset{2l-1}{\gamma_{00}} - \sum_{s=1}^{p} \underset{2l-1}{\overset{s}{S_r}} \overset{s}{\psi}_{,r}.$$

These are additional *dipole* solutions, and we assume that no other dipole expressions are contained in γ_{00}. Again the S_r are functions of

τ only, to be determined later. The $\underset{2l-2}{\gamma_{00}}$ now contain the single pole solutions so as to enforce the integrability of (8.1b). We can easily see what change in $\underset{2l-1}{\gamma_{0m}}$ is induced by (8.18). The answer is, that $\underset{2l-1}{\gamma_{0m}}$ changes into

$$\underset{2l-1}{\gamma_{0m}} - \sum_{s} \left(\underset{2l-2}{\overset{s}{S}_m \overset{s}{\psi}} \right)_{,\underset{1}{0}} . \tag{8.19}$$

Indeed, if the old $\underset{2l-1}{\gamma_{0m}}$ satisfies the original equation (8.1b):

$$\underset{2l-1}{\gamma_{0m}}{}_{,ss} - \underset{2l-1}{\gamma_{0sm}}{}_{,s} = \underset{2l-2}{\gamma_{ms}}{}_{,\underset{1}{0s}} - \underset{2l-2}{\gamma_{00}}{}_{,\underset{1}{m0}} + 2\underset{2l-1}{\Lambda'_{0m}}, \tag{8.20}$$

then $\underset{2l-2}{\gamma_{00}}$, $\underset{2l-1}{\gamma_{0m}}$ *with* the additional expressions written out in (8.18) and (8.19) satisfy the equation too. This is so, because $2\underset{2l-1}{\Lambda'_{0m}}$ being non-linear can contain neither $\underset{2l-2}{\gamma_{00}}$ nor $\underset{2l-1}{\gamma_{0m}}$. Therefore the addition of dipoles does not affect the integrability of (8.1b).

Now the last and decisive step: we replace in (8.1c) $\underset{2l-2}{\gamma_{00}}$, $\underset{2l-1}{\gamma_{0m}}$, by the new expressions according to (8.18) and (8.19) and adjust the S's so that the surface integrals will vanis hidentically. This requires a somewhat more lengthy calculation.

Written out explicitly, equation (8.1c) is:

$$\begin{aligned} &\underset{2l}{\gamma_{mn}}{}_{,ss} - \underset{2l}{\gamma_{ms}}{}_{,ns} - \underset{2l}{\gamma_{ns}}{}_{,ms} + \delta_{mn}\underset{2l}{\gamma_{rs}}{}_{,rs} \\ &= -\underset{2l-1}{\gamma_{0m}}{}_{,\underset{1}{0n}} - \underset{2l-1}{\gamma_{0n}}{}_{,\underset{1}{0m}} + 2\delta_{mn}\underset{2l-1}{\gamma_{0r}}{}_{,\underset{1}{0r}} + \underset{2l-2}{\gamma_{mn}}{}_{,\underset{2}{00}} - \delta_{mn}\underset{2l-2}{\gamma_{00}}{}_{,\underset{2}{00}} + 2\underset{2l}{\Lambda'_{mn}} \\ &= 2\underset{2l}{\Lambda_{mn}}. \end{aligned} \tag{8.21}$$

We introduce into (8.21)

$$\underset{2l-2}{\gamma_{00}} - \sum_{s} \underset{2l-2}{\overset{s}{S}_r \overset{s}{\psi}}{}_{,r}, \tag{8.22}$$

$$\underset{2l-1}{\gamma_{0m}} - \sum_{s} \left(\underset{2l-2}{\overset{s}{S}_m \overset{s}{\psi}} \right)_{,\underset{1}{0}} \tag{8.23}$$

for the old $\underset{2l-2}{\gamma_{00}}, \underset{2l-1}{\gamma_{0m}}$. We now obtain new expressions added to the old $\underset{2l}{\Lambda_{mn}}$. The difficulty is, that now the contributions come not only from the linear expressions, but also from Λ'_{mn} which will contain terms of the type $\underset{2l-2}{\gamma_{00}}\cdot\underset{2}{\gamma_{00}}$. The result of the calculations is given in A.8, and contains many expressions of which we shall here write only the first three which arise from the linear terms (the others, as we shall see, are unimportant). Instead of the old $2\underset{2l}{\Lambda_{mn}}$ we have:

$$2\underset{2l}{\Lambda_{mn}}$$

$$(8.24)\qquad +\sum_s\left(\underset{2l}{\overset{s}{\ddot{S}}_m}\overset{s}{\psi}_{,n}+\underset{2l}{\overset{s}{\ddot{S}}_n}\overset{s}{\psi}_{,m}-\delta_{mn}\underset{2l}{\overset{s}{\ddot{S}}_r}\overset{s}{\psi}_{,r}\right)$$

$$+\ \dots$$

where the dots at the end indicate the omitted expressions. As we are here discussing the problem of surface integrals, we are justified in omitting them because they do not give any contribution to the surface integrals. We see too, that the expressions written out here have a vanishing divergence, and this is true for the omitted terms also. Calculating the surface integrals (A.4), we find that the old surface integral

$$(8.25)\qquad \frac{1}{4\pi}\int^{s} 2\underset{2l}{\Lambda_{mn}}n_n\, dS = \underset{2l}{\overset{s}{C}_m}$$

changes into

$$(8.26)\qquad \underset{2l}{\overset{s}{C}_m}-\underset{2l}{\overset{s}{\ddot{S}}_m}.$$

Therefore it can be made zero, by choosing

$$(8.27)\qquad \underset{2l}{\overset{s}{\ddot{S}}_m} = \underset{2l}{\overset{s}{C}_m}.$$

Thus we can always, by adding dipole solutions in $\underset{2l-2}{\gamma_{00}}$, force the surface integrals to vanish identically.

By proceeding in this way, we accumulate single poles and dipoles, and the additional expressions in γ_{00} are:

$$-\sum_{l}\lambda^{2l-2}\sum_{s=1}^{p}\left(4\underset{2l-2}{\overset{s}{m}}\;\overset{s}{\psi}+\underset{2l-2}{\overset{s}{S_r}}\;\overset{s}{\psi}_{,r}\right). \tag{8.28}$$

We violated our rule of not introducing dipoles. However, this was done for γ_{00} only. We can, at the end of the approximation procedure, annihilate all these additional dipole expressions by taking

$$\sum_{l}\lambda^{2l-2}\underset{2l-2}{\overset{s}{S_r}}=0. \tag{8.29}$$

Differentiating this twice, we obtain, because of (8.27):

$$\sum_{l}\lambda^{2l}\underset{2l}{\overset{s}{\ddot{S}_m}}=\sum_{l}\lambda^{2l}\underset{2l}{\overset{s}{C_m}}=0. \tag{8.30}$$

These are the 3p equations of motion. Thus the motion is determined, if dipole solutions are rejected.

On the other hand, the m's can be calculated from the C_0's according to (8.17). Denoting the total coefficient at $\overset{s}{\psi}$ by $-4\overset{s}{M}$, we have:

$$\overset{s}{M}=\lambda^{2}\underset{2}{\overset{s}{m}}+\lambda^{4}\underset{4}{\overset{s}{m}}+\lambda^{6}\underset{6}{\overset{s}{m}}+\dots \tag{8.31}$$

where $\underset{4}{\overset{s}{m}}$, $\underset{6}{\overset{s}{m}}$, ... are functions of the original constants $\underset{2}{\overset{s}{m}}$ and of known functions of the time.

The equations (8.30) and (8.31) will contain only a finite number of terms depending on the order to which we wish to carry out the actual calculations.

9. On the Choice of a Co-ordinate System

We shall now see that it is possible to simplify our equations through the proper choice of a co-ordinate system. Let us assume that

$$\underset{2l-2}{\overset{*}{\gamma}}_{00},\quad \underset{2l-1}{\overset{*}{\gamma}}_{0m},\quad \underset{2l}{\overset{*}{\gamma}}_{mn} \tag{9.1}$$

are solutions of our system (6.3), where the Φ's and Λ's are defined by (6.4) and (6.5). Then we can show that any

$$(9.2)\qquad \begin{cases} \underset{2l-2}{\gamma_{00}} = \underset{2l-2}{\overset{*}{\gamma}_{00}}, \\[2ex] \underset{2l-1}{\gamma_{0m}} = \underset{2l-1}{\overset{*}{\gamma}_{0m}} + \underset{2l-1}{a_{0,m}}, \\[2ex] \underset{2l}{\gamma_{mn}} = \underset{2l}{\overset{*}{\gamma}_{mn}} + \underset{2l}{a_{m,n}} + \underset{2l}{a_{n,m}} - \delta_{mn}\underset{2l}{a_{r,r}} + \delta_{mn}\underset{2l-1}{a_{0}}{}_{,\underset{1}{0}}, \end{cases}$$

with $\underset{2l-1}{a_0}$, $\underset{2l}{a_m}$ *arbitrary* are also solutions of our equations. This can be shown just by straightforward substitution in (6.4). A simple calculation shows that all the a's vanish from these equations. Thus we can, at each approximation step, impose four conditions upon the field. Let us choose, as is usually done, the following four co-ordinate conditions:

$$(9.3\text{a})\qquad \underset{2l-2}{\gamma_{00}}{}_{,\underset{1}{0}} - \underset{2l-1}{\gamma_{0r,r}} = 0,$$

$$(9.3\text{b})\qquad \underset{2l-1}{\gamma_{0m}}{}_{,\underset{1}{0}} - \underset{2l}{\gamma_{mr,r}} = 0.$$

Indeed, if γ^* do not satisfy such a condition, then a's can be found that ensure it. The equations for the a's are:

$$(9.4\text{a})\qquad \underset{2l-1}{a_{0,rr}} = \underset{2l-2}{\overset{*}{\gamma}_{00}}{}_{,\underset{1}{0}} - \underset{2l-1}{\overset{*}{\gamma}_{0r,r}},$$

$$(9.4\text{b})\qquad \underset{2l}{a_{m,rr}} = \underset{2l-1}{\overset{*}{\gamma}_{0m}}{}_{,\underset{1}{0}} - \underset{2l}{\overset{*}{\gamma}_{mr,r}}.$$

With the co-ordinate condition (9.3) our system of equations is considerably simplified. Equations (6.3) now become:

$$(9.5\text{a})\qquad \underset{2l-2}{\gamma_{00,rr}} = \underset{2l-4}{\gamma_{00}}{}_{,\underset{2}{00}} + 2\underset{2l-2}{\Lambda'_{00}},$$

$$(9.5\text{b})\qquad \underset{2l-1}{\gamma_{0m,rr}} = \underset{2l-3}{\gamma_{0m}}{}_{,\underset{2}{00}} + 2\underset{2l-1}{\Lambda'_{0m}},$$

$$(9.5\text{c})\qquad \underset{2l}{\gamma_{mn,rr}} = \underset{2l-2}{\gamma_{mn}}{}_{,\underset{2}{00}} + 2\underset{2l}{\Lambda'_{mn}},$$

which together with the co-ordinate conditions

(9.6a) $$\underset{2l-2\;1}{\gamma_{00,\,0}} - \underset{2l-1}{\gamma_{0r,\,r}} = 0,$$

(9.6b) $$\underset{2l-1\;1}{\gamma_{0m,\,0}} - \underset{2l}{\gamma_{mr,\,r}} = 0,$$

now form a symmetrical system of equations, where in (9.5) all the known functions on the right-hand side are at least two orders lower than those on the left.

The surface integrals that must vanish and which give the equations of motion are:

(9.7a) $$\int^{s} \Big(\underset{2l-3\;2}{\gamma_{0m,\,00}} - \underset{2l-2\;1}{\gamma_{00,\,0m}} + 2\underset{2l-1}{\Lambda_{0m}}\Big) n_m \, dS = 0,$$

(9.7b) $$\int^{s} \Big(\underset{2l-2\;2}{\gamma_{nm,\,00}} - \underset{2l-1\;1}{\gamma_{n0,\,0m}} + 2\underset{2l}{\Lambda'_{nm}}\Big) n_m \, dS = 0.$$

We can deduce them from our old formulae, using the lemma, or directly, differentiating (9.6), adding to (9.5) and using the lemma.

If, as in Sec. 8, we now introduce dipoles in order to satisfy (9.7b), we do not violate (9.6a).

Sometimes it is more convenient to use other co-ordinate conditions. For example, the one used in the actual calculations is:

(9.8a) $$\underset{2l-2\;1}{\gamma_{00,\,0}} - \underset{2l-1}{\gamma_{0s,\,s}} = 0,$$

(9.8b) $$\underset{2l}{\gamma_{mn,\,n}} = 0.$$

The equations then are:

(9.9a) $$\underset{2l-2}{\gamma_{00,\,rr}} = 2\underset{2l-2}{\Lambda'_{00}},$$

(9.9b) $$\underset{2l-1}{\gamma_{0m,\,rr}} = 2\underset{2l-1}{\Lambda'_{0m}},$$

(9.9c) $$\underset{2l}{\gamma_{mn,\,rr}} = -\underset{2l-1}{\gamma_{0m,\,0n}} - \underset{2l-1\;1}{\gamma_{0n,\,0m}} + \delta_{mn}\underset{2l-2\;2}{\gamma_{00,\,00}} + \underset{2l-2\;2}{\gamma_{mn,\,00}} + 2\underset{2l}{\Lambda'_{mn}}$$
$$= 2\underset{2l}{\Lambda_{mn}}$$

and the surface conditions are:

(9.10a) $$\int^{s} \Big(2\underset{2l-1}{\Lambda'_{0m}} - \underset{2l-2\;1}{\gamma_{00,\,0m}}\Big) n_m \, dS = 0,$$

(9.10b) $$\int^{s} 2\underset{2l}{\Lambda_{mn}} n_m \, dS = 0.$$

The question arises: to what extent does the co-ordinate condition influence the equations of motion? We shall return to this problem in the last section and we shall show that the equations of motion to the sixth order do not depend on the choice of the co-ordinate system.

10. The Newtonian Approximation

We shall discuss now the first three equations for $l = 2$. The equations are:

(10.1) $$\underset{2}{\gamma_{00,\,rr}} = 0,$$

(10.2) $$\underset{3}{\gamma_{0m,\,rr}} = 0,$$

(10.3) $$\underset{4}{\gamma_{nm,\,rr}} = 2\underset{4}{\Lambda_{nm}}.$$

The co-ordinate conditions that we accept are:

(10.4) $$\underset{3}{\gamma_{0r,\,r}} - \underset{2\;\;1}{\gamma_{00,\,0}} = 0.$$

(10.5) $$\underset{4}{\gamma_{mr,\,r}} = 0.$$

The explicit form of $\underset{4}{\Lambda_{mn}}$ is given in A.10.

The character of our entire solution will depend essentially upon the choice of the harmonic function we take as the solution of (10.1). As we are interested in solutions representing particles, we shall write:

(10.6) $$\begin{cases} \underset{2}{\gamma_{00}} = 2\varphi; \quad \varphi = \displaystyle\sum_{s=1}^{p} \Big\{-2\overset{s}{m}\underset{2}{\overset{s}{\psi}}\Big\}, \\ \overset{s}{\psi} = \Big[\Big(x^r - \overset{s}{\xi^r}\Big)\Big(x^r - \overset{s}{\xi^r}\Big)\Big]^{-\frac{1}{2}} = \Big(\overset{s}{r}\Big)^{-1}. \end{cases}$$

From (10.2) we see that $\underset{3}{\gamma}_{0m}$ is a harmonic function too, which must, however, satisfy the co-ordinate condition also. From (10.4) we have:

$$\underset{3}{\gamma}_{0r,\,r} = \underset{2}{\gamma}_{00,\,\underset{1}{0}} = -\sum_{s} \left\{ 4\overset{s}{\underset{2}{m}}\overset{s}{\psi}_{,\,r}\overset{s}{\underset{1}{\dot{\xi}}}{}^{r} \right\}. \tag{10.7}$$

The constant $\overset{s}{\underset{2}{m}}$, which we identify with the gravitational mass of the particles is assumed to be positive. Therefore the exclusion of dipoles, together with the field equations and the co-ordinate condition determine uniquely $\underset{3}{\gamma}_{0n}$:

$$\underset{3}{\gamma}_{0n} = \sum_{s} 4\overset{s}{\underset{2}{m}}\overset{s}{\psi}\overset{s}{\underset{1}{\dot{\xi}}}{}^{n}. \tag{10.8}$$

To this $\underset{3}{\gamma}_{0n}$ we could add, according to (9.2), the gradient of any function and in this way obtain a general solution. But as our entire procedure consists in employing only rational functions of $\left(x^r - \overset{s}{\xi}{}^r\right)$, any such addition would introduce new singularities (not of the character of a single pole), or a non-Galilean field at infinity. Thus we should regard $\underset{3}{\gamma}_{0n}$ in (10.8) as characterizing the problem of particles, regardless of whether we introduce the co-ordinate condition (10.4) or not.

Just for the sake of simplicity, let us now restrict our consideration to *two* particles and write (omitting the indices below $\overset{s}{m}, \varphi, f, g$):

$$\left\{ \begin{array}{l} \varphi = f + g, \\ f = -2\overset{1}{m}\overset{1}{\psi}; \quad g = -2\overset{2}{m}\overset{2}{\psi}, \\ \overset{1}{\xi}{}^r = \eta^r; \quad \overset{2}{\xi}{}^r = \zeta^r. \end{array} \right. \tag{10.9}$$

The next step then, since the surface integral (9.10a) vanishes for $l = 3$, because

$$\int^{s} \left(2\underset{3}{\Lambda}{}'_{0m} - \underset{2}{\gamma}_{00,\,\underset{1}{0m}} \right) n_m \, dS = -\int^{s} \underset{2}{\gamma}_{00,\,\underset{1}{0m}} n_m \, dS = 0, \tag{10.10}$$

is to determine

$$
\begin{cases}
\overset{1}{\underset{4}{C}}{}_m = \dfrac{1}{4\pi}\displaystyle\int^{1} 2\underset{4}{\Lambda}_{mr}n_r\,dS, \\
\overset{2}{\underset{4}{C}}{}_m = \dfrac{1}{4\pi}\displaystyle\int^{2} 2\underset{4}{\Lambda}_{mr}n_r\,dS.
\end{cases} \tag{10.11}
$$

If we wish to finish our approximation procedure here, the equations of motion up to the fourth, or as we shall call it, the Newtonian approximation, are:

$$
\overset{1}{\underset{4}{C}}{}_m = 0; \quad \overset{2}{\underset{4}{C}}{}_m = 0\cdot \tag{10.12}
$$

All we have to do now is to calculate the surface integrals, according to the method outlined in A.4. The result of this particular calculation is given in A.10. It is:

$$
\begin{cases}
\overset{1}{\underset{4}{C}}{}_m(\tau) = 4\overset{1}{m}\left\{\ddot{\eta}^m + \dfrac{1}{2}\tilde{g}_{,m}\right\} = 0, \\
\overset{2}{\underset{4}{C}}{}_m(\tau) = 4\overset{2}{m}\left\{\ddot{\zeta}^m + \dfrac{1}{2}\tilde{f}_{,m}\right\} = 0, \\
\tilde{g}_{,m} = g_{,m} \quad \text{for} \quad x^s = \eta^s, \\
\tilde{f}_{,m} = f_{,m} \quad \text{for} \quad x^s = \zeta^s.
\end{cases} \tag{10.13}
$$

The form (10.13) is actually independent of the variables x^s. In the last equations we see that $\tilde{g}_{,m}$, say, is obtained by differentiating g with respect to x^s and then by replacing x^s by η^s. But the result will be the same if we *first* replace x^s by η^s and *then* differentiate with respect to η^s or ζ^s. Thus:

$$
\begin{cases}
\tilde{g}_{,s} = \dfrac{\partial g(r)}{\partial \eta^s} = -\dfrac{\partial g(r)}{\partial \zeta^s}, \\
g(r) = -\dfrac{2\overset{2}{m}}{r}; \quad r^2 = (\eta^s - \zeta^s)(\eta^s - \zeta^s).
\end{cases} \tag{10.14}
$$

We can, therefore, think of our equations of motion as involving the differentiation of functions depending only on the position of singularities, as is characteristic of the theories based on the concept of action at a distance. Indeed, we see that our equations are precisely the Newtonian equations of motion, deduced here as the first approximation from the field equations. The treatment of p particles (instead of two) does not add any new difficulties if we deal with the Newtonian approximation only.

11. Transition to the Next Approximation

We wish to go now beyond the Newtonian approximation. But then we must calculate $\underset{4}{\gamma_{mn}}$, since $\underset{6}{\Lambda_{mn}}$ depends on $\underset{4}{\gamma_{mn}}$. The characteristic feature of this method is that generally, if we wish to find the equations of motion to the $2l$ approximation (inclusive) then we do not need to calculate $\underset{2l}{\gamma_{mn}}$, because $\underset{2l}{C_m}$ does not contain it. But now, if we wish to go one step further we must find $\underset{4}{\gamma_{mn}}$ for which the equations are:

$$\underset{4}{\gamma_{mn,rr}} = 2\underset{4}{\Lambda_{mn}}. \tag{11.1}$$

This is "the transition step" that we have to take before proceeding to the next approximation. These equations are integrable only if we *do* assume Newtonian motion. Otherwise we would have to add dipoles. Yet if we wish to proceed *only* to the next approximation we may assume Newtonian motion and additional expressions induced by the dipole fields are not necessary.

If in (11.1) we assume Newtonian motion, then (11.1) can be integrated, because the surface integral of $\underset{4}{\Lambda_{mn}}$ vanishes then. But if we do this, we introduce Newtonian motion into $\underset{6}{\Lambda_{mn}}$. This is admissible because any difference between $\underset{6}{\Lambda}$ calculated this way and $\underset{6}{\Lambda}$ calculated with the proper motion is of order $\underset{8}{\Lambda}$. Thus since we do not propose to go beyond $\underset{6}{\Lambda}$ we may ignore the additional dipole fields.

It is for this reason that the previous special calculations in [1] were correct, but the general theory was not.

We shall now solve

$$\underset{4}{\gamma}_{mn,rr} = 2\underset{4}{\Lambda}_{mn} = -\underset{3}{\gamma}_{0m,\,0n} - \underset{1}{\gamma}_{0n,\,0m} + 2\delta_{mn}\underset{3}{\varphi}_{,00} \tag{11.2}$$
$$-2\varphi\varphi_{,mn} - \varphi_{,m}\varphi_{,n} + \tfrac{3}{2}\,\delta_{mn}\varphi_{,s}\varphi_{,s}$$

assuming the Newtonian equation of motion, i.e. (10.13).

We can ignore the dipole expressions because we are interested only in the equations of motion to the next approximation. But, for the same reason, we are interested only in those expressions in $\underset{4}{\gamma}_{mn}$ which give a contribution to the corresponding surface integral of $\underset{6}{\Lambda}_{mn}$.

An inspection of $\underset{6}{\Lambda}_{mn}$ (A.12) shows that we need only the knowledge of $\underset{4}{\gamma}_{mn}$ in the neighbourhood of the singularities, and we may ignore in it the terms which do not become infinite as $\underset{s}{r} \to 0$, since the surface integral due to these terms must vanish (see A.12). On the other hand $\underset{4}{\gamma}_{ss}$ which also appears in $\underset{6}{\Lambda}$ should and will be calculated in the entire space.

In the equation (11.2) we have, on the right-hand side "cross products", that is, products belonging to different singularities. Because of them (11.2) can only be integrated in the neighbourhood of the first singularity, say. The expression arising from the second singularity can be expanded into a power series near the first singularity. Retaining all the expressions that may give some contribution to the surface integral and those only, we have in the neighbourhood of the *first* singularity:

$$(11.3)\quad \begin{cases} \underset{4}{\gamma}_{mn} = \{f[(x^n-\eta^n)\dot{\eta}^m + (x^m-\eta^m)\dot{\eta}^n - \delta_{mn}(x^s-\eta^s)\dot{\eta}^s]\}_{,0} \\ \quad + \{\underset{1}{g}[(x^n-\zeta^n)\dot{\zeta}^m \underset{2}{+} (x^m-\zeta^m)\dot{\zeta}^n - \delta_{mn}(x^s-\zeta^s)\dot{\zeta}^s]\}_{,0} \\ \quad + \tfrac{7}{4}r^2 f_{,m} f_{,n} + \tfrac{7}{4}r^2 g_{,m} g_{,n} \\ \quad - f_{,m}(x^n-\eta^n)\tilde{g} \\ \quad + \alpha_{mn} f + \beta_{mn} g. \end{cases}$$

Here only the expression

$$-f_{,m}(x^n-\eta^n)\tilde{g};\quad \tilde{g}=g \quad \text{for} \quad x^s=\eta^s,$$

is due to the interaction terms. The two last expressions are the additive harmonic functions (dipoles are excluded) and they are determined by the co-ordinate condition

$$\underset{4}{\gamma}_{mr,\,r}=0. \tag{11.4}$$

The result is:

$$\left\{\begin{aligned} \alpha_{mn} &= 2\dot{\eta}^m\dot{\eta}^n+\delta_{mn}\tilde{g},\\ \beta_{mn} &= 2\dot{\xi}^m\dot{\xi}^n+\delta_{mn}\tilde{f},\\ &\tilde{f}=f(r);\quad \tilde{g}=g(r);\quad r^2=(\eta^s-\zeta^s)(\eta^s-\zeta^s). \end{aligned}\right. \tag{11.5}$$

But, let us say once more, that all this is true only if the Newtonian motion is assumed.

Finally, as we mentioned before, $\underset{4}{\gamma}_{rr}$ can be calculated rigorously. The result is:

$$\underset{4}{\gamma}_{rr}=-2\overset{1}{m}\overset{1}{r}_{,00}-2\overset{2}{m}\overset{2}{r}_{,00}+\tfrac{7}{4}\varphi^2+\alpha f+\beta g. \tag{11.6}$$

Here the α and β are determined so that near the singularity (11.6) will be consistent with (11.3) for $m=n=r$. The result is:

$$\left\{\begin{aligned} \alpha &= 2\dot{\eta}^s\dot{\eta}^s+\tfrac{1}{2}\tilde{g},\\ \beta &= 2\dot{\xi}^s\dot{\xi}^s+\tfrac{1}{2}\tilde{f}. \end{aligned}\right. \tag{11.7}$$

Thus our transition steps are accomplished.

12. Beyond the Newtonian Approximation

We write down the next field equations:

$$\underset{4}{\gamma}_{00,\,rr}=2\underset{4}{\Lambda}_{00}=-\tfrac{3}{2}\varphi_{,r}\varphi_{,r}, \tag{12.1a}$$

$$\underset{5}{\gamma}_{0m,\,rr}=2\underset{5}{\Lambda'}_{0m}=\varphi_{,s}\gamma_{0s,\,m}-\varphi_{,sm}\gamma_{0s}-3\varphi_{,0}\varphi_{,m}, \tag{12.1b}$$

$$\underset{6}{\gamma}_{mn,\,rr}=2\underset{6}{\Lambda}_{mn}. \tag{12.1c}$$

The explicit expressions for $\underset{6}{\Lambda}_{mn}$ are quoted in A.12. The solution of (12.1a) is simple:

$$\underset{4}{\gamma}_{00} = -\tfrac{3}{4}\varphi^2 - 4\underset{4}{\overset{1}{m}}\overset{1}{\psi} - 4\underset{4}{\overset{2}{m}}\overset{2}{\psi}. \tag{12.2}$$

As we know from the general theory, the arbitrary harmonic functions have to be determined in such a way as to make (12.1b) self-consistent, that is, the corresponding surface integral must vanish.

The co-ordinate conditions, are here, as before,

$$\underset{5}{\gamma}_{0r,r} - \underset{4}{\gamma}_{00,\underset{1}{0}} = 0, \tag{12.3a}$$

$$\underset{6}{\gamma}_{mr,r} = 0. \tag{12.3b}$$

Because of this, the conditions for solvability of (12.1b, c) are:

$$\frac{1}{4\pi}\int^{s}\left\{2\underset{5}{\Lambda'}_{0m} - \underset{4}{\gamma}_{00,\underset{1}{0}m}\right\} n_m\, dS = 0, \tag{12.4a}$$

$$\frac{1}{4\pi}\int^{s} 2\underset{6}{\Lambda}_{mr} n_r\, dS = 0. \tag{12.4b}$$

We have in (12.4a) the equations that determine $\underset{4}{m}$. The result of evaluating the surface integrals in (12.4a), (see A.12) is:

$$\begin{cases} \underset{4}{\overset{1}{m}} = \dfrac{1}{2}\overset{1}{m}\left\{\dot{\eta}^s\dot{\eta}^s + \dfrac{1}{2}\tilde{g}\right\} = \dfrac{1}{2}\left(\overset{1}{m}\dot{\eta}^s\dot{\eta}^s - \overset{1}{m}\overset{2}{m}\dfrac{1}{r}\right), \\ \underset{4}{\overset{2}{m}} = \dfrac{1}{2}\overset{2}{m}\left\{\dot{\zeta}^s\dot{\zeta}^s + \dfrac{1}{2}\tilde{f}\right\} = \dfrac{1}{2}\left(\overset{2}{m}\dot{\zeta}^s\dot{\zeta}^s - \overset{1}{m}\overset{2}{m}\dfrac{1}{r}\right), \\ \overset{1}{m} = \underset{2}{\overset{1}{m}};\ \overset{2}{m} = \underset{2}{\overset{2}{m}};\ r^2 = (\eta^s - \zeta^s)(\eta^s - \zeta^s). \end{cases} \tag{12.5}$$

The next step, after the self-consistency of (12.1b) has been insured, is to calculate the $\underset{5}{\gamma}_{0s}$. We need them, because they enter into the next surface integral. Including only relevant terms that can influence the surface integral we find near the first singularity:

$$
\text{(12.6)} \quad \begin{cases}
\underset{5}{\overset{1}{\gamma}}_{0m} = -\frac{7}{4} r f_{,m} f_{,r} \dot{\eta}^r + \frac{3}{4} f^2 \dot{\eta}^m \\
\quad + \frac{3}{2}(x^s - \eta^s)(\dot{\eta}^s - \dot{\zeta}^s) f \tilde{g}_{,m} \\
\quad - (x^m - \eta^m) f \tilde{g}_{,s} (\dot{\eta}^s - \dot{\zeta}^s) \\
\quad + \frac{1}{2}(x^s - \eta^s) f_{,m} \dot{\zeta}^s \{\tilde{g} + \tilde{g}_{,r}(x^r - \eta^r)\} \\
\quad + \frac{1}{2}(x^s - \eta^s)\{f \tilde{g}_{,s} \dot{\zeta}^m + f_{,m} \tilde{g} \dot{\zeta}^s\} \\
\quad + \alpha_{0m} f.
\end{cases}
$$

Again α_{0m} is determined from the co-ordinate condition (12.3a) and the result is:

$$
\text{(12.7)} \qquad a_{0m} = -\dot{\eta}^s \dot{\eta}^s \dot{\eta}^m + \tilde{g} \dot{\eta}^m - \tilde{g} \dot{\zeta}^m.
$$

Now the scene is set for the last and most difficult calculation:

$$
\text{(12.8)} \qquad \underset{6}{\overset{1}{C}}_m = \frac{1}{4\pi} \int^1 2 \underset{6}{\Lambda}_{mn} n_m \, dS.
$$

Some remarks about this calculation are made in A.12, and partial results given. We obtain:

$$
\text{(12.9)} \quad \underset{6}{\overset{1}{C}}_m = -4 \overset{1}{m} \overset{2}{m} \left\{ \left[\dot{\eta}^s \dot{\eta}^s + \frac{3}{2} \dot{\zeta}^s \dot{\zeta}^s - 4 \dot{\eta} \dot{\zeta}^s - 4 \frac{\overset{2}{m}}{r} - 5 \frac{\overset{1}{m}}{r} \right] \frac{\partial}{\partial \eta^m} \left(\frac{1}{r} \right) \right.
$$

$$
\left. + [4 \dot{\eta}^s (\dot{\zeta}^m - \dot{\eta}^m) + 3 \dot{\eta}^m \dot{\zeta}^s - 4 \dot{\zeta}^s \dot{\zeta}^m] \frac{\partial}{\partial \eta^s} \left(\frac{1}{r} \right) + \frac{1}{2} \frac{\partial^3 r}{\partial \eta^s \, \partial \eta^r \, \partial \eta^m} \dot{\zeta}^s \dot{\zeta}^r \right\}.
$$

Thus the equation of motion belonging to this stage of approximation is:

$$
\text{(12.10)} \qquad \lambda^4 \underset{4}{\overset{1}{C}}_m + \lambda^6 \underset{6}{\overset{1}{C}}_m = 0.
$$

We can now re-absorb the λ's by substituting new units for τ and $\overset{1}{m}, \overset{2}{m}$:

$$
\text{old } \tau = \lambda \cdot \text{new } \tau; \quad \text{old mass} = \lambda^{-2} \cdot \text{new mass}.
$$

Preserving the old symbols for the new units we have for the equations of motion of the first particle:

$$
\begin{aligned}
\ddot{\eta}^m - \overset{2}{m}\frac{\partial(1/r)}{\partial\eta^m} = \overset{2}{m}\Bigg\{\Bigg[\dot{\eta}^s\dot{\eta}^s + \frac{3}{2}\dot{\zeta}^s\dot{\zeta}^s - 4\dot{\eta}^s\dot{\zeta}^s - 4\frac{\overset{2}{m}}{r} - 5\frac{\overset{1}{m}}{r}\Bigg]\frac{\partial}{\partial\eta^m}(1/r) \\
+ [4\dot{\eta}^s(\dot{\zeta}^m - \dot{\eta}^m) + 3\dot{\eta}^m\dot{\zeta}^s - 4\dot{\zeta}^s\dot{\zeta}^m]\frac{\partial}{\partial\eta^s}(1/r) \\
+ \frac{1}{2}\frac{\partial^3 r}{\partial\eta^s\,\partial\eta^r\,\partial\eta^m}\dot{\zeta}^s\dot{\zeta}^r\Bigg\}.
\end{aligned}
\tag{12.11}
$$

The equations of motion for the other particle are obtained by replacing

$$\overset{1}{m},\ \overset{2}{m},\ \eta,\ \zeta, \quad \text{by} \quad \overset{2}{m},\ \overset{1}{m},\ \zeta,\ \eta,$$

respectively.

These are the equations of motion of two particles. They can be integrated and conclusions concerning perihelion motion of a double star can be drawn from them [5]. The entire method can also be adapted for the case of a charged particle in an electromagnetic field [4].

13. The Equations of Motion and the Co-ordinate Condition

The contents of the last three sections are not new. Its presentation, however, is different than that given before in [1] and [2], since it has been adjusted to the new theory. There is one more question that we wish to answer and which we did not treat before. It is possible to do so only now after the general theory has been perfected. We ask: To what extent do the equations of motion as formulated in (12.11) depend on the particular choice of the coordinate system?

We reject any particular choice of co-ordinate system and write the first two equations:

$$\underset{2}{\Phi}_{00} + 2\underset{2}{\Lambda}_{00} = -\underset{2}{\gamma}_{00,\,rr} = 0, \tag{13.1}$$

$$\underset{3}{\Phi}_{0m} + 2\underset{3}{\Lambda}_{0m} = -\underset{3}{\gamma}_{0m,\,rr} + \underset{3}{\gamma}_{0r,\,mr} - \underset{2}{\gamma}_{00,\,m\underset{1}{0}} = 0. \tag{13.2}$$

We *assume* that we start our approximation procedure with the same $\underset{2}{\gamma_{00}}$ and $\underset{3}{\gamma_{0m}}$ functions as *we did before*. But from now on, while dealing with the rest of the equations we shall look for *general* solutions not restricted by any additional co-ordinate conditions.

Thus the equations that we wish to consider now are:

(13.3a) $$\underset{5}{\Phi}_{mn}+2\underset{4}{\Lambda}_{mn} = 0,$$

(13.3b) $$\underset{4}{\Phi}_{00}+2\underset{4}{\Lambda}_{00} = 0,$$

(13.3c) $$\underset{5}{\Phi}_{0m}+2\underset{5}{\Lambda}_{0m} = 0.$$

In the previous three sections we solved these equations, using special co-ordinate conditions. Let us *now* call the special solutions that we obtained there:

(13.4) $$\underset{4}{\overset{*}{\gamma}}_{mn},\ \underset{4}{\overset{*}{\gamma}}_{00},\ \underset{5}{\overset{*}{\gamma}}_{0m}.$$

Knowing them, as we do, we can find the general solution of (13.3). The procedure is similar to that outlined in Sec. 9, only slightly different, because we have now a set of equations of order $(2l)$, $(2l)$, and $(2l+1)$, whereas before we had a set of order $(2l-2)$, $(2l-1)$, and $(2l)$. But a straightforward substitution shows, that because of the linear expressions in (13.3), (and they alone enter the argument), the general solution of (13.3) is:

(13.5a) $$\underset{4}{\gamma}_{mn} = \underset{4}{\overset{*}{\gamma}}_{mn}+\underset{4}{a}_{m,n}+\underset{4}{a}_{n,m}-\delta_{mn}\underset{4}{a}_{r,r},$$

(13.5b) $$\underset{4}{\gamma}_{00} = \underset{4}{\overset{*}{\gamma}}_{00}+\underset{4}{a}_{r,r},$$

(13.5c) $$\underset{5}{\gamma}_{0m} = \underset{5}{\overset{*}{\gamma}}_{0m}+\underset{5}{a}_{0,m}+\underset{4}{a}_{m,\underset{1}{0}},$$

where a_μ are arbitrary. The question then is: If we substitute these new expressions into the Λ's do we change the integrals

(13.6) $$\int \underset{4}{\Lambda}_{mr}n_r\,dS,\quad \int \underset{5}{\Lambda}_{0r}n_r\,dS,\quad \int \underset{6}{\Lambda}_{mr}n_r\,dS?$$

As far as the first two integrals are concerned the answer is easy; $\underset{4}{\Lambda}$ is not changed; only linear expressions in $\underset{5}{\Lambda}$ are affected, but the surface integral of the additional expressions disappears because of the lemma. But it is different with the third surface integral. In $\underset{6}{\Lambda}$ new terms appear containing the a's. They appear both through the linear and the non-linear expressions. But these additional expressions—quoted in the last appendix—are such that their surface integral vanishes. Thus in the sense explained here the equations of motion do not depend on the choice of the co-ordinate system. This dependence would appear probably in the next approximation steps ($\underset{8}{\Lambda}$), but it does not enter into the surface integral of $\underset{6}{\Lambda}$. This is a satisfying result, because it is difficult to see the meaning of our co-ordinate conditions

$$(13.7)\qquad \begin{cases} \underset{4}{\gamma}_{mr,\,r} = 0, \\ \underset{5}{\gamma}_{0r,\,r} - \underset{4}{\gamma}_{00,\,\underset{1}{0}} = 0, \\ \underset{6}{\gamma}_{mr,\,r} = 0, \end{cases}$$

and it is good to know that our equations of motion are independent of it. This result is general. If we have a system

$$(13.8)\qquad \begin{cases} \underset{2l}{\Phi}_{mn} + 2\underset{2l}{\Lambda}_{mn} = 0, \\ \underset{2l}{\Phi}_{00} + 2\underset{2l}{\Lambda}_{00} = 0, \\ \underset{2l+1}{\Phi}_{0m} + 2\underset{2l+1}{\Lambda}_{0m} = 0, \end{cases}$$

then the surface integral of $\underset{2l+2}{\Lambda_{mn}}$ is independent of the co-ordinate conditions introduced in this particular approximation stage. This is so, because the $\underset{2l}{a}$'s combine with the φ's in the same way in each approximation step.

Appendices

A.2

The field equations are:

$$(A.2, 1)\quad R_{\mu\nu} = -\begin{Bmatrix} \varrho \\ \mu\nu \end{Bmatrix}_{|\varrho} + \begin{Bmatrix} \varrho \\ \mu\varrho \end{Bmatrix}_{|\nu} + \begin{Bmatrix} \varrho \\ \mu\sigma \end{Bmatrix}\begin{Bmatrix} \sigma \\ \varrho\nu \end{Bmatrix} - \begin{Bmatrix} \varrho \\ \mu\nu \end{Bmatrix}\begin{Bmatrix} \sigma \\ \varrho\sigma \end{Bmatrix}.$$

Introducing here the h's as defined in (2.3) and splitting (A.2, 1) into linear and non-linear terms we have

$$(A.2, 2a)\quad R_{00} = -\tfrac{1}{2}h_{00|ss} + h_{0s|0s} - \tfrac{1}{2}h_{ss|00} + L'_{00}$$

$$(A.2, 2b)\quad \begin{aligned} R_{0n} = &-\tfrac{1}{2}h_{0n|ss} + \tfrac{1}{2}h_{0s|ns} \\ &+ \tfrac{1}{2}h_{ns|0s} - \tfrac{1}{2}h_{ss|n0} + L'_{0n} \end{aligned}$$

$$(A.2, 2c)\quad \begin{aligned} R_{mn} = &-\tfrac{1}{2}h_{mn|ss} + \tfrac{1}{2}h_{ms|ns} + \tfrac{1}{2}h_{ns|ms} - \tfrac{1}{2}h_{ss|mn} \\ &+ \tfrac{1}{2}h_{mn|00} - \tfrac{1}{2}h_{m0|n0} - \tfrac{1}{2}h_{n0|m0} \\ &+ \tfrac{1}{2}h_{00|mn} + L'_{mn}. \end{aligned}$$

Here $L'_{\mu\nu}$ are the non-linear expressions. We form now:

$$(A.2, 3)\quad -2(R_{\mu\nu} - \tfrac{1}{2}\eta_{\mu\nu}\eta^{\alpha\beta}R_{\alpha\beta}) = 0.$$

Substituting the γ's for the h's, we see that (A.2, 3) written out is (2.10)–(2.18), where

$$(A.2, 4)\quad \Lambda'_{\mu\nu} = L'_{\mu\nu} - \tfrac{1}{2}\eta_{\mu\nu}\eta^{\alpha\beta}L'_{\alpha\beta}.$$

A.4

In calculating the surface integrals we need to take into account only expressions that go to infinity like r^{-2}, because only such expressions will give finite contributions. Since all the field functions are finite (outside of the singularity), and since the contributions do not depend on the shape of the surface, we may ignore all other expressions. But we have to keep the surface fixed, because in our calculations a complicated expression whose surface integral does not depend on the shape of the surface, is split into partial expressions with non-vanishing divergence. Thus in our calculations the surface is always a two-dimensional "sphere" with radius shrinking to zero. Let us assume, for the sake of simplicity, that the space co-ordinate of the singularity is (0, 0, 0). We shall first give some examples of the surface integrals formed around such a singularity.

Example 1. We calculate:

$$\int^{0} \psi_{,s} n_s \, dS; \quad \psi = r^{-1}; \quad r^2 = x^s x^s;$$

We have:

$$\int^{0} \psi_{,s} n_s \, dS = -\int^{0} \frac{x^s x^s}{r^4} r^2 \sin\theta \, d\theta \, d\varphi = -4\pi.$$

Example 2. We calculate:

$$\int^{0} \psi_{,s} n_r \, dS = -\int^{0} \frac{x^s x^r}{r^4} r^2 \sin\theta \, d\theta \, d\varphi = -\frac{4\pi}{3} \delta_{sr}.$$

Example 3. We calculate:

$$\int^{0} \psi_{,mn} n_n \chi(r) \, dS.$$

To find such a surface integral we expand $\chi(r)$ as a power series in the neighbourhood of the singularity:

$$\chi = \chi(0) + \chi_{,s}(0) x_s + \ldots .$$

The only contribution is from the second expression, that is, we have to calculate:

$$\begin{aligned}\chi_{,s}(0) \int^{0} \psi_{,mn} n_n x^s \, dS &= -\chi_{,s}(0) \int^{0} \frac{x^m x^s}{r^4} r^2 \sin\theta \, d\theta \, d\varphi \\ &\quad + 3\chi_{,s}(0) \int^{0} \frac{x^m x^s}{r^4} r^2 \sin\theta \, d\theta \, d\varphi \\ &= \frac{8\pi}{3} \chi_{,m}(0).\end{aligned}$$

In the course of our calculations we shall have to find more complicated surface integrals and the following table will prove to be useful:

Table of Surface Integrals

I. $$\frac{1}{4\pi}\int^{0} \psi,_{n} n_n \, dS = -1.$$

II. $$\frac{1}{4\pi}\int^{0} \psi,_{s} n_n \, dS = -\frac{1}{3}\,\delta_{sn}.$$

III.. $$\frac{1}{4\pi}\int^{0} x^r \psi,_{ns} n_n = \frac{2}{3}\,\delta_{rs}.$$

IV. $$\frac{1}{4\pi}\int^{0} x^r \psi,_{ms} n_n \, dS = -\frac{1}{15}\,\{2\delta_{rn}\delta_{ms} - 3\delta_{rm}\delta_{ns} - 3\delta_{rs}\delta_{mn}\}.$$

V. $$\frac{1}{4\pi}\int^{0} x^r \psi,_{rs} n_n \, dS = \frac{2}{3}\,\delta_{ns}.$$

VI. $$\frac{1}{4\pi}\int^{0} x^n \psi,_{ms} n_n \, dS = 0.$$

VII. $$\frac{1}{4\pi}\int^{0} x^n x^s \psi,_{mrk} n_n = 0.$$

VIII. $$\frac{1}{4\pi}\int^{0} x^m x^s \psi_{\,nrl} n_n \, dS = \frac{2}{5}\,\delta_{ms}\,\delta_{lr} - \frac{3}{5}\,(\delta_{ml}\,\delta_{rs} + \delta_{mr}\,\delta_{ls}).$$

A.8

The linear terms of (8.26) give the following contribution to $2\underset{2l}{\Lambda}_{mn}$:

(A.8, 1) $$\sum_{s=1}^{p}\left\{\underset{2l-2}{\overset{s}{S}_m}\overset{s}{\psi},_{n} + \underset{2l-2}{\overset{s}{S}_n}\overset{s}{\psi},_{m} - \delta_{mn}\underset{2l-2}{S_r}\psi,_{r}\right\},_{\underset{2}{00}}.$$

The non-linear terms can be found in the following way: Inspecting the terms in $\underset{6}{\Lambda}_{mn}$ (A.12, 3) we see products of $\underset{4}{\gamma_{00}}$ and $\underset{2}{\gamma_{00}}$ or, as it is there called 2φ. Thus, if we put there the expression in (8.18) in place of $\underset{4}{\gamma_{00}}$ and write for brevity:

$$(S_r\psi) = \sum_s \overset{s}{S_r} \underset{2l-2}{\overset{s}{\psi}} \tag{A.8, 2}$$

we get five new terms. Thus with the abbreviation (A.8, 2) we have in *every* approximation the following additional terms:

$$\begin{aligned} &\{(S_m\psi),_n + (S_n\psi),_m - \delta_{mn}(S_r\psi),_r\},_{00} \\ &+\varphi(S_r\psi),_{rmn} + \tfrac{1}{2}\varphi,_n(S_r\psi),_{rm} \\ &+\tfrac{1}{2}\varphi,_m(S_r\psi),_{rn} - \tfrac{3}{2}\,\delta_{mn}\varphi,_s(S_r\psi),_{rs} + \varphi,_{mn}(S_r\psi),_r. \end{aligned} \tag{A.8, 3}$$

Only three of the linear terms give us a contribution to the surface integral. It is more difficult to see that the non-linear terms do not give any contribution, since it requires some knowledge of how to deal with surface integrals which is outlined in A.4, and which we shall here assume. We can write the non-linear terms in (A.8, 3), in the following way:

$$\begin{aligned} &\{\varphi,_{mn}(S_r\psi),_r - \{\varphi,_{mr}(S_n\psi)\},_r \\ &+\varphi,_{mn}(S_r\psi),_r \\ &+\tfrac{1}{2}\{\varphi,_n(S_r\psi),_m\},_r - \tfrac{1}{2}\{\varphi,_r(S_n\psi),_m\},_r \\ &+\tfrac{1}{2}\varphi,_r(S_n\psi),_{mr} \\ &-\tfrac{3}{2}\delta_{mn}\varphi,_r(S_s\psi),_{sr}. \end{aligned} \tag{A.8, 4}$$

These are the non-linear expressions, and their divergence vanishes because φ is a harmonic function. The expressions written out in pairs in (A.8, 4) do not give any contribution to the surface integrals, because of our lemma in Sec. 3. Thus the only contribution could come from the terms:

$$\tfrac{3}{2}\varphi,_s S_n\psi,_{ms} - \tfrac{3}{2}\,\delta_{mn}\varphi,_s S_r\psi,_{rs}. \tag{A.8, 5}$$

Here only the "cross products" could give contributions and we find with the help of the table in A.4, that the result is zero.

A.10

In the $l = 2$ approximation we have:

$$
\text{(A.10, 1)} \quad
\begin{cases}
\underset{2}{\gamma}_{00} = 2\varphi = 2\jmath + 2g, \\
\underset{3}{\gamma}_{0n} = -2f\dot{\eta}^n - 2g\dot{\zeta}^n = \underset{3}{h}_{0n}, \\
\underset{2}{h}_{00} = \varphi = \jmath + g, \\
\underset{2}{h}^{00} = -h_{00} = -\varphi \\
\underset{3}{h}^{0n} = \underset{3}{h}_{0n} = \underset{3}{\gamma}_{0n}, \\
\underset{2}{h}_{mn} = -\underset{2}{h}^{mn} = \delta_{mn}\varphi.
\end{cases}
$$

A straightforward calculation gives:

$$
\text{(A.10, 2)} \quad
\begin{cases}
2\underset{2}{\Lambda}_{00} = 0, \\
2\underset{3}{\Lambda}_{0m} = -\underset{2}{\gamma}_{00,\, m\underset{1}{0}}, \\
2\underset{4}{\Lambda}_{mn} = -\underset{3}{\gamma}_{0m,\, \underset{1}{0}n} - \underset{3}{\gamma}_{0n,\, \underset{1}{0}m} + 2\delta_{mn}\varphi_{,\, \underset{2}{00}} \\
\qquad\qquad -2\varphi\varphi_{,mn} - \varphi_{,m}\varphi_{,n} + \frac{3}{2}\delta_{mn}\varphi_{,s}\varphi_{,s}.
\end{cases}
$$

The contributions to the surface integrals are (for the first singularity):

$$
\begin{aligned}
-\gamma_{0m,\,0n} &\to 4\overset{1}{m}\ddot{\eta}^m \cdot 4\pi, \\
-\eta_{0n,\,0m} &\to \tfrac{4}{3}\overset{1}{m}\ddot{\eta}^m \cdot 4\pi, \\
2\delta_{mn}\varphi_{,\,00} &\to -\tfrac{4}{3}\overset{1}{m}\ddot{\eta}^m \cdot 4\pi, \\
-2\varphi\varphi_{,\,mn} &\to \tfrac{8}{3}\overset{1}{m}\tilde{g} \cdot 4\pi, \\
-\varphi_{,m}\varphi_{,n} &\to -\tfrac{8}{3}\overset{1}{m}\tilde{g} \cdot 4\pi, \\
\tfrac{3}{2}\delta_{mn}\varphi_{,s}\varphi_{,s} &\to 2\overset{1}{m}\tilde{g}_{,m} \cdot 4\pi.
\end{aligned}
$$

$$
\left(\overset{1}{m} = \underset{2}{\overset{1}{m}}\right).
$$

A.12

A straightforward calculation of $\underset{4}{\Lambda}, \underset{5}{\Lambda}, \underset{6}{\Lambda}$ gives

$$\text{(A.12, 1)}\quad 2\underset{4}{\Lambda}_{00} = -\tfrac{3}{2}\varphi_{,s}\varphi_{,s},$$

$$\text{(A.12, 2)}\quad 2\underset{5}{\Lambda}'_{0m} = \varphi_{,s}\underset{3}{\gamma}_{0s,m} - \varphi_{,sm}\underset{3}{\gamma}_{0s} - 3\varphi_{,0}\varphi_{,m},$$

$$\begin{aligned}
2\underset{6}{\Lambda}_{mn} = {} & -\underset{5}{\gamma}_{0m,0n} - \underset{5}{\gamma}_{0n,0m} + \delta_{mn}\underset{4}{\gamma}_{00,00} + \underset{4}{\gamma}_{mn,00} - \varphi\underset{4}{\gamma}_{00,mn} \\
& - \varphi\underset{4}{\gamma}_{ss,mn} - \varphi_{,mn}\underset{4}{\gamma}_{00} - \varphi_{,mn}\underset{4}{\gamma}_{ss} + \varphi_{,ms}\underset{4}{\gamma}_{ns} \\
& + \varphi_{,ns}\underset{4}{\gamma}_{ms} - \delta_{mn}\varphi_{,sr}\underset{4}{\gamma}_{sr} - 2\varphi_{,s}\underset{4}{\gamma}_{mn,s} + \varphi_{,s}\underset{4}{\gamma}_{ms,n} \\
& + \varphi_{,s}\underset{4}{\gamma}_{ns,m} - \tfrac{1}{2}\varphi_{,m}\underset{4}{\gamma}_{ss,n} - \tfrac{1}{2}\varphi_{,n}\underset{4}{\gamma}_{ss,m} - \tfrac{1}{2}\varphi_{,n}\underset{4}{\gamma}_{00,m} \\
& - \tfrac{1}{2}\varphi_{,m}\underset{4}{\gamma}_{00,n} + \tfrac{3}{2}\delta_{mn}\varphi_{,s}\underset{4}{\gamma}_{rr,s} + \tfrac{3}{2}\delta_{mn}\varphi_{,s}\underset{4}{\gamma}_{00,s} \\
& - \underset{3}{\gamma}_{0s}\underset{3}{\gamma}_{0n,ms} - \underset{3}{\gamma}_{0s}\underset{3}{\gamma}_{0m,ns} + 2\underset{3}{\gamma}_{0s}\underset{3}{\gamma}_{0s,mn} \\
& + \tfrac{1}{2}\delta_{mn}\underset{3}{\gamma}_{0s,r}\underset{3}{\gamma}_{0r,s} - \tfrac{3}{2}\delta_{mn}\underset{3}{\gamma}_{0s,r}\underset{3}{\gamma}_{0s,r} + \underset{3}{\gamma}_{0s,m}\underset{3}{\gamma}_{0s,n} \\
& + \underset{3}{\gamma}_{0m,s}\underset{3}{\gamma}_{0n,s} - \varphi_{,0n}\underset{3}{\gamma}_{0m} - \varphi_{,0m}\underset{3}{\gamma}_{0n} + 2\delta_{mn}\underset{3}{\gamma}_{0s}\varphi_{,0s} \\
& - \varphi_{,0}\underset{3}{\gamma}_{0m,n} - \varphi_{,0}\underset{3}{\gamma}_{0n,m} - \varphi_{,n}\underset{3}{\gamma}_{0m,0} - \varphi_{,m}\underset{3}{\gamma}_{0n,0} \\
& + 2\varphi\underset{3}{\gamma}_{0m,0n} + 2\varphi\underset{3}{\gamma}_{0n,0m} - 2\delta_{mn}\varphi\varphi_{,00} \\
& + 2\varphi\varphi\varphi_{,mn} - \varphi\varphi_{,m}\varphi_{,n} + \tfrac{3}{2}\delta_{mm}\varphi\varphi_{,s}\varphi_{,s} \\
\text{(A.12, 3)}\qquad & + \tfrac{1}{2}\delta_{mn}\varphi_{,0}\varphi_{,0}.
\end{aligned}$$

The surface integral (12.4a) for $s = 1$ is, because of (12.2), and (12.1b):

$$\frac{1}{4\pi}\int\left(\varphi_{,r}\underset{3}{\gamma}_{0r,m} - \varphi_{,rm}\underset{3}{\gamma}_{0n} - \frac{3}{2}\varphi_{,0}\varphi_{,m} + \frac{3}{2}\varphi\varphi_{,0m} + 4\left(\underset{4}{\overset{1}{m}\psi}\right)_{0m}\right)n_m\, dS = 0.$$

The contributions of these five expressions are respectively:

$$(1) \rightarrow -\frac{4\overset{1}{m}}{3}\tilde{g}_{,s}\dot{\zeta}^s - 4\overset{1}{m}\tilde{g}_{,s}\dot{\eta}^s,$$

$$(2) \to -\frac{8\overset{1}{m}}{3}\tilde{g}_{,s}\dot{\xi}^s,$$

$$(3) \to \quad 3\overset{1}{m}\tilde{g}_{,s}+\overset{1}{m}\tilde{g}_{,s}\dot{\eta}^s,$$

$$(4) \to \quad 2\overset{1}{m}\tilde{g}_{,s}\dot{\eta}^s,$$

$$(5) \to -4\underset{4}{\overset{1}{\dot{m}}}.$$

Therefore:

$$-4\underset{4}{\overset{1}{\dot{m}}} = \overset{1}{m}\tilde{g}_{,s}\dot{\xi}^s+\overset{1}{m}\tilde{g}_{,s}\dot{\eta}^s = 2\overset{1}{m}\tilde{g}_{,s}\dot{\eta}^s-\overset{1}{m}\tilde{g}_{,s}\dot{\eta}^s+\overset{1}{m}\tilde{g}_{,s}\zeta^s$$
$$= -\overset{1}{m}(2\dot{\eta}^s\dot{\eta}^s+\tilde{g})_{,0}.$$

From the last equation (12.5) follows immediately.

The last step is to calculate the surface integrals due to $\underset{6}{\Lambda}$. Here a skilful use of the lemma may save the calculation of many surface integrals. Indeed, $2\underset{6}{\Lambda}_{mn}$ can be written in the following form:

$$
\begin{aligned}
2\Lambda_{mn} = {} & \Big(\varphi_{,n}\underset{4}{\gamma}_{sm}-\varphi_{,s}\underset{4}{\gamma}_{nm}\Big)_{,s}+\Big(\varphi\underset{4}{\gamma}_{ms,n}-\varphi\underset{4}{\gamma}_{mn,s}\Big)_{,s}\\
&+\Big(\delta_{ms}\varphi_{,r}\underset{4}{\gamma}_{rn}-\delta_{mn}\varphi_{,r}\underset{4}{\gamma}_{rs}\Big)_{,s}+\Big(\delta_{mn}\varphi_{,s}\underset{4}{\gamma}_{rr}-\delta_{ms}\varphi_{,n}\underset{4}{\gamma}_{rr}\Big)_{,s}\\
&+\tfrac{1}{1}\Big(\delta_{mn}\varphi\underset{4}{\gamma}_{rr,s}-\delta_{ms}\varphi\underset{4}{\gamma}_{rr,n}\Big)_{,s}+\Big(\delta_{mn}\underset{5}{\gamma}_{0s,0}-\delta_{ms}\underset{5}{\gamma}_{0n,0}\Big)_{,s}\\
&+\tfrac{1}{2}\Big(\underset{3}{\gamma}_{0s,m}\underset{3}{\gamma}_{0n}-\underset{3}{\gamma}_{0n,m}\underset{3}{\gamma}_{0s}\Big)_{,s}+\Big(\delta_{mn}\varphi_{,0}\underset{3}{\gamma}_{0s}-\delta_{ms}\psi_{,0}\underset{3}{\gamma}_{0n}\Big)_{,s}\\
&+\Big(\underset{3}{\gamma}_{0n}\underset{3}{\gamma}_{0m,s}-\underset{3}{\gamma}_{0s}\underset{3}{\gamma}_{0m,n}\Big)_{,s}+\tfrac{1}{2}\Big(\delta_{mn}\underset{3}{\gamma}_{0s,r}\underset{3}{\gamma}_{0r}-\delta_{ms}\underset{3}{\gamma}_{0n,r}\underset{3}{\gamma}_{0r}\Big)_{,s}\\
&+\Big(\delta_{ms}\underset{3}{\gamma}_{0r,n}\underset{3}{\gamma}_{0r}-\delta_{mn}\underset{3}{\gamma}_{0r,s}\underset{3}{\gamma}_{0r}\Big)_{,s}\\
&-\underset{5}{\gamma}_{0m,0n}+\underset{4}{\gamma}_{mn,00}+\underset{3}{\gamma}_{0s}\underset{3}{\gamma}_{0s\,mn} && [\alpha_1+\alpha_2+\alpha_3]\\
&-\tfrac{1}{2}\delta_{mn}\underset{3}{\gamma}_{0s,r}\underset{3}{\gamma}_{0s,r}-\Big(\varphi_{,n}\underset{3}{\gamma}_{0m}\Big)_{,0} && [\alpha_4+\alpha_5]\\
&-\Big(\varphi_{,m}\underset{3}{\gamma}_{0n}\Big)_{,0}+\Big(\varphi\underset{3}{\gamma}_{0m,n}\Big)_{,0}+\Big(\varphi\underset{3}{\gamma}_{0n,m}\Big)_{,0} && [\alpha_6+\alpha_7+\alpha_8]\\
&-\tfrac{3}{2}\delta_{mn}\varphi_{,0}\varphi_{,0}-\tfrac{1}{2}\varphi\underset{4}{\gamma}_{ss,mn} && [\alpha_9+\alpha_{10}]
\end{aligned}
$$

TABLE OF SURFACE INTEGRALS FOR $\int_6^{1} \Lambda_{nts} n_s \, dS$

No.	Expression	α_1	α_2	α_3	α_4	α_5	α_6	α_7	α_8	α_9	α_{10}	α_{11}	α_{12}	α_{13}	α_{14}	α_{15}	α_{16}	α_{17}	Result	Remarks
1	$\overset{1}{m} g_{,s} \dot{\eta}^s \dot{\eta}^m$	$-\frac{16}{3}$				$-\frac{4}{3}$		$-\frac{8}{3}$	$-\frac{4}{15}$		$\frac{8}{15}$	$\frac{4}{15}$				$\frac{4}{5}$			-8	$\tilde{g}_{,s} = -2\overset{2}{m} \frac{\partial \frac{1}{r}}{\partial \eta^s}$
2	$\overset{1}{m} \tilde{g} \ddot{\eta}^m$	-2						-4	$-\frac{4}{3}$			$-\frac{29}{3}$	3	$\frac{11}{3}$	$-\frac{5}{3}$	$\frac{2}{3}$	$\frac{32}{5}$	$-\frac{22}{3}$	-8	$\tilde{g} = -\frac{2\overset{2}{m}}{r}$; $\ddot{\eta}^m = -\frac{1}{2}\tilde{g}_{,m}$
3	$\overset{1}{m} \tilde{g}_{,m} \dot{\eta}^s \dot{\eta}^s$	1					$-\frac{4}{3}$	$-\frac{4}{5}$			$\frac{8}{5}$	$\frac{4}{5}$	1	$\frac{1}{3}$	$-\frac{1}{3}$	$-\frac{4}{15}$			2	$\tilde{g}_{,m} = -2\overset{1}{m} \ddot{\eta}^m$
4	$\overset{1}{m} \tilde{g}_{,m} \dot{\zeta}^s \dot{\zeta}^s$											2	$\frac{1}{3}$	1	$-\frac{1}{3}$				3	
5	$\overset{1}{m} \tilde{g}_{,m} \tilde{f}$	$\frac{4}{3}$				2	$\frac{2}{3}$					$\frac{1}{2}$	$\frac{1}{6}$	$\frac{1}{2}$	$-\frac{1}{6}$				5	$\tilde{g}_{,m}\tilde{f} = \tilde{g}\tilde{f}_{,m}$; $\tilde{f} = -\frac{2\overset{1}{m}}{r}$
6	$\overset{1}{m}\overset{2}{m} \tilde{r}_{,00m}$											-2							-2	$\tilde{r}_{,00m} = \left(\overset{2}{r}_{,00m}\right)$ for $x^s = \eta^s$ *
7	$\overset{1}{m} \tilde{g}_{,s} \zeta^m \dot{\eta}^s$	$\frac{16}{5}$				$\frac{8}{3}$	$\frac{4}{5}$	$\frac{4}{3}$											8	
8	$\overset{1}{m} \tilde{g}_{,s} \dot{\zeta}^s \dot{\eta}^m$	$\frac{16}{5}$					$-\frac{8}{15}$	4	$\frac{4}{3}$	-2									6	
9	$\overset{1}{m} \tilde{g}_{,m} \dot{\eta}^s \dot{\zeta}^s$	$-\frac{32}{15}$		$-\frac{16}{3}$	$-\frac{8}{3}$		$\frac{4}{5}$		$\frac{4}{3}$										-8	
10	$\overset{1}{m} \tilde{g}_{,s} \dot{\zeta}^s \dot{\zeta}^m$	$-\frac{8}{3}$				-4	$-\frac{4}{3}$												-8	

$$^*\tilde{r}_{,00m} = \frac{\partial^3 r}{\partial \eta^s \partial \eta^r \partial \eta^m} \dot{\zeta}^s \dot{\zeta}^r, \text{ as } \frac{\partial^3 r}{\partial \eta^s \partial \eta^m} \frac{\partial \frac{1}{r}}{\partial \eta^s} = 0.$$

$$+\tfrac{1}{2}\varphi_{,n}\underset{4}{\gamma}_{ss,m}+\tfrac{1}{2}\varphi_{,n}\underset{4}{\gamma}_{00,m} \qquad [\alpha_{11}+\alpha_{12}]$$

$$+\tfrac{1}{2}\varphi_{,m}\underset{4}{\gamma}_{00,n}-\tfrac{1}{2}\delta_{mn}\underset{4}{\gamma}_{,s}\varphi_{00,s} \qquad [\alpha_{13}+\alpha_{14}]$$

$$-\varphi\varphi_{,00}\delta_{mn}-2\varphi\varphi_{,m}\varphi_{,n} \qquad [\alpha_{15}+\alpha_{16}]$$

$$+\tfrac{11}{4}\varphi\varphi_{,s}\varphi_{,s}\delta_{mn}. \qquad [\alpha_{17}]$$

Because of the lemma we have to find now the surface integrals of only 17 expressions denoted successively by $\alpha_1, \alpha_2, \ldots, \alpha_{17}$. The result of this calculation is summarized in the table. Only ten types of expressions (or their equivalents) appear in the result. The table tells us what is the contribution of each of the α's to the final result. The only α that does not give a contribution is $\alpha_2 = \underset{4}{\gamma}_{mn,00}$.

A.13

The additional expressions in $\underset{6}{\Lambda}_{mn}$ induced through rejection of the coordinate condition are:

$$\begin{aligned}
&2(\delta_{mn}\alpha_{0,r0}-\delta_{mr}\alpha_{0,n0})_{,r}\\
+\;&(\varphi_{,m}\alpha_{n,r}-\varphi_{,m}\alpha_{r,n})_{,r}\\
+\;&(\varphi_{,n}\alpha_{m,r}-\varphi_{,r}\alpha_{m,n})_{,r}\\
+\;&(\varphi_{,n}\alpha_{s,m}-\varphi_{,s}\alpha_{n,m})_{,s}\\
-&2(\delta_{mn}\varphi_{,s}\alpha_{s,r}-\delta_{mr}\varphi_{,s}\alpha_{s,n})_{,r}\\
+&2(\delta_{mn}\varphi_{,s}\alpha_{r,r}-\delta_{ms}\varphi_{,n}\alpha_{r,r})_{,s}\,.
\end{aligned}$$

They are written in such a way that the vanishing of each line is evident, because of the lemma.

References

[1] A. EINSTEIN, L. INFELD and B. HOFFMANN, *Ann. of Math.*, vol. 39, 1(1938) 66.
[2] A. EINSTEIN and L. INFELD, *Ann. of Math.*, vol. 41, 2 (1940) 455.
[3] L. INFELD, *Phys. Rev.*, vol. 53 (1938) 836.
[4] L. INFELD and P. R. WALLACE, *Phys. Rev.*, vol. 57 (1940) 797.
[5] H. P. ROBERTSON, *Ann. of Math.*, vol. 39, 1 (1938) 101.

Institute for Advanced Study
University of Toronto

NOTES ON EXTRACT 6

A SOMEWHAT different view of relativity from Einstein's (which yet agrees in nearly all important details of calculation) is set out in this paper by Fock. This version has been corrected by the author. Fock's book *The Theory of Space, Time and Gravitation*, has been published in English at the Pergamon Press in 1959 (first edition) and in 1964 (second edition). The analysis of the notion of relativity has been pursued in this book and also in Fock's papers "The principles of relativity and of equivalence in the Einsteinian gravitation theory" (*Kgl. Norske Videnskabers Selskabs Forhandlinger*, vol. 36, N2 4–5, Frondheinm, 1963) and "Les principes physiques de la théorie de gravitation d'Einstein", *Ann. Inst. Henri Poincaré*, vol. V, No. 3, p. 205, Paris, 1966.)

EXTRACT 6[†]

Three Lectures on Relativity Theory

V. FOCK

Physical Institute, University of Leningrad, U.S.S.R.
(Delivered to the Colloquium of the Institute for Theoretical Physics, Copenhagen, Denmark, February 18, 20, 22, 1957.)

First Lecture: on Homogeneity, Covariance, and Relativity

In my first lecture, I try to elucidate some general notions connected with relativity theory. I speak on homogeneity, covariance, and relativity. My considerations are of a very simple nature but, nevertheless, I hope that they may be of interest, because simple notions are often the most difficult ones.

If we consider the geometrical aspect of the theory of space and time, this theory naturally divides into the theory of homogeneous (uniform) space–time and that of the nonhomogeneous (nonuniform) space–time. The former may be called Galilean space and the latter the Riemannian or Einsteinian space. (I sometimes use the word space instead of space–time.)

The property of space–time of being homogeneous means that (a) there are no privileged points in space and in times (b) there are no privileged directions, and (c) there are no privileged inertial frames (that all frames moving uniformly and in a straight line with respect to one another are on the same footing).

The uniformity of space and time manifests itself in the existence of the Lorentz group. In particular, the equality of points in space and

† *Rev. Mod. Phys.* **29**, 325 (1957).

time corresponds to the possibility of a displacement, the equality of directions corresponds to that of spatial rotations, and the equality of inertial frames corresponds to a special Lorentz transformation. The displacements contain four parameters, the rotation three (the three angles), and the transformation to a moving frame also three (the three components of velocity). This gives together ten parameters—the maximum possible number, if we do not take into account scale transformations $x' = \lambda x$.

The statement that the Lorentz transformation leaves invariant the expression for the square of the line element is to be understood in the following sense.

If one writes ds^2 as

$$ds^2 = dx_0^2 - dx_1^2 - dx_2^2 - dx_3^2$$

or

$$ds^2 = \eta_{\mu\nu}\, dx_\mu\, dx_\nu$$

then, after the transformation from (x) to (x'), we have

$$ds^2 = \eta_{\mu\nu}\, dx'_\mu\, dx'_\nu$$

with *the same* matrix

$$||\eta_{\mu\nu}|| = \begin{Bmatrix} 1 & 0 & 0 & 0 \\ 0 & -1 & 0 & 0 \\ 0 & 0 & -1 & 0 \\ 0 & 0 & 0 & -1 \end{Bmatrix}.$$

In studying the properties of homogeneous (uniform) space–time, the use of Galilean coordinates is convenient, but not essential. The property of space–time of being uniform may be as well expressed in general coordinates.

Let the substitution

$$x'_\mu = f_\mu(x_0 x_1 x_2 x_3) \equiv f_\mu(x)$$

be performed in the expression for ds^2. Then,

$$ds^2 = g_{\mu\nu}(x)\, dx_\mu\, dx_\nu$$

changes into

$$ds^2 = g'_{\mu\nu}(x')\, dx'_\mu\, dx'_\nu$$

so that

$$g'_{\mu\nu}(x') = g_{\alpha\beta} \frac{\partial x_\alpha}{\partial x'_\mu} \frac{\partial x_\beta}{\partial x'_\nu}. \tag{1}$$

If the mathematical form of the functions $g'_{\mu\nu}$ is the same as that of the $g_{\mu\nu}$, that is, if

$$g'_{\mu\nu}(x) = g_{\mu\nu}(x), \tag{2}$$

then the space admits a transformation group.

For an infinitesimal transformation

$$x'_\alpha = x_\alpha + \eta^\alpha(x);$$

this leads to

$$\nabla^\mu \eta^\nu + \nabla^\nu \eta^\mu = 0,$$

and these equations are completely integrable if

$$R_{\mu\nu,\,\alpha\beta} = K(g_{\nu\alpha}g_{\mu\beta} - g_{\mu\alpha}g_{\nu\beta}),$$

that is, for a space of constant curvature. Galilean space corresponds to vanishing curvature

$$R_{\mu\nu,\,\alpha\beta} = 0.$$

What I wish to stress is that the properties of the uniform Galilean space–time can be expressed in a generally covariant manner. On the other hand, the Einsteinian gravitation theory supposes the space–time to be nonuniform. It is just this fundamental assumption, and not the general covariance of equations, that distinguishes the gravitation theory from the theory of the Galilean space–time.

This distinction has not been sufficiently understood, or in any case not sufficiently stressed, by many physicists and, paradoxical as it may seem, by Einstein himself, although the French mathematician Cartan has drawn attention to it many years ago. Einstein called both theories relativity theories. But what is relativity? This word has been misused. It is natural to connect the notion relativity with uniformity of space and time. The uniformity of Galilean space with respect to positions, directions, and nonaccelerated motions may be as well termed as relativity of positions, of directions, and of nonaccelerated motions. That is the true content of Einstein's principle of relativity of 1905. Use of the word relativity in this sense is quite legitimate.

But, if one does this, if one connects relativity with uniformity, the relativity has nothing to do with general covariance, that is, with covariance in which (1) is true but (2) is not necessarily satisfied. This means also that, in the theory of nonuniform space–time, there is no principle of relativity. The generalization of the theory which consists in replacement of a uniform space–time by a nonuniform one means a restriction and not a generalization of relativity. If one uses the word relativity consistently, then the general principle of relativity is nonsensical.

In saying this, I do not want to introduce any doubt as to the validity of the wonderful Einsteinian gravitation theory, but only to stress the inconsistency of the use of the name "general relativity" when applied to gravitation theory.

Einstein himself proposed for his theory the name "general relativity", because the transformations considered in this theory are more general than the Lorentz transformations. But he omitted to state that, in the case of ordinary relativity, one has to consider transformations for which (2) must also be fulfilled, while, in the case of the so-called general relativity, this equation does not have to be taken into account. Thus, in the "general" theory, Einstein uses the word relativity simply as covariance, while in the "special" theory, the same word relativity is used as uniformity. Since covariance has nothing to do with uniformity, there arises a confusion which is very harmful to the understanding of Einstein's theory. If one uses the word relativity in both senses, then one has to admit such statements as "in general relativity there is no relativity" or "the Lagrangian form of nonrelativistic equations of motion satisfies the requirements of general relativity", etc.

This confusion is more harmful than it would seem at first glance. It leads to statements like "rotation is relative" which are obviously false, because the distinction between a geodetic and a nongeodetic is absolute and not relative.

The general covariance of equations has been considered for a long time as a specific property of the Einsteinian gravitation theory, by which it is distinguished from other physical theories. But later on, it was recognized that the covariance by itself cannot lead to any physical consequences. The true key to Einstein's discovery and the most difficult step was the limitation of the functions describing the gravitational field to geometrical ones (to the $g_{\mu\nu}$'s). Historically, the covariance requirement played a great part also, but this is because it was combined with other requirements, such as simplicity and beauty of the theory.

Nevertheless, the covariance requirement is still considered in a somewhat mystical way, as something prohibiting the use of well-defined coordinate frames, like Galilean coordinates in uniform space–time. The existence of Galilean coordinate frames is a characteristic of the inherent properties of the uniform space–time of the "special" theory. Likewise, there may be in "general" theory coordinate frames distinguished by some remarkable properties and characteristic of the kind of the nonuniform space–time considered.

In what follows, I wish to draw your attention to the fact that, for a rather general class of problems of gravitation theory, there exist such coordinate systems that may be considered as generalizations of ordinary inertial systems. I mean not the local geodetic system valid in the vicinity of a point and of an instant of time, but the nonlocal generalization of the inertial frames of reference, valid throughout space.

In order to investigate whether such systems exist, it is necessary to make definite assumptions as to the physical system considered and as to the properties of space–time as a whole. This is necessary because of the nonlocal character of the problem, that requires a solution of Einstein's gravitational equations with conditions at infinity.

In the case of an isolated system of masses, it is natural to consider the system as embedded in a Galilean space–time. In a Galilean space–time, the following *theorem* holds.

Let

$$\Box\psi = \frac{1}{c^2}\frac{\partial^2\psi}{\partial t^2} - \left(\frac{\partial^2\psi}{\partial x^2} + \frac{\partial^2\psi}{\partial y^2} + \frac{\partial^2\psi}{\partial z^2}\right).$$

If ψ satisfies the wave equation $\Box\psi = 0$ and is finite everywhere and tends to zero at infinity like $1/r$, as well as its derivatives, and if in addition the radiation condition

$$\lim_{r\to\infty}\left\{\frac{\partial(r\psi)}{\partial r} + \frac{1}{c}\frac{\partial(r\psi)}{\partial t}\right\} = 0$$

is satisfied for all values of the time t, *then ψ vanishes identically*. The radiation condition states that only outgoing waves are allowed.

A similar theorem may be proved in the case that $\Box\psi$ refers to a static Einsteinian space–time which is Galilean at infinity. It is to be supposed that the theorem holds also for a nonstatic Einsteinian space-time, though a formal proof may be difficult.

Let the space–time be such that the aforestated theorem is valid. Then, one can introduce auxiliary conditions for the coordinates in such a way that they behave like Galilean coordinates and are determined like them throughout the space–time (a Lorentz transformation remains of course arbitrary).

The auxiliary conditions are of the form $\Box x_\nu = 0$; $\nu = 0, 1, 2, 3$. But we have

$$\Box\psi = g^{\mu\nu}\frac{\partial^2\psi}{\partial x_\mu\,\partial x_\nu} - \Gamma^\nu\frac{\partial\psi}{\partial x_\nu},$$

where

$$\Gamma^\nu = -\frac{1}{(-g)^{\frac{1}{2}}}\frac{\partial \mathfrak{g}^{\mu\nu}}{\partial x_\mu}; \qquad \mathfrak{g}^{\mu\nu} = (-g)^{\frac{1}{2}}g^{\mu\nu}.$$

Consequently, the condition is equivalent to

$$\partial \mathfrak{g}^{\mu\nu}/\partial x_\mu = 0.$$

Let the coordinates x_μ satisfy this condition. To find the most general form

$$x'_\alpha = f^\alpha(x_0x_1x_2x_3)$$

we put

$$f^{\alpha} = a_{\alpha} + e_{\beta} a_{\alpha\beta} x_{\beta} + \eta^{\alpha}. \quad (e_0 = 1; \quad e_1 = e_2 = e_3 = -1).$$

The linear part of this is a Lorentz transformation. Now, η^{α} must satisfy the wave equation $\Box \eta^{\alpha} = 0$, since f^{α} and the linear part satisfy it. Further, η^{α} must vanish at infinity (because the transformation must reduce there to a Lorentz transformation) and also η^{α} must satisfy the radiation condition (this follows from the radiation condition for the $g^{\mu\nu}$'s):

$$\eta^{\alpha} = \text{outgoing wave at infinity.}$$

But the conditions imposed upon η^{α} are so stringent that, according to the theorem, η^{α} = zero everywhere. Thus, the whole arbitrariness of the coordinates resides in the Lorentz transformation.

We thus come to the conclusion that, in the case of an isolated system of masses, there is no essential difference in the coordinate question, between the so-called general and so-called special relativity theory. In both cases, arbitrary coordinates are admissible, since the equations are, or may be, written covariantly with respect to general transformations. But, in both cases, auxiliary conditions may be imposed upon the coordinates in such a way that only a Lorentz transformation remains arbitrary.

The coordinates so defined—I call them harmonic—are particularly adapted to the solution of Einstein's equations, and all the solutions that I shall discuss in the following lectures are obtained in these coordinates. But the value of the harmonic coordinates resides not only in their practical importance, but also in the fact that they help us to understand the general features of gravitation theory. Their existence shows that the usual sharp distinction between the coordinate problem in special and in general theory is somewhat artificial. In both theories, coordinates exist that are determined to a Lorentz transformation, but in both theories any other coordinate system may be used.

NOTES ON EXTRACT 7

ONE of the most influential ideas in giving a new direction to general relativity is contained in this paper of Pirani. It seeks to answer the problem: What is the appropriate definition of radiation in gravitation theory?, and its answer is largely that adopted later.

EXTRACT 7†

Invariant Formulation of Gravitational Radiation Theory

F. A. E. PIRANI

Department of Mathematics, King's College, Strand, London, England
[Received October 18, 1956]

In this paper, gravitational radiation is defined invariantly within the framework of general relativity theory. The definition is arrived at by assuming (*a*) that gravitational radiation is characterized by the Riemann tensor, and (*b*) that it is propagated with the fundamental velocity. Therefore a gravitational wave front should appear as a discontinuity in the Riemann tensor across a null 3-surface; the possible form of this discontinuity is here calculated from Lichnerowicz's continuity conditions.

The concept of an observer who follows the gravitational field is defined in terms of the eigenbivectors of the Riemann tensor. It is shown that the 4-velocity of this observer is timelike for one of Petrov's three canonical types of Riemann tensor, but null for the other two types. The first type is identified with the absence of radiation, the other two with its presence. This constitutes the definition. It is shown that the difference between the no-radiation type and one of the radiation types can be made to correspond to the discontinuity possible across a null 3-surface; this demonstrates the consistency of the wave front and following-the-field concepts.

A covariant approximation to the canonical energy-momentum pseudo-tensor is defined, using normal coordinates, which are given a physical interpretation. It is shown that when gravitational radiation is present, the approximate gravitational energy-flux cannot be removed by a local Lorentz transformation, which supports the definition of radiation.

It is proved that, as would be demanded of a sensible definition, there can be no gravitational radiation present in a region of empty space–time where the metric is static.

† *Phys. Rev.* **105,** 1089 (1957).

1. Introduction

The investigation of gravitational radiation in general relativity theory is hampered by the lack of an invariant definition of that concept. The presence of gravitational radiation must be distinguishable, mathematically, from a peculiar choice of the coordinate system, and physically, from a peculiar motion of the observer. In a covariant, nonlinear theory, the definition should not, if the concept of radiation has any real validity, depend on the weakness of fields or on special coordinate conditions. An invariant definition is proposed in this paper.

This definition is given in terms of the Riemann tensor. Just as it is the Riemann tensor which indicates a genuine gravitational field in the first place, so

(*A*) It is the Riemann tensor which characterizes the presence of radiation.

Physically, this is because the Riemann tensor describes the variations in the gravitational field from event to event in space–time. In accordance with the principle of equivalence, only the variations in the field, and not the field itself, can produce any real physical effects. The question is: what sort of variations in the field should be classified as gravitational radiation?

To answer this question, one must first of all decide which attributes of radiation, a concept until now familiar largely through electromagnetic theory, may be assumed to apply also to the gravitational case. In making the present definition, it will be assumed that its essential attribute is:

(*B*) In empty space–time, gravitational radiation is propagated with the fundamental velocity.

The two assumptions (*A*) and (*B*) serve to characterize gravitational radiation completely. Two main arguments are developed in the following sections to support the definition; these arguments depend respectively on the following consequences of (*A*) and (*B*):

(*C*) A gravitational wave-front manifests itself as a discontinuity in the Riemann tensor across a null 3-surface.

(*D*) The motion of an observer following the gravitational field is determined by the Riemann tensor. In the presence of gravitational radiation, such an observer would have to move with the fundamental velocity in order to keep up with the field.

These ideas will now be developed in more detail. In connection with (*A*), one may investigate the variations in the gravitational field directly by writing down the equation of geodesic deviation. This equation gives the variation in the field between neighboring space–time events in terms of the Riemann tensor.[1] The physical effects so represented are set out in detail in Sec. 2.

Assumption (*B*) is supported by very general considerations, as well as some specific ones, like the result of Lichnerowicz[2] that the characteristic surfaces of Einstein's equations are null 3-surfaces. Lichnerowicz starts from continuity conditions which are sufficient to ensure that the equations have physically unique solutions in empty space–time.

In Sec. 2, Lichnerowicz's conditions will be used to determine what discontinuity in the Riemann tensor is permissible across a null 3-surface. In accordance with statement (*C*) above, one would expect to find such a discontinuity whenever a source of gravitational radiation was switched on or off.

The idea of an observer following the field, introduced in statement (*D*), is one already well known in the ordinary Maxwell–Lorentz electromagnetic theory, and not difficult to generalize to gravitational field theory. In the Maxwell–Lorentz theory, an observer is said to be following the field if he moves so that in his restframe the Poynting vector vanishes. He therefore observes no flux of field energy. If this idea is restated covariantly (but still in the Maxwell–Lorentz theory)

[1] F. A. E. Pirani, *Helv. Phys. Acta* (to be published); and *Acta Phys. Polon.* (to be published).

[2] A. Lichnerowicz, *Théories relativistes de la gravitation et de l'électromagnétisme* (Masson et Cie, Paris, 1955), p. 33.

in terms of the energy-tensor of the field, it is found that an observer may always follow the field by acquiring a suitable 4-velocity, unless it is a null field (self-conjugate field), in which case it would be necessary to acquire the fundamental velocity in order to make the energy flux vanish. This is because in a null field **E** and **H** are perpendicular and of equal magnitude in every Lorentz frame. Plane waves and spherical waves are common examples of null fields.[3] From the point of view adopted here, only null fields will be counted as radiation.

The idea of following the field, expressed in such terms, does not admit an immediate covariant generalization to the case of the gravitational field, for, because of the principle of equivalence, there is no covariant gravitational field energy-tensor. The generalization will be achieved by considering the geometrical properties of the two fields. It will be found that in each case certain eigenvectors of the field can be defined. In the electromagnetic case it turns out that an observer following the field has the timelike eigenvector for 4-velocity; when the field is a null field this timelike vector collapses onto the null cone,[3] and it is this which is characteristic of the presence of radiation.

In the same way, a timelike eigenvector may in general be defined, in terms of the Riemann tensor, for the gravitational field in empty space-time. This vector is interpreted as the 4-velocity of an observer following the field, and in some fields this vector collapses onto the null cone. As in the electromagnetic case, this situation is identified with the presence of radiation.

These timelike eigenvectors can be given a further physical significance in both gravitational and electromagnetic fields. For example, in a non-null electromagnetic field, the 4-velocity which follows the field yields, for a given field, an extreme magnitude for the Lorentz force. Similarly, the (more complicated) physical effects of the gravitational field also reach extreme values for an observer having the 4-velocity which follows the field. This will be discussed in detail in Sec. 4.

[3] For a detailed discussion of the geometrical and algebraic properties of the electromagnetic field, see J. L. Synge, *Relativity: the Special Theory* (North-Holland Publishing Company, Amsterdam, 1956), Chap. IX.

The eigenvectors of the gravitational field are defined in Sec. 3, with the aid of Petrov's classification[4] of empty space–time Riemann tensors into canonical types. The elegant geometrical-algebraic techniques developed by Ruse[5] and others supply the basis for this classification, which yields two types with radiation present and one with no radiation. It will be shown that the difference between the non-radiation type and one of the radiation types can be made, by a suitable alignment of axes, to correspond exactly to the discontinuity in the Riemann tensor across a wave front permitted by Lichnerowicz's conditions. This will demonstrate the consistency of the idea (Sec. 2) of a gravitational wave front with the idea (Sec. 3) that an observer following the gravitational radiation field must have the fundamental velocity.

This geometrical approach is necessary because there is no covariant gravitational energy-momentum tensor in Einstein's theory. This lack is to be expected, because of the principle of equivalence. In Lorentz-invariant field theories the energy-momentum tensor depends on the field strengths, but these have locally no absolute significance for the gravitational field. The canonical energy-momentum pseudotensor[6] t^{ν}_{μ} is a quadratic function of the field strengths (Christoffel symbols) and satisfies a conservation law, but it is not covariant—it can in fact be made to vanish entirely along an arbitrary open curve in space–time. Nevertheless, t^{ν}_{μ} has been used in definitions of gravitational radiation by various authors,[7] but always in weak field approximations and under physically obscure coordinate conditions.

It might be possible to construct covariant, and therefore physically significant, expressions out of t^{ν}_{μ} over extended regions of space–time, into which only the variations in the gravitational field, not the

[4] A. Z. Petrov, *Sci. Not. Kazan State Univ.* **114,** 55 (1954).

[5] H. S. Ruse, *Proc. Roy. Soc. (Edinburgh)* **62,** 64 (1944); *Quart. J. Math. (Oxford)* **17,** 1 (1946); *Proc. London Math. Soc.* **50,** 75 (1947).

[6] Range and summation conventions: lower case Greek indices 0, 1, 2, 3; lower case Latin indices 1, 2, 3.

[7] E.g., L. Landau and E. Lifshitz, *The Classical Theory of Fields* (Addison-Wesley Publishing Company, Inc., Cambridge, 1951); J. N. Goldberg, *Phys. Rev.* **99,** 1873 (1955).

field itself, could enter, but no one seems to have succeeded in doing this, or even to have attempted it. The difficulties could be resolved if one could reformulate the weak-field approximation method in a covariant way valid in an extended region.

The alternative to this, which is adopted in Sec. 4, is to develop a covariant local approximation method valid in arbitrarily strong fields but only in small regions of space–time. This is done by introducing normal coordinates,[8] which are given a physical interpretation. The energy-momentum pseudotensor t_μ^ν vanishes at the origin of normal coordinates but not in a finite neighborhood of the origin, and may be expanded in a power series with tensor coefficients which are functions of the Riemann tensor and its covariant derivatives. By averaging over a small 2-region, one may construct a covariant approximation $\overline{t_\mu^\nu}$ to the mean energy-momentum pseudotensor in the region. It is found that when no gravitational radiation is present, there exist observers (with suitable 4-velocities) who observe no gravitational energy flux, but that when gravitational radiation is present, such observers cannot be found. This corresponds exactly to the electromagnetic field case, and supports the definition of gravitational radiation.

Various examples are discussed in Sec. 5, and some deficiencies of the present approach are mentioned in Sec. 6.

2. Nature of a Gravitational Wave Front

The nature of a gravitational wave front will now be investigated, by finding the discontinuity in the Riemann tensor permissible across a null 3-surface. The calculation is based on Lichnerowicz's continuity conditions,[9] which are conditions on the metric tensor and its derivatives sufficient to ensure that Einstein's equations *in vacuo*,[10]

$$G_{\mu\nu} = 0, \tag{2.1}$$

[8] B. Riemann, *Göttingen Abhandl.* **13,** 1 (1862); see O. Veblen, *Invariants of Quadratic Differential Forms* (Cambridge University Press, Cambridge, 1927), for a lucid exposition.

[9] Reference 2, Chap. I.

[10] The cosmological term is disregarded throughout this paper. It could be restored to the work without any difficulty, but at some cost in conciseness.

are physically unique. These conditions are essentially the same as those found by O'Brien and Synge[11] from assumptions about the finiteness of certain quantities in the boundary layer between two regions of a continuous medium.

Lichnerowicz postulates that space–time can be divided up into overlapping regions, in each of which there exists a coordinate system such that (i) the metric tensor $g_{\mu\nu}$ is continuous, (ii) the first partial derivatives[12] $g_{\mu\nu,\varrho}$ are continuous, (iii) the second and third partial derivatives of $g_{\mu\nu}$ are piecewise continuous. Space–time is assumed be a Riemannian manifold with certain differentiability properties which do not affect the present argument.

Lichnerowicz's analysis is developed by taking in one of the regions a coordinate system such that, say, the surface $S: x^0 = 0$ is a surface of discontinuity of the gravitational field. Then according to the postulates, the coordinate system can be chosen so that all the $g_{\mu\nu}$, all the $g_{\mu\nu,\varrho}$, and all the $g_{\mu\nu,\varrho\sigma}$ with the possible exception of $g_{\mu\nu,00}$ are continuous across S. Now the covariant components of the Riemann tensor are[13]

$$R_{\varrho\sigma\mu\nu} = [\sigma\nu, \varrho]_{,\mu} - [\sigma\mu, \varrho]_{,\nu} + \Gamma^{\pi}_{\sigma\mu}[\varrho\nu, \pi] - \Gamma^{\pi}_{\sigma\nu}[\varrho\mu, \pi]. \tag{2.2}$$

If the first two terms of this are written out in full, it may be seen that the only components of $R_{\varrho\sigma\mu\nu}$ which admit discontinuities across S are those with just two indices 0, one in each of the pairs $\varrho\sigma$, $\mu\nu$. This is the same as a result of O'Brien and Synge.[14]

Results of this kind may be put into covariant form by transforming to a local Minkowskian coordinate system and then introducing a *tetrad* (orthonormal frame, quadruped, Vierbein, 4-nuple) of orthogonal unit vectors directed along the axes of the local Minkowskian system. Tensor equations may then be rewritten as scalar equations by contracting with tetrad vectors, and the special coordinate system discarded.

[11] O'Brien and J. L. Synge, *Comm. Dublin Inst.* A, No. 9 (1952).

[12] Comma denotes partial differentiation: $g_{\mu\nu,\varrho} \equiv \partial g_{\mu\nu}/\partial x^{\varrho}$.

[13] $(\sigma_{\mu\nu}) = \frac{1}{2}(g_{\nu\varrho,\sigma} + g_{\varrho\sigma,\nu} - g_{\sigma\nu,\varrho})$ and $\Gamma \equiv g^{\pi}_{\sigma\nu}\,[\sigma\nu, \varrho]$ are Christoffel symbols of the first and second kinds respectively.

[14] Reference 11, Eq. (5.12).

Physically, the timelike tetrad vector is identified as the 4-velocity of an observer, making measurements at the event in question, who uses space axes having the directions of the three spacelike tetrad vectors. The scalars formed by contracting any tensor with tetrad vectors are just the physical components of the tensor, measured by this observer.[15]

For the present purpose, the 3-surface S is to be a null 3-surface. It is not convenient to have any of the x^μ as a null coordinate; therefore one writes, say, $\xi = 0$ in place of $x^0 = 0$ for the equation of S. One may by a linear transformation introduce at any point P of S local Minkowskian coordinates such that $ds^2 = \eta_{\mu\nu}\, dx^\mu\, dx^\nu$, where[16]

$$\eta_{\mu\nu} = \operatorname{diag}(1, -1, -1, -1),$$

and in a finite neighborhood of P, one may take $\xi = 2^{-\frac{1}{2}}(x^0 - x^1)$. If also $\zeta = 2^{-\frac{1}{2}}(x^0 + x^1)$ in this neighborhood, then at P,

$$ds^2 = 2\, d\xi\, d\zeta - (dx^2)^2 - (dx^3)^2.$$

Then if Δ denotes amount of discontinuity across S, Lichnerowicz's conditions require that at P,

$$\Delta(g_{\mu\nu}) = 0,$$
$$\Delta(g_{\mu\nu,\sigma}) = \Delta(\partial g_{\mu\nu}/\partial\xi) = \Delta(\partial g_{\mu\nu}/\partial\zeta) = 0,$$
$$\Delta(\partial^2 g_{\mu\nu}/\partial\xi\, \partial\zeta) = \Delta(\partial^2 g_{\mu\nu}/\partial\zeta^2) = 0,$$

but

$$\Delta(\partial^2 g_{\mu\nu}/\partial\xi^2) = a_{\mu\nu},$$

where $a_{\mu\nu}$ are any numbers. The possible discontinuities in $R_{\mu\nu\varrho\sigma}$ are now easily found from (2.2). A straightforward calculation shows that the only a's contributing to the Riemann tensor in empty space–time are

$$-a_{22} = a_{33} = \sigma \quad \text{and} \quad a_{23} = \phi, \tag{2.3}$$

[15] For details, see the second paper of reference 1.

[16] Units are chosen, throughout, so that $c = 1$.

where σ and ϕ are arbitrary. These correspond exactly to the two types of "transverse-transverse" waves found in the linear approximation theory of gravitational radiation.[17]

All the terms in ϕ may be reduced to zero by a rotation of axes through angle $\tan^{-1}(\phi/\sigma)$ in the 23-plane.

The resulting discontinuity in the Riemann tensor will now be written in covariant form by introducing a tetrad of unit vectors[18] λ^{μ}_{α}, which at P are directed along the coordinate axes, so that at P, $\lambda^{\alpha}_{\mu} = \delta^{\alpha}_{\mu}$. On account of the orthonormality, it is true everywhere that

$$g_{\mu\nu}\lambda^{\mu}_{\alpha}\lambda^{\nu}_{\beta} = \eta_{\alpha\beta}.$$

Now define

$$\lambda^{\alpha\mu} \equiv \eta^{\alpha\beta}\lambda^{\mu}_{\beta},$$

so that $\lambda^{0\mu} = \lambda^{\mu}_{0}$, $\lambda^{\alpha\mu} = -\lambda^{\mu}_{\alpha}$. Then it is not difficult to prove that

$$\eta_{\alpha\beta}\lambda^{\alpha\mu}\lambda^{\beta\nu} = \lambda^{\alpha\mu}\lambda^{\alpha}_{\nu} = g^{\mu\nu}, \quad \lambda^{\alpha}_{\mu}\lambda^{\nu}_{\alpha} = \delta^{\nu}_{\mu},$$

and so forth. It is convenient to abbreviate

$$\lambda^{\mu}_{0} = \lambda^{0\mu} \equiv \lambda^{\mu}_{\beta}.$$

This notation differs slightly from that of Eisenhart,[19] who uses indicators instead of $\eta_{\alpha\beta}$.

It is convenient to introduce a simple 6-dimensional formalism for discussing the Riemann tensor and other bivector-tensors. The 6-dimensional pseudo-Euclidean space (the Klein space) is introduced whose vectors are just the bivectors (skew tensors) in the local tangent

[17] A. S. Eddington, *Mathematical Theory of Relativity* (Cambridge University Press, New York, 1924), second edition, p. 247.

[18] Here μ is a vector index and α a label distinguishing the four vectors. The Greek letters α, β, γ, δ, ε and Latin letters a, b, c, d, e will be used only for labels, but shall satisfy the same range and summation conventions (Greek 0, 1, 2, 3; Latin 1, 2, 3) as ordinary vector and tensor indices μ, ν, ϱ, σ, τ, ... and m, n, p, r, ... From now on, an index given a particular value will be understood to be a label index, not a tensor index. This notation obviates the necessity of bracketing indices.

[19] L. P. Eisenhart, *Riemannian Geometry* (Princeton University Press, Princeton, 1949), Chap. 3.

Minkowski space defined by the tetrad $\lambda^{\mu}_{\alpha\cdot}$. The rule for going over to the 6-dimensional formalism is the following:

If $H_{\alpha\beta}$ are the physical components, with respect to a given tetrad at a given space–time event P, of any skew tensor $H_{\mu\nu}$, then the corresponding 6-vector in the 6-space at P is[20] H_A, got by relabeling the suffixes $\alpha\beta$ according to the scheme

$$\begin{array}{lcccccc} \alpha\beta: & 23 & 31 & 12 & 10 & 20 & 30, \\ A: & 1 & 2 & 3 & 4 & 5 & 6. \end{array} \tag{2.4}$$

Accordingly, any bivector-tensor corresponds to a symmetric tensor in the 6-space. The physical components $R_{\alpha\beta\gamma\delta}$ of the Riemann tensor, for example, go over to the components of a symmetric 6-tensor R_{AB}, each of the suffix pairs $\alpha\beta$, $\gamma\delta$ being relabeled according to the scheme (2.4).

In order that the raising and lowering of indices in the 6-space should correspond to the raising and lowering of index pairs in the 4-space of physical components, the metric tensor of the 6-space must be chosen to be

$$\eta_{AB} = \operatorname{diag}(1, 1, 1, -1, -1, -1) \tag{2.5}$$

which corresponds to the bivector-tensor $\eta_{\alpha\gamma}\eta_{\beta\delta}-\eta_{\alpha\delta}\eta_{\beta\gamma}$.

The discontinuity in the Riemann tensor at any event on the null 3-surface S in empty space–time may be calculated straightforwardly from (2.2) and (2.3) and written in terms of R_{AB}. It turns out to be

$$\Delta R_{AB} = \begin{bmatrix} \cdot & \cdot & \cdot & \cdot & \cdot & \cdot \\ \cdot & -\sigma & -\phi & \cdot & -\phi & \sigma \\ \cdot & -\phi & \sigma & \cdot & \sigma & \phi \\ \cdot & \cdot & \cdot & \cdot & \cdot & \cdot \\ \cdot & -\phi & \sigma & \cdot & \sigma & \phi \\ \cdot & \sigma & \phi & \cdot & \phi & -\sigma \end{bmatrix} \tag{2.6}$$

Here σ and ϕ are arbitrary numbers, but the terms in ϕ may be eliminated by a rotation, as stated above. It will be shown in Sec. 3

[20] Upper case letters A, B, C range and sum over 1, 2, . . ., 6.

that the difference between the no-radiation and one of the radiation canonical types of Riemann tensor, referred to a suitably oriented tetrad, is precisely the array of σ's in (2.6), which supports the interpretation of (2.6) as the discontinuity across a gravitational wave front.

The physical effects of the discontinuities (2.6) may be studied in terms of the equation of geodesic deviation[21]

$$\delta^2\eta^\mu/\delta\tau^2 + R^\mu_{\nu\varrho\sigma}\lambda^\nu\eta^\varrho\lambda^\sigma = 0, \tag{2.7}$$

which describes the relative acceleration of two neighboring (spherically symmetric) test particles.[1] In Eq. (2.7), $\lambda^\mu = dx^\mu/d\tau$ is the unit tangent vector to the geodesic world-line C of one of the particles, τ is proper time along C, and η^μ is the orthogonal displacement vector to the (neighboring) world-line of the other particle. To reach this physical interpretation directly, one has only to refer (2.7) to a tetrad comprising λ^μ, which is the 4-velocity of the particle with world-line C, and three spacelike vectors λ^μ_α orthogonal to and parallelly propagated along C. Then (2.7) becomes

$$d^2X^a/d\tau^2 + K^a_b(\tau)X^b = 0, \tag{2.8}$$

where $X^a = \eta^\mu\lambda^a_\mu$ are the physical components of the displacement vector (X^0 vanishes) and

$$K^a_b = R^a_{0b0} \tag{2.9}$$

are some of the physical components of the Riemann tensor. In the Newtonian equation corresponding to (2.8), X^a are the coordinates of the second particle relative to the first, and $K^a_b = \partial^2V/\partial x^a\partial x^b$, where V is the ordinary Newtonian gravitational potential. Thus $-K^a_bX^b$ is to be identified as the relative acceleration of two particles with relative coordinates X^b, arising from the difference in gravitational field between the particles.

It follows that as the gravitational wave front described by (2.6)

[21] J. L. Synge and A. Schild, *Tensor Calculus* (University of Toronto Press, Toronto, 1949), p. 93.

passes the pair of test particles, there will be a discontinuity

$$\Delta K^a_b = \begin{bmatrix} \cdot & \cdot & \cdot \\ \cdot & -\sigma & \phi \\ \cdot & \phi & \sigma \end{bmatrix} \tag{2.10}$$

across the wave front. The tetrads to which (2.6) and (2.8) are referred have been (and always may be) chosen to coincide at the space–time event where the wave front passes the particles. It can be seen that the discontinuity in the relative acceleration depends on the relative position of the particles, and in particular that there is no discontinuity if the two particles are aligned in the direction of propagation of the wave front, which is the 1-direction.

This result represents in an invariant manner the transverse character of gravitational radiation. Two particles lying in the 23-plane (which is perpendicular to the direction of propagation) will suffer a discontinuous change in relative acceleration. If, for example, the 2-axis is chosen so that $\phi = 0$, and the line joining the particles makes an angle θ with this axis, then according to (2.10) the change in the relative acceleration will take place in a direction making an angle $-\theta$ with the same axis.

3. Canonical Forms for the Riemann tensor

In this section, the idea of *following the gravitational field* is made precise. This is done by fairly straightforward generalization from the case of the electromagnetic field. In that case, eigenvectors for the field are defined by the equations

$$T^{\nu}_{\mu}\xi_{\nu} = \lambda\xi_{\mu}, \tag{3.1}$$

where T^{ν}_{μ} is the electromagnetic energy tensor. It is found[3] that in a general field both timelike, spacelike, and null eigenvectors exist; these lie in two orthogonal 2-spaces but are otherwise undetermined. In a null field, on the other hand, there is no timelike eigenvector; all the eigenvectors are spacelike, except for one which is null, and all lie in a 3-space tangent to the null cone along the direction of the null eigenvector.

The sense in which an observer follows the electromagnetic field comes most easily out of consideration of the Poynting vector. Rather than introduce a particular Lorentz frame, one may define a Poynting 4-vector in a covariant way, to be

$$P_\varrho = (\delta^\mu_\varrho - v_\varrho v^\mu) T_{\mu\nu} v^\nu, \tag{3.2}$$

where v^μ is the 4-velocity of the observer measuring the field. This is easily seen to reduce to the usual definition when v^μ lies along the time axis in a local Lorentz frame. Since

$$P_\varrho v^\varrho = 0, \tag{3.3}$$

P_ϱ must be spacelike. If n_ϱ is the 4-normal to any small 2-surface Σ carried along by the observer with velocity v^ϱ, then the electromagnetic energy flux across Σ is

$$P_\varrho n^\varrho = T_{\mu\nu} v^\mu n^\nu. \tag{3.4}$$

Now one says that such and such an observer is following the electromagnetic field if he measures zero energy flux across all 2-surfaces which he carried along, however oriented. By (3.3) and (3.4) this can occur only if

$$P_\varrho = 0, \tag{3.5}$$

which implies that

$$T_{\varrho\nu} v^\nu = (T_{\mu\nu} v^\mu v^\nu) v_\varrho, \tag{3.6}$$

so that v_ϱ must be an eigenvector of $T_{\mu\nu}$. This establishes the connection between the concept of following the electromagnetic field and the eigenvectors of the electromagnetic energy tensor. As mentioned above, a null field has *no* timelike eigenvector, so that the Poynting vector will not vanish for any observer with a finite velocity. The energy flow in a null field cannot be abolished by a Lorentz transformation. A null field has one null eigenvector, say ξ^μ, belonging to the eigenvalue zero, so that

$$T_{\mu\nu}\xi^{\nu} = 0. \tag{3.7}$$

Thus an "observer" moving with the speed of light in the direction of ξ^{ν} (which is essentially the propagation vector) would observe no energy flux past him.

In the gravitational case, there is no energy-momentum tensor of the gravitational field itself (the pseudotensor is discussed in Sec. 4), but in accord with the arguments developed in Sec. 1, one may seek in the geometrical structure of the Riemann tensor a definition of "following the field" analogous to that developed in the electromagnetic case. The definition is naturally more complicated, because the Riemann tensor is a more complicated object than the Maxwell energy tensor. The definition is made in two stages. First of all, eigenvectors (skew tensors) are defined for the Riemann tensor. By using Petrov's canonical forms,[4] these eigenbivectors may be written down explicitly for the three algebraically distinct types of Riemann tensor in empty space–time. The eigenbivectors correspond geometrically to 2-spaces, or pairs of 2-spaces, in space–time. The intersections of these 2-spaces with one another define a number of 4-vectors (assumed normalized if they are not null), which will be referred to as *Riemann principal vectors.*

> An observer with a timelike Riemann principal vector as 4-velocity is said to be following the gravitational field.

It turns out that for two of the three types of Riemann tensor, this timelike principal vector collapses onto the null cone. The occurrence of these types of Riemann tensor is identified with the presence of gravitational radiation.

The eigenbivectors $P_{\mu\nu}$ of the Riemann tensor are defined by the equation

$$R_{\mu\nu\varrho\sigma}P^{\varrho\sigma} = \lambda P_{\mu\nu}, \tag{3.8}$$

or

$$R_{AB}P^{B} = \lambda P_{A}$$

in the 6-dimensional formalism introduced in Sec. 2.

Now Petrov[4] has shown that by a suitable choice of the reference tetrad at any event in empty space–time, one may reduce the Riemann tensor to a canonical form of one of the following three types:

Type I:

$$(3.9) \qquad R_{AB} = \begin{bmatrix} \alpha_1 & \cdot & \cdot & \beta_1 & \cdot & \cdot \\ \cdot & \alpha_2 & \cdot & \cdot & \beta_2 & \cdot \\ \cdot & \cdot & \alpha_3 & \cdot & \cdot & \beta_3 \\ \beta_1 & \cdot & \cdot & -\alpha_1 & \cdot & \cdot \\ \cdot & \beta_2 & \cdot & \cdot & -\alpha_2 & \cdot \\ \cdot & \cdot & \beta_3 & \cdot & \cdot & -\alpha_3 \end{bmatrix}.$$

$$\begin{bmatrix} \sum_{k=1}^{3} \alpha_k = 0 \\ \sum_{k=1}^{3} \beta_k = 0 \end{bmatrix}$$

Type II:

$$(3.10) \quad R_{AB} = \begin{bmatrix} -2\alpha & \cdot & \cdot & -2\beta & \cdot & \cdot \\ \cdot & \alpha-\sigma & \cdot & \cdot & \beta & \sigma \\ \cdot & \cdot & \alpha+\sigma & \cdot & \sigma & \beta \\ -2\beta & \cdot & \cdot & +2\alpha & \cdot & \cdot \\ \cdot & \beta & \sigma & \cdot & -(\alpha-\sigma) & \cdot \\ \cdot & \sigma & \beta & \cdot & \cdot & -(\alpha+\sigma) \end{bmatrix}.$$

Type III:

$$(3.11) \qquad \begin{bmatrix} \cdot & -\sigma & \cdot & \cdot & \cdot & \sigma \\ -\sigma & \cdot & \cdot & \cdot & \cdot & \cdot \\ \cdot & \cdot & \cdot & \sigma & \cdot & \cdot \\ \cdot & \cdot & \sigma & \cdot & \sigma & \cdot \\ \cdot & \cdot & \cdot & \sigma & \cdot & \cdot \\ \sigma & \cdot & \cdot & \cdot & \cdot & \cdot \end{bmatrix}.$$

In Type I, the reference tetrad yielding the canonical form is in general fully determined; accidental equality between different α's or β's may introduce some freedom. In Types II and III, the reference tetrad is determined only up to a Lorentz rotation in the 10-plane and a spatial rotation in the 23-plane. The α's and β's are scalar invariants of the

Riemann tensor, but the value of σ depends on the choice of axes in the 10-plane.

These forms of R_{AB} are determined, first, by the limitation of transformations in the 6-space to those generated by changes of tetrad (i.e., by real Lorentz transformations, including rotations) in space–time, and secondly, by the nonsymmetry of R^A_B (equivalently, by the indefinite character of the metric η_{AB}). As a result, the elementary divisors of R^A_B need not be simple, and Types II and III result when they are not.

The eigenbivectors of R_{AB}, defined by (3.8), are easily found from (3.9)–(3.11); they are either simple bivectors, dual in pairs or of the form $P_A = S_A \pm i{}^0S_A$, where S_A is simple and ${}^0S_A$ is its dual.[22]

Thus each eigenbivector P^A of R_{AB} defines a pair of orthogonal 2-spaces. The P^A are readily found from (3.9)–(3.11) to be (conveniently normalized) the following:

Type I: Six independent eigenbivectors:

If $\beta_1 = 0$, $P^A = \delta^A_1$ and $P^A = \delta^A_4$ (dual pair);

if $\beta_1 \neq 0$, $P^A = \delta^A_1 \pm i\delta^A_4$,

If $\beta_2 = 0$, $P^A = \delta^A_2$ and $P^A = \delta^A_5$ (dual pair);

if $\beta_2 \neq 0$, $P^A = \delta^A_2 \pm i\delta^A_5$,

If $\beta_3 = 0$, $P^A = \delta^A_3$ and $P^A = \delta^A_6$ (dual pair);

if $\beta_3 \neq 0$, $P^A = \delta^A_3 \pm i\delta^A_6$

Type II: Four independent eigenbivectors:

If $\beta = 0$, $P^A = \delta^A_1$ and $P^A = \delta^A_4$ (dual pair), and

$P^A = \delta^A_2 - \delta^A_6$ and $P^A = \delta^A_3 + \delta^A_5$ (dual pair).

If $\beta \neq 0$, $P^A = \delta^A_1 \pm i\delta^A_4$, and $P^A = \delta^A_2 - \delta^A_6 \pm i(\delta^A_3 + \delta^A_5)$.

[22] A simple bivector may be characterized by det $(S_{\mu\nu}) = 0$; it may always be written in the form $S_{\mu\nu} = X_\mu Y_\nu - X_\nu Y_\mu$, and defines a 2-space in space–time. The dual of any bivector $P_{\mu\nu}$ is

$${}^0P_{\mu\nu} = \tfrac{1}{2} g_{\mu\varrho} g_{\nu\sigma} e^{\varrho\sigma\tau\pi} P_{\tau\pi} ({}^0P_A = \tfrac{1}{2} g_{AB} e^{BC} P_C),$$

where $e^{\varrho\sigma\tau\pi} = \pm\sqrt{(-g)}$ is the alternating tensor, which will be understood to take the negative sign when $\varrho\sigma\tau\pi$ are in the order 0123. The dual of a simple bivector $S_{\mu\nu}$ defines a 2-space orthogonal to that defined by $S_{\mu\nu}$ itself.

Type III: Two independent eigenbivectors:

$$P^A = \delta_2^A - \delta_6^A, \quad \text{and} \quad P^A = \delta_3^A + \delta_5^A.$$

The different pairs of 2-spaces represented by these simple bivectors are not orthogonal. Their intersections yield the Riemann principal vectors r^α, which are (conveniently normalized) as follows:

Type I: $r^\alpha = \delta_1^\alpha,\ \delta_2^\alpha,\ \delta_3^\alpha,\ \delta_0^\alpha$. The Riemann principal vectors are just the vectors of the reference tetrad. One is timelike, three spacelike.

Type II: $r^\alpha = \delta_0^\alpha - \delta_1^\alpha,\ \delta_2^\alpha,\ \delta_3^\alpha$. The first is null, the others spacelike. Because of the freedom of rotation in the 23-plane, the last two may be replaced by any linear combination of themselves.

Type III: $r^\alpha = \delta_0^\alpha - \delta_1^\alpha$. There is only one Riemann principal vector, and it is a null vector.

According to the criteria set out earlier, gravitational radiation is now defined as follows:

> At any event in empty space–time, gravitational radiation is present if the Riemann tensor is of Type II or Type III, but not if it is of Type I.

Now it will be noticed that the σ's in (3.10) appear in the same positions and with the same signs as the σ's in (2.6), which exhibits the discontinuities permissible across a null 3-surface. This correspondence has of course been achieved in part by orienting the two reference tetrads so that the 1-direction is picked out for asymmetrical treatment in each case. However, it has also the physical significance that the discontinuities possible across a gravitational wave front, according to Lichnerowicz's conditions, are just what are required for the transition from a space–time region without gravitational radiation to one with gravitational radiation, according to the definition just proposed. This is of course precisely what one would wish, to show the compatibility of the two approaches to the problem.

It will be noticed also that the transition from Type I to Type II reduces the number of independent α's and β's from two each to one each. This implies some additional symmetry in the radiation field, which may at first sight be surprising. However, a physical interpretation which at once suggests itself is that because of the nonlinearity of the field (that is, because the gravitational field effectively enters its own source-function), gravitational waves without any kind of symmetry would interfere with themselves to the extent of destruction.[22a]

The physical effects of gravitational waves may be investigated by using the equation of relative acceleration (2.8), in exactly the same way as the effects of discontinuities were investigated in Sec. 2. The difference between Type I and Type II space–times shows up clearly if one examines the behavior of test particles moving with velocities different from that specified by the timelike tetrad vector. As an example, consider the effect on K^a_b [defined by (2.9)] of the local Lorentz transformation defined at an event by

$$\begin{cases} \bar{\lambda}^0_\mu = \lambda^0_\mu \cosh\theta + \lambda^1_\mu \sinh\theta, \\ \bar{\lambda}^1_\mu = \lambda^0_\mu \sinh\theta + \lambda^1_\mu \cosh\theta, \\ \bar{\lambda}^2_\mu = \lambda^2_\mu, \quad \bar{\lambda}^3_\mu = \lambda^3_\mu, \end{cases} \tag{3.12}$$

where the unbarred tetrad is that to which the canonical forms (3.9) and (3.10) are referred. Then the comparative values of K_{ab} (omitting K_{1b}, which are unaltered) are

Type I:

$$K_{22} = -(\alpha-\sigma), \quad \bar{K}_{22} = -(\alpha-\sigma\cosh 2\theta),$$
$$K_{23} = 0, \quad \bar{K}_{23} = \tfrac{1}{2}(\beta_3-\beta_2)\sinh 2\theta,$$
$$K_{33} = -(\alpha+\sigma), \quad \bar{K}_{33} = -(\alpha+\sigma\cosh 2\theta).$$

Type II:

$$K_{22} = -(\alpha-\sigma), \quad \bar{K}_{22} = -(\alpha-\sigma e^{-2\theta}),$$
$$K_{23} = 0, \quad \bar{K}_{23} = 0,$$
$$K_{33} = -(\alpha+\sigma), \quad \bar{K}_{33} = -(\alpha+\sigma e^{-2\theta}).$$

[22a] *Note added in proof.*—The remaining discussion, where it refers to particular canonical types, is restricted to Types I and II. The absence of scalar invariants in Type III suggests that space–times of this type would represent radiation without sources, but the interpretation of this is not obvious, and further consideration of it is left to a subsequent paper. The writer knows of no example of a Type III space–time; he would be grateful if new examples of empty space–time metrics of any type were sent to him for study.

Here the barred K's are those referred to the tetrad $\bar{\lambda}^{\alpha}_{\mu}$, and in Type I, α_2 and α_3 have been replaced by $\alpha = \frac{1}{2}(\alpha_2+\alpha_3)$ and $\sigma = \frac{1}{2}(\alpha_3-\alpha_2)$, for ready comparison with Type II.

An essential difference between the types is represented by the fact that in Type I, the changes in K_{22} and K_{33} go as θ^2, for small θ, while in Type II, they go as θ. The Type I changes are essentially a special-relativistic effect, in the sense of "being of the same kind as familiar effects such as the Lorentz contraction", but the Type II changes are characteristic of a non-Lorentzian phenomenon. The first-order change in K_{23}, another instance of non-Lorentzian behavior, suggests that the β's may be connected with the rotational properties of the field.

For strong Lorentz transformations (large θ), the Type I K's become large in absolute value for both signs of θ, but the Type II K's approach finite limits for large positive θ, so that in Type I the K's have extreme values for $\theta = 0$, while in Type II the extreme values are approached only as $\theta \to \infty$, that is, as the observer's velocity approaches the fundamental velocity in the direction of propagation of the radiation.

4. Reduction of the Energy-momentum Pseudotensor

As is well known, one may convert the covariant conservation equations[23]

$$T^{\nu}_{\mu;\,\nu} = 0$$

into the form

$$\left\{(-g)^{\frac{1}{2}}T^{\nu}_{\mu}+(1/\varkappa)t^{\nu}_{\mu}\right\}_{,\,\nu} = 0$$

by introducing the canonical energy-momentum pseudotensor t^{ν}_{μ}. This fact, and the canonical origin of t^{ν}_{μ}, lead one to identify it as the "energy" momentum pseudotensor of the gravitational field". The physical argument is, roughly, that the deviations of space–time from flatness introduce additional terms into the conservation equations, and that these deviations are consequences of the existence of the gravitational field. All would be well, were it not that t^{ν}_{μ} depends

[23] A semicolon denotes a covariant derivative.

on idiosyncrasies in the choice of coordinates as well as on actual physical phenomena. The nonhomogeneous transformation properties of t_μ^ν make it impossible to construct any scalar quantities out of it, at least in a direct way, and so its physical interpretation must be suspect, because of this essential dependence on the coordinate system. It is hard to see how one can attach any physical meaning to t_μ^ν unless one can first attach a physical meaning to the coordinate system. The same difficulty would arise with vectors and tensors, except that one can construct scalars (e.g., physical components) out of them by contracting with other vectors or tensors, and these scalars are of course independent of the coordinate system.

The usual procedure in dealing with t_μ^ν is to make weak-field approximations and to assume mathematically convenient coordinate conditions. These methods are controversial and their physical significance obscure. Difficulties in dealing with t_μ^ν might anyhow have been expected from consideration of the principle of equivalence. Since the gravitational field can be abolished at an event by a coordinate transformation (in the sense that the $\Gamma_{\mu\nu}^\varrho$ can be made to vanish), the gravitational energy, momentum and stress at an event can readily be understood to be as ephemeral as the coordinate system. The energy of the field resides not in its value at a single event, but in its variation from one event to another. It is not surprising that one cannot abolish t_μ^ν throughout any finite 2-surface in a general space–time. However, a mean value $\bar{t}_\mu^\nu$ may be defined over the 2-surface of a small 3-volume, and by a suitable physical prescription of the coordinate system, such a definition can be made covariant.

The coordinates to be prescribed are well known to mathematicians under the name normal coordinates,[8] and have been used in general relativity theory before,[24, 25] but it is nevertheless desirable to give some physical justification for this choice.

The choice of coordinate system depends on the physical situation involved. For many purposes it is enough to specify at an event a tetrad

[24] G. D. Birkhoff, *Relativity and Modern Physics* (Harvard University Press, Cambridge, 1923).

[25] T. Y. Thomas, *Phil. Mag.* **48,**, 1056 (1924).

of unit vectors, or the corresponding local Minkowskian coordinate axes, representing the 4-velocity of an observer and rectangular Cartesian axes in his local instantaneous 3-space. The essential thing is that it should be possible in principle to identify the chosen system with one which could be used by an observer in the given physical situation. Recently,[1] the writer compared the behavior of test particles in a gravitational field in the general relativity theory and in the Newtonian theory (see Sec. 2 above). In that case it was appropriate to introduce local Cartesian coordinate systems in the instantaneous 3-spaces along one of the particle world-lines, and the coordinate systems at different events were related by parallelly propagating along the world-line the tetrad vectors representing the coordinate axes. As might be expected, it was found that this mode of propagation led to a description of gravitational phenomena most closely resembling that obtained in the Newtonian theory from the use of ordinary Newtonian inertial frames. However, the whole formalism, being designed for a comparison with the Newtonian theory, was essentially nonrelativistic.

The present case is rather different. The formalism just described is appropriate to the discussion of dynamical effects, as in the discussion following Eq. (2.8) above, but the whole idea of the energy–momentum tensor is essentially a relativistic one, developed largely within the framework of a relativistic theory—the Maxwell theory—and it would be inappropriate to develop the same idea in general relativity theory, regarded now specifically as a field theory of gravitation, except in a relativistic manner. Therefore what is required is a convenient 4-dimensional analog of the Minkowskian inertial systems of special relativity, but one defined more completely than by a tetrad of unit vectors. Some loss of general covariance is inevitable, and the whole aim is anyhow a little artificial, the idea being to relate the novel concept of gravitational radiation developed here to a conventional idea of radiation developed specifically for electromagnetic theory—although it must be admitted that the discussion of the Poynting vector at the beginning of Sec. 3 applies also to flows of other sorts of energy.

Having, then, the aim of investigating the energy-momentum pseudotensor by analogy with Lorentz-invariant field theories, it is appropriate to choose a coordinate system which approximates to a Minkowskian system. In the weak-field approximation method this is done by considering a metric which deviates slightly from the Minkowski metric at sufficiently large distances from material particles. The conceptual difficulties which arise in the use of that method can be ascribed to the lack of a covariant formulation of the weak-field approximation. The alternative adopted here is a local approximation method capable of invariant formulation: the introduction, in the neighborhood of any chosen space–time event, of a normal coordinate system, which approximates to Minkowskian inertial system in a mathematically and physically well-defined way.

The physical interpretation of normal coordinates comes out of their exact correspondence to Minkowskian coordinates in one particular respect, namely the measurement of interval. This is best explained by summarizing the relevant properties of such a coordinate system, which are the following:

Normal coordinates x^μ can always be chosen so that at any chosen space–time event O,

(4.1) (i) $$x^\mu = 0,$$

(4.2) (ii) $$g_{\mu\nu} = \eta_{\mu\nu},$$

(4.3) (iii) $$\Gamma^\varrho_{\mu\nu} = g_{\mu\nu,\varrho} = 0,$$

(4.4) (iv) $$g_{\mu\nu,\varrho\sigma} = \tfrac{1}{3}(R_{\varrho\mu\nu\sigma}+R_{\varrho\nu\mu\sigma}),$$

and

(v) at every point P in the neighborhood of O,

(4.5) $$x^\mu = up^\mu_0,$$

where

(4.6) $$p^\mu_0 = dx^\mu/du$$

is a vector tangent at O to the geodesic OP, and

(a) if OP is timelike, u is the proper time τ from O to P;

(b) if OP is spacelike, u is the properdistance s from O to P;

(c) if OP is null, u is a preferred parameter in terms of which the equation of the null geodesic OP takes the form

$$\frac{d^2x^\mu}{du^2}+\Gamma^\mu_{\nu\varrho}\frac{dx^\nu}{du}\frac{dx^\varrho}{du}=0,$$

(this defines u up to a linear transformation on the null geodesics through O. The origin of u is chosen to be at O, and x^μ defined by (4.5) does not depend on the scale of u);

(vi) the normal coordinate systems at O are connected to one another by homogeneous Lorentz transformations at O.

It is clear from examination of these properties that an observer who assigns coordinates in the neighborhood of a given event O by theodolite measurements at O and interval measurements from O as if space-time were flat, will assign normal coordinates. Thus the employment of normal coordinates exploits to the full the locally Minkowskian properties of a Riemannian space. In order to connect this to previous work, it is convenient again to introduce a tetrad of unit vectors directed along the coordinate axes.

It is property (iv) which supplies the key to a covariant expression for the energy-momentum pseudotensor t_μ. The latter is homogeneous quadratic in the $g_{\mu\nu,\,\sigma}$, and so if it is expanded in a power series about the origin of normal coordinates, the first nonvanishing term has an invariant coefficient, a function of $R_{\mu\nu\varrho\sigma}$. By taking an average over a small 2-sphere, an invariant average expression is obtained.[25a] The details of the calculation are as follows: The energy-momentum pseudotensor is defined by

$$t^\nu_\mu = L\delta^\nu_\mu - g_{\varrho\sigma,\,\mu}\frac{\partial L}{\partial g_{\varrho\sigma,\,\nu}}, \tag{4.7}$$

where

$$L=(-g)^{\frac{1}{2}}g^{\mu\nu}[\Gamma^\varrho_{\mu\nu}\Gamma^\sigma_{\varrho\sigma}-\Gamma^\varrho_{\mu\sigma}\Gamma^\sigma_{\nu\varrho}] \tag{4.8}$$

[25a] *Note added in proof.*—This is perhaps rather an unusual definition of average, being in effect

$$t^\varphi_\psi = \lim_{r\to 0}(4\pi r^4)^{-1}\int t^\varphi_\psi\, d^2S.$$

is the first-order Lagrangian for the field. A straightforward calculation yields for L this explicit expression in terms of the $g_{\mu\nu,\varrho}$:

(4.9) $$L = \tfrac{1}{8}(-g)^{\frac{1}{2}} S^{\pi\tau\mu\nu\varrho\sigma} g_{\pi\tau,\sigma} g_{\mu\nu,\varrho},$$

where

(4.10) $$S^{\pi\tau\mu\nu\varrho\sigma} = S^{\mu\nu\pi\tau\sigma\varrho} = g^{\pi\tau} U^{\mu\nu\varrho\sigma} + g^{\varrho\sigma} U^{\mu\nu\lambda\pi} - g^{\pi\varrho} U^{\mu\nu\tau\sigma} + g^{\tau\varrho} U^{\mu\nu\pi\sigma},$$
$$U^{\mu\nu\varrho\sigma} = g^{\mu\varrho} g^{\nu\sigma} + g^{\mu\sigma} g^{\nu\varrho} - g^{\mu\nu} g^{\varrho\sigma}.$$

From (4.7), (4.8), and (4.9) one may write t_μ^ν explicitly in terms of the $g_{\mu\nu,\varrho}$:

(4.11) $$t_\psi^\phi = \tfrac{1}{8}(\delta_\tau^\varrho \delta_\psi^\phi - 2\delta_\psi^\varrho \delta_\tau^\phi)(-g)^{\frac{1}{2}} S^{\pi\varkappa\mu\nu\tau\sigma} g_{\pi\varkappa,\sigma} g_{\mu\nu,\varrho}.$$

Differentiating (4.11) twice with respect to x^ξ and setting $g_{\mu\nu,\varrho} = 0$ in agreement with (4.3), one obtains

(4.12) $$t_\psi^\phi = 0, \quad t_{\psi,\xi}^\phi = 0,$$

(4.13) $$t_{\psi,\xi\eta}^\phi = \tfrac{1}{8}(\delta_\tau^\varrho \delta_\psi^\phi - 2\delta_\psi^\varrho \delta_\tau^\phi)(-g)^{\frac{1}{2}} S^{\pi\varkappa\nu\alpha\tau\sigma}(g_{\pi\varkappa,\sigma\xi} g_{\nu\mu,\varrho\eta} + g_{\pi\varkappa,\sigma\eta} g_{\mu\nu,\varrho\xi}).$$

Now making use of (4.4) and the field equations for empty space–time (2.1), one finds after a straightforward calculation

(4.14) $$t_{\psi,\xi\eta}^\phi = \tfrac{1}{9}(\delta_\tau^\varrho \delta_\psi^\phi - 2\delta_\psi^\varrho \delta_\tau^\phi)(\delta_\xi^\varkappa \delta_\eta^\lambda + \delta_\eta^\varkappa \delta_\xi^\lambda)(R_\varkappa^{\tau\mu\nu} + R_\varkappa^{\tau\nu\mu}) R_{\varrho\mu\nu\lambda}.$$

It follows from (4.9) that the mean value of t_ψ^ϕ over the surface of a small sphere about O in the 3-space $t = 0$ will be

(4.15) $$\bar{t}_\psi^\phi = \tfrac{1}{6} t_{\psi,ss}^\phi.$$

Substituting from (4.14) into (4.15) and introducing the unit vector $\lambda^\varrho = \delta_0^\varrho$ directed along the time axis, one obtains

(4.16) $$\bar{t}_\psi^\phi = (1/27)(\delta_\tau^\varrho \delta_\psi^\phi - 2\delta_\psi^\varrho \delta_\tau^\phi)(\lambda^\varkappa \lambda^\lambda - g^{\varkappa\lambda})(R_\varkappa^{\tau\mu\nu} + R_\varkappa^{\tau\nu\mu}) R_{\varrho\mu\nu\lambda}.$$

This covariant expression is to be interpreted as the approximate mean gravitational energy-momentum tensor determined by an observer with 4-velocity $\lambda^\varkappa$ by measurements in his instantaneous 3-space. It will be observed that $\bar{t}_{\varkappa\psi} = g_{\varkappa\phi} \bar{t}_\psi^\phi$ is a symmetric tensor.

A straightforward but lengthy calculation yields the value of $\bar{t}^{\phi}_{\psi}$ for the Riemann tensor in canonical form. It is of importance to compare the physical components for Types I and II. One finds for Type I

$$\bar{t}^{\beta}_{\alpha} = (1/27)\Big[2\Big\{\sum_{k=1}^{3} \alpha_k^2(\delta^0_\alpha\delta^\beta_0 + 2\delta^k_\alpha\delta^\beta_k + 3\delta^{k+1}_\alpha\delta^\beta_{k+1} + 3\delta^{k+2}_\alpha\delta^\beta_{k+2})\Big\} - 9\delta^\beta_\alpha \sum_{k=1}^{3} \alpha_k^2\Big]. \tag{4.17}$$

It will be noticed that all the off-diagonal terms vanish. It follows that an observer measuring these physical components in his rest frame observes no gravitational energy flow. On the other hand, for Type II one finds

$$\begin{aligned}\bar{t}^{\beta}_{\alpha} = (1/27)[\alpha^2(&-42\delta^\beta_\alpha + 16\delta^1_\alpha\delta^\beta_1 + 22\delta^2_\alpha\delta^\beta_2 \\ &+ 22\delta^3_\alpha\delta^\beta_3) + 4\alpha\sigma(\delta^2_\alpha\delta^\beta_2 - \delta^3_\alpha\delta^\beta_3) \\ &+ 8\sigma^2(\delta^0_\alpha + \delta^1_\alpha)(\delta^\beta_0 - \delta^\beta_1)].\end{aligned} \tag{4.18}$$

This is of the form

$$\bar{t}^{\beta}_{\alpha} = \text{a diagonal part} + (8/27)\sigma^2\xi_\alpha\xi^\beta,$$

where $\xi_\alpha = \delta^0_\alpha + \delta^1_\alpha$ is a null vector in the direction of propagation of the radiation represented by Type II. This part of $\bar{t}^{\beta}_{\alpha}$ is of exactly the same form as the energy tensor of an electromagnetic null field, and so should be identified as that part arising purely from gravitational radiation. The terms in α^2, on the other hand, are to be associated with the nonradiative part of the field, and the terms in $\alpha\sigma$ with the interaction between the two parts of the field. An observer measuring a Type II field will, according to this definition of $\bar{t}^{\nu}_{\mu}$, observe gravitational energy flow in the 1-direction.

These results lend plausibility to the definition of gravitational radiation proposed in Sec. 3. If one accepts the energy-momentum pseudotensor as a respectable part of Einstein's theory, then the calculations in this section show that when, according to the proposed definition, gravitational radiation is present, there must be an energy flux through a small 2-surface.

5. Examples

It would not be satisfactory if empty static space–time regions could admit the presence of radiation; that they cannot is shown by the following rather clumsy proof.

A static space–time region, rigorously defined, is one in which there is an everywhere-timelike group of motions of the region into itself (apart from boundaries) whose generators form a normal congruence.

It follows that if the timelike tetrad vector λ^{μ} is taken to be tangent to the generators, then

$$\gamma_{0ab} = 0, \tag{5.1}$$

where $\gamma_{0ab} = \lambda_{\mu:\,\nu}\lambda^{\mu}_{a}\lambda^{\nu}_{b}$ are some of the Ricci rotation coefficients.[19] A standard formula[19] then at once gives

$$R_{0abc} = 0. \tag{5.2}$$

Then a rotation of the spacelike tetrad vectors will diagonalize the symmetric 3-tensor R_{0a0b}, and it follows from the field equations (2.1) that R_{abcd} must be simultaneously diagonalized. Hence the Riemann tensor is now in Type I canonical form, and so no gravitational radiation is present. It follows from a result of Taub[26] that there can be no plane gravitational waves filling all space–time.

The simplest empty space–time gravitational field is the Schwarzschild field. Taking the metric in the form

$$ds^2 = \left(1-\frac{2m}{r}\right)dt^2-\left(1-\frac{2m}{r}\right)^{-1}dr^2-r^2(d\theta^2+\sin^2\theta\, d\phi^2), \tag{5.3}$$

and labeling $r\theta\phi t$ in the order 1230, one finds, with tetrad vectors directed along the coordinate axes, a Type I Riemann tensor already in canonical form with

$$-\tfrac{1}{2}\alpha_1 = \alpha_2 = \alpha_3 = m/r^3, \quad \beta_k = 0. \tag{5.4}$$

The Riemann principal vectors are not fully determined, however,

[26] A. H. Taub, *Ann. Math.* **53,** 472 (1951).

because of the symmetries of the field, which show up in the equality of α_2 and α_1. According to Birkhoff's theorem,[27] there can be no spherical waves, since the Schwarzschild field is the only spherically symmetric empty space–time solution of Einstein's Eq. (2.1).

The cylindrically symmetric metric introduced by Rosen,[28] in discussing cylindrical waves,

$$ds^2 = e^{2\gamma-2\psi}(dt^2-d\varrho^2)-e^{-2\psi}\varrho^2\,d\phi^2-e^{2\psi}\,dz^2,$$

$$\psi = \psi(\varrho, t), \quad \gamma = \gamma(\varrho, t), \tag{5.5}$$

is of Type II, with

$$\sigma = \frac{\partial^2\psi}{\partial\varrho\partial t}+5\frac{\partial\psi}{\partial\varrho}\frac{\partial\psi}{\partial t}+\frac{\partial\psi}{\partial t}\frac{\partial\gamma}{\partial\varrho}-3\frac{\partial\psi}{\partial\varrho}\frac{\partial\gamma}{\partial t}, \tag{5.6}$$

as may readily be found by taking tetrad vectors along the coordinate axes. Radiation will be present unless the above expression for σ vanishes.

6. Discussion

The definition proposed in this paper provides an unambiguous local criterion of the presence of gravitational radiation, but it suffers from several defects. In the first place it counts as radiation only those gravitational disturbances which are propagated with the fundamental velocity. If it should turn out to be desirable that phenomena propagated with lower velocities be classified as radiation, then they would not be included under this definition. In particular, standing waves are not included. However, the analysis points to a new and powerful tool for the investigation of gravitational fields in general, namely the scalar invariants α_k and β_k.[29]

[27] Reference 24, p. 253.

[28] N. Rosen, *Bull. Research Council Israel* **3,** 328 (1954).

[29] That these might become significant had already been emphasized by several workers in the formal talks and the discussion at the Berne Conference on Relativity Theory, July, 1955 (unpublished).

In the second place, the definition is a local geometric-algebraic one, and does not reveal at all how the properties of the radiation may vary along the path of propagation. This hiatus can be filled, at least formally, by introducing Petrov's canonical forms (3.9)–(3.11) into the conservation law for the matter-free gravitational field:

$$R^{\mu}_{\nu\varrho\sigma;\mu} = 0, \tag{6.1}$$

which may readily be deduced from the Bianchi identities and the field equations (2.1). The resulting equations, which bear a striking similarity to the ordinary conservation laws for a medium with density and pressures, will be discussed in a subsequent paper.

Another defect of the present discussion is that it gives no indication of what secular changes may occur in radiating matter. Suppose for example that a Schwarzschild particle is disturbed from static spherical symmetry by an internal agency, radiates for some time, and finally is restored to static spherical symmetry. Is its total mass necessarily the same as before? This and similar problems required investigation. Also the status of the scalar invariants of the Riemann tensor in the Einstein, Infeld, and Hoffmann approximation theory deserves clarification, and may be hoped to assist in resolving the annoying ambiguities of interpretation which beset that theory.

I am much indebted to H. Bondi for a remark which stimulated this research, and for many discussions, and to L. Bass for suggesting a valuable improvement in presentation.

NOTES ON EXTRACT 8

THE notion of radiation clarified in Extract 7 is taken up in detail in this paper. It is shown explicitly that gravitational radiation from an isolated system can carry away mass. In finishing one line of research definitively the paper, in section C, opens another when it considers the asymptotic form of the coordinate system and the corresponding transformation group leaving it invariant.

EXTRACT 8†

Gravitational Waves in General Relativity: VII. Waves from Axisymmetric Isolated Systems

H. BONDI, F.R.S., M. G. J. VAN DER BURG

AND A. W. K. METZNER

[*Received 8 January, 1962—Revised 2 April, 1962*]

This paper is divided into four parts. In part A, some general considerations about gravitational radiation are followed by a treatment of the scalar wave equation in the manner later to be applied to Einstein's field equations.

In part B, a co-ordinate system is specified which is suitable for investigation of outgoing gravitational waves from an isolated axi-symmetric reflexion-symmetric system. The metric is expanded in negative powers of a suitably defined radial co-ordinate r, and the vacuum field equations are investigated in detail. It is shown that the flow of information to infinity is controlled by a single function of two variables called the *news function*. Together with initial conditions specified on a light cone, this function fully defines the behaviour of the system. No constraints of any kind are encountered.

In part C, the transformations leaving the metric in the chosen form are determined. An investigation of the corresponding transformations in Minkowski space suggests that no generality is lost by assuming that the transformations, like the metric, may be expanded in negative powers of r.

In part D, the mass of the system is defined in a way which in static metrics agrees with the usual definition. The principal result of the paper is then deduced, namely, that *the mass of a system is constant if and only if there is no news; if there is news, the mass decreases monotonically so long as it continues*. The linear approximation is next discussed, chiefly for its heuristic value, and employed in the analysis of a receiver for gravitational waves. Sandwich waves are constructed, and certain non-radiative but non-static solutions are discussed. This part concludes with a tentative classification of time-dependent solutions of the types considered.

† *Proc. Roy. Soc.* A, **269**, 21 (1962).

PART A. GENERAL CONSIDERATIONS

H. BONDI, F.R.S.

1. Introduction

A great deal of work has been done on gravitational waves. In the first instance the linearized theory has been developed extensively, but it seems doubtful whether its results can be fully trusted. The non-linearity of the gravitational field is one of its most characteristic properties, and it is likely that at least some of the crucial properties of the field show themselves only through the non-linear terms. Moreover, it is never entirely clear whether solutions derived by the usual method of linear approximation necessarily correspond in every case to exact solutions, or whether there might be spurious linear solutions which are not in any sense approximations to exact ones. Next, although a good deal is known about exact plane and cylindrical wave solutions, it is doubtful whether these necessarily display the most important characteristics of physically significant waves, that is, of waves from bounded sources. General relativity is a peculiarly complete theory and may not give sensible solutions for situations too far removed from what is physically reasonable. The simplest field due to a finite source is spherically symmetrical, but Birkhoff's theorem shows that a spherically symmetrical empty-space field is necessarily static. Therefore there cannot be truly spherically symmetrical waves, and thus any description of radiation from a finite system must necessarily involve three co-ordinates significantly. This enormously complicates the mathematical difficulties and thus we have to make use of methods of approximation.

2. Causality

The equations of general relativity, like those of most other wave theories, are symmetrical in time. The choice of the retarded solution is as arbitrary in the gravitational case as any other, but whereas in electromagnetic theory it is the direct appeal to experience that forces us to the retarded solution, no such appeal is possible in the gravitational case. All we can say is that, if our usual notions of causality and the flow of time are not to be upset by gravitational waves, if we are to suppose, in other words, that even the most carefully constructed gravitational receiver would not enable us to look into the future, then we are forced to prescribe purely retarded solutions in this theory as well.

The boundary conditions adopted, therefore, are that we have an isolated material system in an empty space that tends to flatness at infinity, where only outgoing waves are present, and we examine the changes of space which are determined by changes in the material system enclosed.

It might be argued that a closed material system cannot undergo any change if it has been isolated for sufficiently long, but this is incorrect. For the system may have an equation of state containing the time explicitly. Moreover, this time dependence may contain a random element and so produce motions in the system that could not have been forecast from outside. This lack of the possibility of forecasting is an important and characteristic point. It is well known that the solutions of hyperbolic equations, such as we are dealing with in general relativity, need not be analytic. On the contrary, it is typical of hyperbolic systems that non-analytic behaviour can be propagated, though only along characteristics which are the wave fronts defined by the system of equations. It will, of course, be realized that an analytic function of time is one whose entire future can be forecast from an arbitrarily small section of time, whereas a non-analytic function is one whose future is undetermined. We shall, therefore, expect to find in our work that the behaviour of the system can be described by functions that need not be analytic and can thus contain the effects of the possible "time-bomb" character of the system enclosed.

3. The Loss of Mass

The symmetry properties of gravitational waves are restricted by conservation laws. The conservation of mass effectively prohibits purely spherically symmetrical waves and, similarly, the conservation of momentum prohibits waves of dipole symmetry. The lowest kind of symmetry which we can associate with gravitational waves is that of a quadrupole. However, the significance of this result is substantially reduced by the non-linearity of the equations. In a linear theory the absence of a purely spherically symmetrical mode implies that there can be no spherically symmetrical component of any wave motion at all, but this is not so in a non-linear theory; on the contrary, perhaps the most important character of gravitational waves concerns just this. For a wave to be a wave in any real physical sense it must convey energy: accordingly, an outgoing wave must diminish the energy of the source and, therefore, its mass.

Contemplate now a transmitter quiescent for a semi-infinite period (so that during this time we have a static situation), then emitting by moving in a suitable way for a finite period, with the field eventually returning again to a static situation. If the waves are real physical waves, i.e. if they carry energy, then in the final situation the transmitter must have less mass than in the initial situation. But the mass is the spherically symmetrical part of the gravitational field and therefore a diminution in mass means a change in the spherically symmetrical component. No change can be expected if the whole situation is purely spherically symmetrical throughout by virtue of Birkhoff's theorem. However, if the field was initially spherically symmetrical and is eventually spherically symmetrical, yet there is an intermediate non-spherically symmetrical wave-emitting period, a change of mass may occur, that is, a change in the coefficient of the r^{-1} term in the static solution. Thus we would expect the higher terms to react back through the non-linearities and to produce an effect on the spherically symmetrical term which represents the mass and thus the energy of the source. The situation is in marked contrast to the electromagnetic case, for there the one thing that cannot be radiated away at all is source strength, that is, charge. In the gravitational case the source strength must diminish if the wave carries energy. The case here mentioned of a transmitter initially and eventually quiescent is the case that will receive most attention in the rest of this paper, because in this case, and in this case alone, we are concerned with initial and final situations that are static and therefore well understood.

The loss of energy, that is the loss of mass, is immediately connected with the problem of the availability to receivers of the radiated energy. This raises the question of what constitutes a receiver for gravitational waves and how much energy it can absorb from such a wave. The matter is discussed more fully in Section D of this paper, but it may perhaps be worth pointing out now that in electromagnetic theory we are familiar with the distinction between near-field transfer as, say, in the case of electromagnetic induction of energy, and radiative wave transfer of energy. The distinction between these two types of transfer is normally clear-cut in the electromagnetic case, though even there difficulties can occur (Bondi, 1961). We shall see that these difficulties are very much greater in the gravitational case.

4. Huygens's Principle and the Change of Wave Form

Different kinds of waves show a number of different properties and one of the most important of these is whether or not they adhere to Huygens's principle. This means, briefly, whether after the end of excitation the wave rings on ("has a tail"), or whether with the end of excitation there is an end to the wave motion, propagated throughout space with the fundamental velocity. It is well known that the ordinary d'Alembert wave equation with an odd number of spatial dimensions satisfies Huygens's principle but that in an even number of spatial dimensions or with suitable additive terms the wave equation does not satisfy this principle, and therefore its solutions then possess a tail. Whether gravitational waves have tails is one of the questions that will be investigated in the course of this paper.

5. Method of Treatment

The extreme complexity of the field equation for empty space makes it clear that a method of expansion should be used to examine the problem. The method that will be used here is that of expansion in negative powers of a suitably defined radius. This seems to be a very suitable method for a wave problem, and the difficulties that have previously stood in the way of such an analysis are avoided by the choice of a suitable system of co-ordinates. The problem of convergence is, naturally, always a very real problem in such a method. However, arguments will be given suggesting that the difficulties arising from this can be contained. At the same time it must be emphasized that many of the strange features that appear in the course of the work are due to this method rather than inherent in the equations. It may therefore be useful to consider here the ordinary scalar wave equation using the same methods as will later be applied to the gravitational wave problem.[1] These methods are not the easiest methods for dealing with the scalar wave equation, but they seem to be the most promising for the gravitational wave case. Consider then

$$\partial^2 Q/\partial t^2 = \nabla^2 Q. \tag{1}$$

Separate now the part Q_n of Q proportional to the surface harmonic S_n and introduce a null variable u by the relation $u = t-r$. The wave equation (1) now takes the form

$$2\left(\frac{\partial^2 Q_n}{\partial r\,\partial u}+\frac{1}{r}\frac{\partial Q_n}{\partial u}\right) = \frac{\partial^2 Q_n}{\partial r^2}+\frac{2}{r}\frac{\partial Q_n}{\partial r}-\frac{n(n+1)}{r^2}Q_n. \tag{2}$$

We attempt to find solutions of this equation by an expansion in negative powers of r

$$Q_n = \sum_k \overset{k}{L}_n(u) r^{-k-1}. \tag{3}$$

Substitution readily yields the recurrence relation

$$2(k+1)\frac{\mathrm{d}\overset{k+1}{L}_n}{\mathrm{d}u} = (n-k)(n+k+1)\overset{k}{L}_n. \tag{4}$$

The class of solutions of (1) that we expect to be able to represent in form (3) is essentially the class of outward travelling waves, for with general inward travelling waves we would expect to find arbitrary functions of $t+r$, that is of $u+2r$, which only in the rarest cases would admit expansion in powers of r^{-1}. Indeed the Sommerfeld condition may be taken to be virtually equivalent to the validity of the expansion (3).

The recurrence relation (4) shows

(i) $\overset{0}{L}_n$ is the lowest $\overset{k}{L}_n$ occurring, so that we have a satisfactory expansion diminishing to zero at infinity.

(ii) Since $\mathrm{d}\overset{n+1}{L}_n/\mathrm{d}u = 0$, the set of $\overset{k}{L}_n$ is divided into two disconnected parts, one from $\overset{0}{L}_n$ to $\overset{n}{L}_n$, the other from $\overset{n+1}{L}_n$ onwards.

(iii) $\overset{n}{L}_n$ is the only non zero term if we impose the condition that Q_n is static.

Consider again the combination of all the surface harmonics S_n. Calling $\overset{n}{L}_n$ the nth moment M_n of the generating distribution, a static system is therefore fully described by all its moments. Next, suppose the system to vary in time, with each of the moments a given function of the time or, rather, of the variable u. Again confining attention to the part proportional to $S_n (n \geqslant 1)$, we have

$$\overset{n}{L}_n = M_n(u) \text{ (given)}.$$

Then, by (4)

$$(5) \quad \overset{n-1}{L}_n = \frac{\mathrm{d}M_n}{\mathrm{d}u}, \quad \overset{n-2}{L}_n = \frac{n-1}{2n-1}\frac{\mathrm{d}^2M_n}{\mathrm{d}u^2}, \quad \ldots, \quad \overset{0}{L}_n = \frac{2^n n!}{(2n)!}\frac{\mathrm{d}^nM_n}{\mathrm{d}u^n}.$$

The coefficient of r^{-1} in the expansion of Q, namely $\overset{0}{L} = \sum_{1}^{\infty} \overset{0}{L}_n S_n$, will be called the coefficient of the *radiative part* of the field. A field for which all the $\overset{0}{L}_n$ vanish is called non-radiative. In addition to static

[1] This approach is considered in detail by Friedlander (1962).

fields, all those fields for which M_n varies like a polynomial of degree not exceeding $(n-1)$ will also be non-radiative.

Suppose now that a field is originally static, is then radiative for a finite period, and finally is non-radiative again (a so-called sandwich wave). Therefore $\overset{0}{L}$ differs from zero only for this finite period of time. If a distant observer registers $\overset{0}{L}$ and knows that before the wave period the system was static, then he can decide by applying (5) whether the final state is static or one of the more general non-radiative solutions. For the wave period will lead from a static situation to a static one only if not merely $\overset{0}{L}_n$ vanishes initially and finally, but also $\overset{1}{L}_n, \overset{2}{L}_n, \ldots, \overset{n-1}{L}_n$ vanish initially and finally.

Therefore the $(n-1)$ additional conditions (wave period $a \leqslant u \leqslant b$)

$$\int_a^b \overset{0}{L}_n(u)\,du = 0, \quad \int_a^b du \int_a^u \overset{0}{L}_n(u')\,du' = \int_a^b (b-u)\overset{0}{L}_n(u)\,du = 0, \; \ldots,$$

$$(6) \quad \int_a^b du \int_a^u du' \ldots \int_a^{u''} \overset{0}{L}_n(u''')\,du''' = \int_a^b \frac{(b-u)^{k-1}}{(k-1)!}\overset{0}{L}_n(u)\,du = 0$$

$$(k = 1, 2, \ldots, n-1)$$

must be satisfied. This is an important result that will be required later, and so it may be worth stressing how this should be applied to the complete solution involving all S_n. Then a distant observer aware only of the complete coefficient $\overset{0}{L}$ of the entire $1/r$ term must apply the following necessary and sufficient conditions for a sandwich wave to lead from a configuration known to be static to a static final configuration:

(i) The S_0 part (i.e. spherically symmetrical part) of $\overset{0}{L}$ is independent of u for $u \geqslant b$.

(ii) The S_1 part (dipole part) of $\overset{0}{L}$ vanishes for $u \geqslant b$.

(iii) The S_2 part (quadrupole part) of $\overset{0}{L}$ vanishes for $u \geqslant b$ and its integral with u from a to b vanishes.

(iv) The S_n part of $\overset{0}{L}$ vanishes for $u \geqslant b$ and the $(n-1)$ conditions (6) apply to its integral over the radiative period (a, b).

Unless all these conditions are satisfied (note that n apply to the S_n part for $n \geqslant 1$) the final state will not be static, even if still non-radiative.

If we consider a more complicated, and especially a non-linear, hyperbolic equation, as we shall do later in this paper, then even if an analysis of this type, with expansion in negative powers of the radius (though not in surface harmonics) is possible, the right-hand side of the equation corresponding to (4) will be vastly more complicated. However, it turns out, in the case of the gravitational field in empty space, that $d\overset{k+1}{L}/du$ is determined entirely by $\overset{0}{L}, \overset{1}{L}, \ldots, \overset{k}{L}$, and their angle derivatives. It need not cause surprise then to find an infinite set of conditions for the field to go from a static state to a static state via a radiative interlude, namely, the set corresponding to (i) to (iv), but with the various S_n unseparated. This situation suggests therefore that Huygens's principle applies to gravitational waves so that after the excitation a completely static situation sets in. However, this argument is no proof, and it is quite possible that there are "tails" in the gravitational case. Perhaps one can pick up a hint of how these tails arise by returning to the linear case (equation (4)) and considering the series beyond $\overset{n}{L}_n$. It is clear that as we go along the series each term will introduce a new arbitrary constant, since each coefficient enters for the first time through its derivative. A simple calculation and summation of the series occurring show that each of these arbitrary constants is multiplied by an expression of the form

$$(7) \qquad \frac{S_n}{r^{n+p}} \frac{1}{u+2r} \left[\text{polynomial in } \frac{u}{u+2r} \text{ of degree } (n+p-1)\right],$$

where p is a positive integer. The process of summation and the final result show that the series occurring converge only over a limited range; in fact, r must be less than $\frac{1}{2}u$. However, it is ascertained by direct substitution that the sum (7) is everywhere a solution of equation (2). We may, if we wish, look at expression (7) as a particular type of combination of incoming and outgoing waves. It is clear that every one of expressions (7) for fixed values of the radius tends to zero as u becomes very large, that is to say, these terms all represent declining modes.

We shall not further investigate the meaning of terms of type (7), but conclude from this heuristic argument that equations of type (4) may generate expressions that tend to zero as $u \to \infty$ which might in the non-linear case represent tails. Moreover, they make it plausible that in general we may obtain useful series for u/r less than some positive constant.

PART B. A SUITABLE CO-ORDINATE SYSTEM

H. Bondi and M. G. J. van der Burg

1. The Character of the Metric

In most investigations of specific systems in general relativity the work can be simplified considerably by a suitable choice of co-ordinate system. In the case of radiation from an isolated system, we are interested in the behaviour of the gravitational field at large distances. As it seems unlikely that a solution in closed form can be found, the main aim in the choice of the system of co-ordinates should be to make an expansion appropriate to large distances as simple as possible. Investigators have often been hindered by the appearance of terms in $\log r$ which prevent expansion in negative powers of r. It is highly desirable so to define the co-ordinates that such terms do not appear. There seem to be two distinct ways in which logarithmic terms arise. In the Schwarzschild and similar solutions, in the usual form, the equation for the radial null geodesics contains a logarithm of r. By using a co-ordinate constant along such radial null geodesics, the appearance of such a term can be avoided. Secondly, it is clear that gravitational waves spread with distance from the source. If the area of intersection of any bundle of these null geodesics with the surfaces of equal phase varies like r^2, then the decay of fields with distance may be expected to occur with suitable powers of $1/r$. If these areas varied like a more complicated function of r, as usually happens, one would expect this function, which often includes logarithmic terms, to appear also in the variation of the strength of the wave with distance.

The co-ordinate system that we shall use is designed to avoid both these possible sources of logarithmic terms. It will first be defined in an intuitive manner; later the definition will be made mathematically more precise. Throughout this paper we shall suppose the 4-space to be axially symmetrical and also reflexion symmetrical. We do not think it likely that this restriction, which materially simplifies the analysis, is too severe for the study of gravitational waves. From the axial symmetry the azimuth angle ϕ is readily defined in an invariant manner. Suppose we now put a source of light at a point O on the axis of symmetry and surround it by a small sphere on which we can produce the azimuth co-ordinate ϕ together with a co-latitude θ and a time co-ordinate u. We then define the u, θ, ϕ co-ordinates of an arbitrary event E to be the u, θ, ϕ co-ordinates of the event at which the light ray OE intersects the small sphere. In other words, along an outward radial light ray the three co-ordinates u, θ, ϕ are constant. If we wish to write down the metric for such a system of co-ordinates (in which the part referring to the azimuth angle ϕ appears separately) then we know that since only the co-ordinate r varies along a light ray, the term g_{11} of the metric tensor must vanish, the four co-ordinates u, r, θ, ϕ being denoted by 0, 1, 2, 3 in that order. Moreover, we must have

$$\Gamma^0_{11} = \Gamma^2_{11} = 0. \tag{8}$$

Owing to the restriction $g_{11} = 0$ this implies

$$g^{00}g_{01,1}+g^{02}g_{12,1} = g^{02}g_{01,1}+g^{22}g_{12,1} = 0. \tag{9}$$

Since neither g nor g_{33} can vanish, the equations are equivalent to

$$g_{12}(g_{12}g_{01,1}-g_{01}g_{12,1}) = g_{01}(g_{12}g_{01,1}-g_{01}g_{12,1}) = 0. \tag{10}$$

This implies either that $g_{01} = 0$, or that $g_{12} = 0$, or that g_{12}/g_{01} is independent of r. If $g_{01} = 0$, it may be shown that $g_{00} < 0$ (for otherwise the signature is incorrect), and thus that u is not a time-line co-ordinate, contrary to its definition. Therefore we reject the possibility $g_{01} = 0$. If g_{12}/g_{01} is independent of r then there exists a function $\tilde{u}(u, \theta)$ such that, with a suitable $\lambda(u, \theta)$,

$$d\tilde{u} = \lambda[du + g_{12}\, d\theta/g_{01}]. \tag{11}$$

Replacing u by $\tilde{u}$ reduces g_{12} to zero. Accordingly we have arrived at a metric distinguished by the condition

$$g_{11} = g_{12} = 0. \tag{12}$$

Although we have defined the u, θ, ϕ co-ordinatives fairly closely now, the radial co-ordinate r remains entirely indeterminate and can be replaced by any function of r, u and θ without changing the character of the metric or, indeed, the coefficients g_{22} and g_{33}. We accordingly define the co-ordinate r by the condition

$$r^4 \sin^2 \theta = g_{22} g_{33} \tag{13}$$

which ensures that the area of the surface element $u =$ constant $r =$ constant is in fact $r^2 \sin \theta \, d\theta \, d\varrho$. We can now write the metric in the form[2]

$$\begin{aligned} ds^2 = {} & (Vr^{-1}\, e^{2\beta} - U^2 r^2 e^{2\gamma})\, du^2 + 2e^{2\beta}\, du\, dr \\ & + 2Ur^2 e^{2\gamma}\, du\, d\theta - r^2(e^{2\gamma}\, d\theta^2 + e^{-2\gamma} \sin^2 \theta\, d\phi^2). \end{aligned} \tag{14}$$

The peculiar form of the first coefficient is chosen for later convenience. The four functions U, V, β, γ, are functions of u, r and θ. All our investigations will be based on this form of the metric.

There is still considerable freedom in setting up such a metric, but this is somewhat reduced by the following consideration. In our problem, space tends to flatness at infinity. We infer from this that tetrad vectors may be chosen such that the physical components of the curvature tensor tend to zero at infinity. Accordingly, the mutual

[2] First given in Bondi (1960).

accelerations in a swarm of freely moving point particles at large distances can be made arbitrarily small by having the swarm sufficiently far away. We now choose such a swarm of particles so far away that the effect of the accelerations can be neglected for a period sufficient for our investigations, and attach the u and θ co-ordinates to this swarm of particles instead of to the small sphere. This ensures that at sufficiently large distances u is a time-like co-ordinate—that is, the coefficient of du^2 remains positive however large r is taken to be.[3]

In any metric in polar co-ordinate form conditions must be imposed in the neighbourhood of the polar axis ($\sin\theta = 0$) in order to insure regularity there. It must be possible to choose a Minkowskian tangent metric which in turn implies that, for a small circle around the axis, the ratio of circumference to radius equals 2π to the second order in the radius, and also that the metric has no kink as the axis is crossed. In our case these conditions imply that, as $\sin\theta \to 0$,

$$V,\ \beta,\ U/\sin\theta,\ \gamma/\sin^2\theta \tag{15}$$

each equals a function of $\cos\theta$ regular at $\cos\theta = \pm 1$. The metric chosen is sufficiently general and chosen by such a physical reasoning that we may suppose it to be valid for sufficiently large distances from an isolated material system. In other words, while co-ordinate patches are the rule rather than the exception, we may reasonably expect that all the space sufficiently far from an isolated system may be covered by one patch, and that this may be expressed in the form of the co-ordinate system given. We next consider the field equations, then impose an outgoing wave condition, and finally, using both these, we further restrict the co-ordinated system.

[3] The significance of this specification may be understood by considering what happens in flat space if the co-ordinate θ is fixed on a small sphere surrounding the origin, but is chosen to be a function of time. Then the "searchlight beam" corresponding to a particular value of θ moves about in space. At large distances points with the same r, θ, ϕ co-ordinates, but different u co-ordinates will, owing to the motion of the "searchlight beam", have a space-like rather than a time-like separation.

2. The Structure of the Field Equations

It is advantageous to consider the relations between the field equations arising from the Bianchi identities before considering the field equations themselves. First, note that the contravariant fundamental tensor for our metric (14) is given by

$$(16)\quad g^{\mu\nu}=\begin{pmatrix} 0 & e^{-2\beta} & 0 & 0 \\ e^{-2\beta} & -Ve^{-2\beta}r^{-1} & Ue^{-2\beta} & 0 \\ 0 & Ue^{-2\beta} & -e^{-2\gamma}r^{-2} & 0 \\ 0 & 0 & 0 & -e^{2\gamma}r^{-2}\sin^{-2}\theta \end{pmatrix}.$$

A list of the three-index symbols is given in the appendix to this paper. For the moment we require only the result

$$(17)\qquad g^{\alpha\varepsilon}\Gamma^{0}_{\alpha\varepsilon}=-2e^{-2\beta}r^{-1}.$$

The contracted Bianchi identities are

$$(18)\qquad g^{\alpha\varepsilon}(R_{\mu\alpha}-\tfrac{1}{2}g_{\mu\alpha}R)_{;\,\varepsilon}=g^{\alpha\varepsilon}(R_{\mu\alpha,\,\varepsilon}-\tfrac{1}{2}R_{\alpha\varepsilon,\,\mu}-\Gamma^{\delta}_{\alpha\varepsilon}R_{\mu\delta})=0.$$

Suppose now that

$$(19)\qquad R_{11}=R_{12}=R_{22}=R_{33}=0.$$

Then the Bianchi identities reduce to

$$(20)\quad\begin{cases} \mu=0: & g^{01}R_{00,1}+g^{11}R_{01,1}+g^{12}(R_{01,2}+R_{0,21}) \\ & \quad +g^{22}R_{02,2}-g^{\alpha\varepsilon}\Gamma^{0}_{\alpha\varepsilon}R_{00}-g^{\alpha\varepsilon}\Gamma^{1}_{\alpha\varepsilon}R_{01}-g^{\alpha\varepsilon}\Gamma^{2}_{\alpha\varepsilon}R_{02}=0, \\ \mu=1: & -g^{\alpha\varepsilon}\Gamma^{0}_{\alpha\varepsilon}R_{01}=0, \\ \mu=2: & g^{01}(R_{02,1}-R_{01,2})-g^{\alpha\varepsilon}\Gamma^{0}_{\alpha\varepsilon}R_{02}=0. \end{cases}$$

We see that the four equations (19) are independent. They will be referred to as "the main equations". Next, we see from the second of equations (20) that R_{01} vanishes as a consequence of the main equations. Using this result and also equation (17), the last of equations (20) may therefore be written

$$(21)\qquad e^{-2\beta}\left(\frac{\partial}{\partial r}+\frac{2}{r}\right)R_{02}=r^{-2}e^{-2\beta}\frac{\partial}{\partial r}(r^{2}R_{02})=0.$$

Accordingly, as a consequence of the main equations (19) only, R_{02} is of the form $f(u, \theta)\, r^{-2}$. Therefore, if R_{02} vanishes for some value of r and all values of u and θ, it will vanish for all values of r. If this condition is satisfied, the first of equations (20) also reduces to the form (21). What we have said about R_{02} then applies equally to R_{00}. Thus, a complete set of field equations resolves into the four main equations (19), the trivial equation $R_{01} = 0$, and the two supplementary conditions, as we shall call them, which only imply that R_{00} and R_{02} vanish for some finite value of r. If R_{00} and R_{02} can be expanded in powers of r then the supplementary conditions merely state that the coefficients of the r^{-2} terms vanish.

3. The Main Equations

We now write the main equations (19) in the following form:

$$0 = R_{11} = -4[\beta_1 - \tfrac{1}{2} r\gamma_1^2] r^{-1}. \tag{22}$$

$$\begin{aligned} 0 = -2r^2 R_{12} = [4r^4 e^{2(\gamma-\beta)} U_1]_1 - 2r^2[\beta_{12} - \gamma_{12} + 2\gamma_1\gamma_2 \\ -2\beta_2 r^{-1} - 2\gamma_1 \cot\theta]. \end{aligned} \tag{23}$$

$$\begin{aligned} 0 = R_{22} e^{2(\beta-\gamma)} - r_2 R_3^3 e^{2\beta} = 2V_1 + \tfrac{1}{2} r^4 e^{2(\gamma-\beta)} U_1^2 \\ -r^2 U_{12} - 4rU_2 - r^2 U_1 \cot\theta - 4rU\cot\theta \\ +2e^{2(\beta-\gamma)}[-1 - (3\gamma_2 - \beta_2)\cot\theta - \gamma_{22} + \beta_{22} + \beta_2^2 + 2\gamma_2(\gamma_2 - \beta_2)]. \end{aligned} \tag{24}$$

$$\begin{aligned} 0 = -R_3^3 e^{2\beta} r^2 = 2r(r\gamma)_{01} + (1 - r\gamma_1)V_1 - (r\gamma_{11} + \gamma_1)V \\ -r(1 - r\gamma_1)U_2 - r^2(\cot\theta - \gamma_2)U_1 \\ +r(2r\gamma_{12} + 2\gamma_2 + r\gamma_1 \cot\theta - 3\cot\theta)U \\ +e^{2(\beta-\gamma)}[-1 - (3\gamma_2 - 2\beta_2)\cot\theta - \gamma_{22} + 2\gamma_2(\gamma_2 - \beta_2)]. \end{aligned} \tag{25}$$

Note that equations (22) to (24) involve only differentiation in the hypersurface $u = \text{const.}$ (hypersurface equations), while (25), the "standard equation", contains a derivative with respect to u. Consider now the structure of these equations without, at first, worrying about the functions of integration. If, for some value of u, γ is given, equation (22) will determine β. Equation (23) will then determine U, and equation (24) will give V. From equation (25), the u-derivative of γ may then be deduced. Thus the function γ may be found at the next instant of u, and then we can again go through the whole cycle. In other words, given γ for some value of u, the future follows, apart from the question of functions of integration, each of which is an arbitrary function of u and θ, but independent of r. We can count them easily enough. Equation (22) determines β apart from an additive function $H(u, \theta)$. In equation (23), two such functions of integration occur, one of them being an addition $-6N(u, \theta)$ to $r^4 e^{2(\gamma-\beta)} U_1$, the other an addition $L(u, \theta)$ to U itself. In equation (24), a function $-2M(u, \theta)$ may be added to V. Finally, equation (25) determines γ_0 apart from a term $c_0(u, \theta)r^{-1}$ which goes out when the first term is formed. Thus γ contains a term $c(u, \theta)r^{-1}$ where $c_0(u, \theta)$ is not determined by (24).

There are, accordingly, a total of five such functions, and we can now re-state what we found before by saying that, given γ for one value of u, and given the five functions $H(u, \theta)$, $N(u, \theta)$, $L(u, \theta)$, $M(u, \theta)$, $c(u, \theta)$, the entire development is determined by the four main equations. We shall see below that, in certain circumstances, co-ordinate considerations serve to eliminate two of these five functions, and that the supplementary conditions yield two relations between the three surviving functions.

The significance of this structure of the equations is readily understood. If we know the situation for a particular value of u = constant (which represents a light cone opening out into the future) then we know everything about incoming waves owing to our knowledge of the various functions for all radii. If the system is to do anything new, then the information about this must be contained in the functions of u and θ that appear as functions of integration in our analysis. Thus we see that all the news there is appears in these functions which, as has been indicated, can be reduced to a single "news function".

4. The Outgoing Wave Condition

If we wish to adopt the principle of causality—in other words, if we wish to eliminate inward travelling waves—we have to apply a suitable condition to γ, etc. This can be done in various ways. We may suppose, for instance, that at one value of u and for a little while beforehand, the entire system was static, that is, without any radiation whatsoever. In that case, the system can be described by the well-known Weyl metric, and it may be shown (see appendix) that, in this case, γ and the other variables have the form of power series in negative powers of r. Alternatively, we may say that the absence of inward flowing radiation is equivalent to the condition that γ (and the other variables) should be of the form, roughly speaking,

$$\gamma = \frac{f(t-r)}{r} + \frac{g(t-r)}{r^2} + \ldots \tag{26}$$

which is equivalent to what we said before. Condition (26) is essentially equivalent to Sommerfeld's radiation condition which, in our case, means that

$$[\partial(r\gamma)/\partial r]_{u=\text{const.}} \to 0$$

as $r \to \infty$. Next consider the main equations in the light of (26). If γ has a term proportional to r^{-1}, together with other terms tending to zero more rapidly at infinity, then it follows from equation (22) that β remains bounded as $r \to \infty$. Similarly equation (23) then shows that U remains bounded as $r \to \infty$, and in fact $\lim U = L(u, \theta)$. Substituting this into equation (24) and integrating one finds that the leading term in V is proportional to r^2. Accordingly, the coefficient g_{00} of the metric is dominated by the second term, which is bound to make it negative for sufficiently large values of r. This is contrary to the way in which the co-ordinate system was defined according to which g_{00} had to remain positive for arbitrarily large values of r. Therefore $L(u, \theta) = 0$.[4] This disposes of one of our functions of integration.

[4] It might be suggested that a more general form of γ should be used with a leading term $Q(u, \theta)$ independent of r. However, Q does not affect the argument just given that $L = 0$ and, if $L = 0$, (14) proves that $Q_0 = 0$. The elementary transformation $\tan \frac{1}{2}\bar{\theta} = \exp \int e^{4Q} \sin^{-1} \theta \, d\theta$, with a consequent adjustment of r, then reduces Q to zero.

On this basis, the leading terms of the various functions are given by

$$(27)\quad \begin{cases} \gamma = c(u, \theta)r^{-1}+\ \dots, \\ U = 2H_2 e^{2H} r^{-1}+\ \dots, \\ \qquad\qquad \beta = H(u, \theta)-\frac{1}{4}c^2 r^{-2}+\ \dots, \\ \qquad\qquad V = re^{2H}[1+2H_2 \cot\theta+4H_2^2+2H_{22}]+\ \dots \end{cases}$$

As a final restriction on the co-ordinate system, we reduce H to zero by a co-ordinate transformation

$$(28)\quad \begin{cases} u = \overset{0}{a}(\bar{u}, \bar{\theta})+\overset{1}{a}(\bar{u}, \bar{\theta})\bar{r}^{-1}+\ \dots, \\ r = \bar{r}+\overset{0}{\varrho}(\bar{u}, \bar{\theta})+\ \dots, \\ \theta = \bar{\theta}+\overset{1}{g}(\bar{u}, \bar{\theta})\bar{r}^{-1}+\ \dots. \end{cases}$$

In order to preserve the character of the metric,

$$(29)\quad \begin{cases} \bar{g}_{11} = 0: \quad 2\overset{1}{a}e^{2H} = -\left(\overset{1}{g}\right)^2, \\ \bar{g}_{12} = 0: \quad \overset{0}{a}_{\bar{\theta}}e^{2H}+2\overset{1}{g} = 0, \\ \bar{g}_{22}\bar{g}_{33} = \bar{r}^4 \sin^2\bar{\theta}: \quad 2\overset{0}{\varrho}+\overset{1}{g}_{\bar{\theta}}+\overset{1}{g}\cot\bar{\theta} = \overset{0}{a}_{\bar{\theta}}H_2 e^{2H}, \\ \bar{g}_{01} = \overset{0}{a}_{\bar{u}}e^{2H}+O(\bar{r}^{-1}). \end{cases}$$

For an arbitrary $\overset{0}{a}$ the second of these equations defines $\overset{1}{g}$, the first $\overset{1}{a}$, the third one $\overset{0}{\varrho}$, and similarly the higher-order equations determine the higher-order coefficients. Then $\bar{H} = 0$ if $\overset{0}{a}$ is chosen so that

$$(30)\quad \overset{0}{a}_{\bar{u}} = \exp(-2H).$$

Thus a suitable transformation of type (28) reduces H to zero.

It may be advantageous to recapitulate briefly how successive restrictions have been imposed on the co-ordinate system:

(i) $du = d\theta = d\phi = 0$ is an outgoing light-ray. This implies $g_{11} = 0$.

(ii) u is time-like, so that $g_{00} > 0$, leading to $g_{12} = 0$.

(iii) Area of 2-surface u = const. r = const. restricted to equal $4\pi r^2$, used to define r and to fix $g_{22}g_{33} = r^4 \sin^2\theta$.

(iv) Field equations for empty space imposed, functions H, N, L, M, c isolated.

(v) Outgoing wave condition imposed. With (ii) this implies $L = 0$.

(vi) H reduced to zero by suitable transformation.

The range of possible transformations of co-ordinate systems satisfying (i) to (vi) is investigated in part C of this paper. It should be noted that in general the three surviving functions of integration c, M, N, cannot be reduced to zero.

If the form (26) for γ is substituted into the main equations it turns out that the other variables do not satisfy the radiation condition unless the r^{-2} term in γ vanishes. (This occurs also in the static case when the Weyl metric is translated into our form.) Carrying out this expansion[5] and substituting into the main equations we obtain the following relations:

(31) If $$\gamma = c(u,\theta)r^{-1} + [C(u,\theta) - \tfrac{1}{6}c^3]r^{-3} + \ldots,$$

(32) $$\text{then } U = -(c_2 + 2c\cot\theta)r^{-2} + [2N(u,\theta) + 3cc_2 + 4c^2\cot\theta]r^{-3} + \tfrac{1}{2}(3C_2 + 6C\cot\theta - 6cN - 8c^2c_2 - 8c^3\cot\theta)r^{-4} + \ldots,$$

(33) $$V = r - 2M(u,\theta) - [N_2 + N\cot\theta - c_2^2 - 4cc_2\cot\theta - \tfrac{1}{2}c^2(1 + 8\cot^2\theta)]r^{-1} - \tfrac{1}{2}[C_{22} + 3C_2\cot\theta - 2C + 6N(c_2 + 2c\cot\theta) + 8c(c^2 + 3cc_2 + 2c^2\cot^2\theta)]r^{-2} + \ldots,$$

(34) $$4C_0 = 2c^2c_0 + 2cM + N\cot\theta - N_2.$$

Thus the form of γ is preserved and the development of the system is fully determined from initial conditions provided the functions c, N, M are known.

[5] No assumption of analyticity is implied. It is probably sufficient if the remainder after the first few terms vanishes suitably at infinity.

5. The Supplementary Conditions

The full form of the equations $R_{02} = 0$ and $R_{00} = 0$, which is exceedingly complicated, is given in the appendix. It is no way obvious how, in general, the substitution of the main equations reduces the supplementary conditions to the inverse square-law term. However, on the basis of the expansions given above, the supplementary conditions reduce enormously. The sole surviving terms are of the r^{-2} form as, of course, they should be, and involve only relations between the three functions c, M and N, namely

$$M_0 = -c_0^2 + \tfrac{1}{2}(c_{22} + 3c_2 \cot\theta - 2c)_0, \tag{35}$$

$$-3N_0 = M_2 + 3cc_{02} + 4cc_0 \cot\theta + c_0 c_2. \tag{36}$$

Thus we see that if M and N are given for one value of u, and c is given as a function of u and θ, the entire situation is fully determined. In other words, the flow of information in the system is entirely controlled by the single function c. In fact, as will be seen later, the whole character of the developments can be read off from equations (35) and (36). This is particularly satisfactory because these are in no way approximate equations, but are exact relations valid supposing only the series expansions to be valid. In order to identify M and N to some extent at least, we consider now static and therefore well understood metrics.

6. The Static Case

It is well known that the empty space axially symmetric static metric can always be reduced to Weyl's form

$$ds^2 = e^{2\psi}\, dt^2 - e^{-2\psi}[e^2\sigma(d\varrho^2 + dz^2) + \varrho^2\, d\phi^2], \tag{37}$$

(38) where

$$\frac{\partial^2 \psi}{\partial \varrho^2} + \frac{1}{\varrho}\frac{\partial \psi}{\partial \varrho} + \frac{\partial^2 \psi}{\partial z^2} = 0,$$

and ψ determines σ by a relation not required here. It follows from (38) that if the metric is Minkowskian at infinity then

$$\psi = -\sum_{n=0}^{\infty} A^{(n)} R^{-n-1} P_n(\cos\Theta), \quad \varrho = R\sin\Theta, \quad z = R\cos\Theta. \tag{39}$$

Here $A^{(0)}$ is the mass m of the system, $A^{(1)}$ is the dipole moment D, and

(40) $$A^{(2)} = Q+\tfrac{1}{3}m^3,$$

where Q is the quadrupole moment, defined so as to vanish for the Weyl form of the Schwarzschild line element. A rather lengthy and tedious transformation is required to connect the Weyl metric with our metric.[6] Therefore only the result will be quoted here, in a form chosen to fit in with the transformation equations given in part C. These equations show that c is an arbitrary function of θ in the static case, given in the general case as compared with the case $c = 0$, in terms of the transformation function α by

(41) $$c = -\tfrac{1}{2}\alpha_{22}+\tfrac{1}{2}\alpha_2 \cot \theta.$$

Then we have

(42) $$\begin{cases} M = m, \\ N = D \sin \theta - m\alpha_2, \\ C = \tfrac{1}{2}Q \sin^2 \theta - \alpha_2 D \sin \theta + \tfrac{1}{2}m\alpha_2^2. \end{cases}$$

Equations (42) allow us to interpret the principal functions occuring in our metric in the static case. The function $M(u, \theta)$ will now be named the *mass aspect*, and the first of equations (42) shows that in the static case the mass aspect equals the mass of the system and is accordingly independent of both its arguments. The functions N and C are seen to be closely related to the dipole and quadrupole moments respectively.

7. The Curvature Tensor

In order to investigate the character of the solutions, it is necessary to know the behaviour of the curvature tensor, and, to be able to interpret it in physical terms, a tetrad formulation is required. Accordingly, we need a system of orthogonal unit vectors, one time-like, and three space-like. Moreover, to prevent spurious effects, we choose the unique set of vectors that, by parallel transport along the null

[6] This is given in appendix 4.

geodesics $du = d\theta = d\phi = 0$, turns at infinity into

$$T^{\mu} = (1, 0, 0, 0), \qquad R^{\mu} = (-1, 1, 0, 0),$$
$$S^{\mu} = \left(0, 0, \frac{1}{r}, 0\right), \quad P^{\mu} = \left(0, 0, 0, \frac{1}{r \sin\theta}\right).$$

To find this set at a general point define

$$q = \int_{r}^{\infty} [\tfrac{1}{2} r e^{\gamma} U_1 + \beta_2 r^{-1} e^{2\beta-\gamma}]\, dr$$
$$= (c_2 + 2c \cot\theta)\, r^{-1} - [\tfrac{3}{2}N + 2c(c_2 + c\cot\theta)] r^{-2} + \dots \tag{43}$$

The set can be shown to be

$$T^{\mu} = [1, \tfrac{1}{2}(1+q^2)e^{-2\beta} - \tfrac{1}{2}Vr^{-1},\ U + qe^{-\gamma}r^{-1}, 0]$$
$$= [1, Mr^{-1} + \tfrac{1}{2}(N_2 + N\cot\theta)r^{-2} + \dots,\ \tfrac{1}{2}Nr^{-3} + \dots, 0],$$
$$R^{\mu} = [-1, \tfrac{1}{2}(1-q^2)e^{-2\beta} + \tfrac{1}{2}Vr^{-1},\ -U - qe^{-\gamma}r^{-1}, 0]$$
$$= [-1, 1 - Mr^{-1} - \tfrac{1}{4}(2N_2 + 2N\cot\theta - c^2)r^{-2} + \dots,$$
$$-\tfrac{1}{2}Nr^{-3} + \dots, 0],$$
$$S^{\mu} = [0, qe^{-2\beta},\ e^{-\gamma}r^{-1}, 0]$$
$$= [0, (c_2 + 2c\cot\theta)r^{-1} - \tfrac{1}{2}(3N + 4c(c_2 + c\cot\theta))\, r^{-2} + \dots,$$
$$r^{-1} - cr^{-2} + \tfrac{1}{2}c^2 r^{-3} + \dots, 0],$$
$$P^{\mu} = [0, 0, 0, e^{\gamma} r^{-1} \sin^{-1}\theta]$$
$$= [0, 0, 0, r^{-1}\sin^{-1}\theta(1 + cr^{-1} + \tfrac{1}{2}c^2 r^{-2} + \dots)]. \tag{44}$$

Owing to the axial symmetry, the number of free components of the curvature tensor is reduced to 12. The surviving components have been worked out, with the use of the power series expansions for the field variables. Again it turns out that, to the approximation worked, several components coincide. We are then left with

$$\left\{\begin{aligned} R_{(TTSS)} &= -R_{(TTPP)} = -R_{(TRSS)} = R_{(TRPP)} = R_{(RRSS)} \\ &= -R_{(RRPP)} \\ &= -c_{00}r^{-1} + \tfrac{1}{2}[c_{022} + c_{02}\cot\theta - 2c_0(1 + 2\cot^2\theta)]r^{-2} \dots +, \\ R_{(TTRS)} &= -R_{(TRSR)} \\ &= -R_{(TSPP)} = R_{(RSPP)} = -(c_{02} + 2c_0\cot\theta)r^{-2} + \dots \\ R_{(TTRR)} &= -R_{(SSPP)} = -2(M + cc_0)r^{-3} + \dots . \end{aligned}\right. \tag{45}$$

PART C. PERMISSIBLE CO-ORDINATE TRANSFORMATIONS

A. W. K. METZNER

1. Evaluation of the Transformations by Power Series

The purpose of this part is to discuss the permissible transformations of co-ordinates that will leave unchanged the character of the metric discussed in part B. By this we mean not only that the form of the metric should remain the same but also that U, β and γ should tend to zero at infinity. It will be assumed throughout that not only the metric but also the co-ordinate transformations can be expanded in powers of r. Thus, expressing the ordinary unbarred co-ordinates in terms of barred ones, we have

$$(46) \qquad \begin{cases} u = a(\bar{u}, \bar{\theta})\bar{r} + \overset{0}{a}(\bar{u}, \bar{\theta}) + \overset{1}{a}(\bar{u}, \bar{\theta})\bar{r}^{-1} + \dots, \\ r = K(\bar{u}, \bar{\theta})\bar{r} + \overset{0}{\varrho}(\bar{u}, \bar{\theta}) + \overset{1}{\varrho}(\bar{u}, \bar{\theta})\bar{r}^{-1} + \dots, \\ \theta = g(\bar{u}, \bar{\theta})\bar{r} + \overset{0}{g}(\bar{u}, \bar{\theta}) + \overset{1}{g}(\bar{u}, \bar{\theta})\bar{r}^{-1} + \dots . \end{cases}$$

Evaluating the components of the new barred metric tensor in terms of the old one by means of tensor transformations, we find

$$\bar{g}_{11} = K^2 g^2 \bar{r}^2 + \dots = 0.$$

Since clearly $K \neq 0$, we must have $g = 0$. With this

$$\bar{g}_{11} = a^2 + 2aK + O(\bar{r}^{-1}) = 0.$$

Thus $a = 0$ or $a = -2K$. The second possibility merely corresponds to a reversal of time, that is, to a consideration of advanced rather than retarded solutions. We reject this, and take the first possibility. Continuing the analysis we have

$$\bar{g}_{00} = K^2 \left(\partial \overset{0}{g} / \partial \bar{u}\right)^2 \bar{r}^2 + O(\bar{r}) = 1 + \dots .$$

Hence

$$\overset{0}{g} = \overset{0}{g}(\bar{\theta}).$$

Next

$$\bar{g}_{22} = K^2\left(d\overset{0}{g}/\partial\bar{\theta}\right)^2 \bar{r}^2 + O(\bar{r}) = \bar{r}^2 + O(\bar{r}).$$

Thus

$$K\left(d\overset{0}{g}/d\bar{\theta}\right) = \pm 1.$$

Furthermore

$$\bar{g}_{22}\bar{g}_{33} = g_{22}g_{33}\left(d\overset{0}{g}/d\bar{\theta}\right)^2 + O(\bar{r}^3) = K^4\bar{r}^4 \sin^2 \overset{0}{g} + O(\bar{r}^3) = \bar{r}^4 \sin^2 \bar{\theta}.$$

(47) From the previous result $\dfrac{d\overset{0}{g}}{d\bar{\theta}} = \pm \dfrac{\sin \overset{0}{g}}{\sin \bar{\theta}}$.

The minus signs are trivial alternatives and thus

(48) $$\tan\left(\tfrac{1}{2}\overset{0}{g}\right) = e^{-\nu} \tan\left(\tfrac{1}{2}\bar{\theta}\right) \qquad (\nu = \text{const.}),$$

(49) $$K = \cosh \nu + \cos \bar{\theta} \sinh \nu.$$

Finally, from these results,

$$\bar{g}_{01} = K\frac{\partial \overset{0}{a}}{\partial \bar{u}} + \ \dots = 1 + \ \dots .$$

(50) Thus

$$\overset{0}{a} = \bar{u}/K(\bar{\theta}) + \alpha(\bar{\theta}),$$

where α is an arbitrary function of its argument. Continuing in this manner to compare coefficients, it turns out that no further freedom exists. Thus the entire range of transformations possible is described by the single constant ν introduced in equation (48), together with the single function α of a single variable introduced in equation (50). What do this constant and this function represent? The constant ν enters the function K whose form immediately suggests that we are dealing with aberration. At large distances, where space is effectively flat, the ν transformation is in fact fully equivalent to a Lorentz transformation corresponding to motion along the axis of symmetry with velocity $-\tanh \nu$. Therefore we can regard the K transformation as a generalized Lorentz transformation, that is to say, as a uniform motion of the material system relative to the fixture of our system of co-ordinates at large distances. The significance of the transformation introduced in equation (50) is a little less obvious. It will be recalled that our system of co-ordinates was fixed in the end by tying down light rays at infinity. In doing this we have dropped the restriction that all the light rays for every angle should originate at the same position. We can imagine, as it were, a different lamp for each angular co-ordinate. In order to keep to the various orthogonality conditions implied by our system of co-ordinates, we cannot introduce a very great deal of freedom there but a kind of static deviation. We can fix, as it were, a light for each angle at one time, and then the motions of these different lights are determined. In this way we account for the function of one variable introduced in equation (50).

The equations (48) to (50) may be applied to find the other terms in expansion (46) and may also be used in order to evaluate in the new co-ordinates the functions defining the metric that were introduced in part B. They tend to be somewhat complicated formulae and so are listed in the appendix.

2. Rigorous Transformations of Minkowski Space

The assumption that the power series (46) represent all possible transformations is open to criticism. No direct proof of this assumption is possible but it is made plausible by the fact, demonstrated below, that *all* transformations for flat space are of this form, together with the assumption made in part B that the coefficients of the general metric can be expanded in power series.

We may write the metric of flat space in the form

$$ds^2 = d\bar{u}^2 + 2\, d\bar{u}\, d\bar{r} - \bar{r}^2(d\bar{\theta}^2 + \sin^2 \bar{\theta}\, d\phi^2),$$

and ask for the most general transformation of this into co-ordinates (u, r, θ, ϕ) giving a metric as specified in part B. We first note that in "mixed" co-ordinates $(u, \bar{r}, \theta, \phi)$ the metric still has $g_{11} = g_{12} = 0$. Thus taking $\bar{u} = \bar{u}(u, \bar{r}, \theta)$, etc., we have

$$(\bar{u}_{\bar{r}} + 1)^2 = 1 + \bar{r}^2 \bar{\theta}_{\bar{r}}^2,$$

$$u_\theta(\bar{u}_{\bar{r}} + 1) = \bar{r}^2 \bar{\theta}_{\bar{r}} \bar{\theta}_\theta.$$

Solving for $\bar{u}_{\bar{r}}$ and $\bar{u}_{\theta}$, we find that the compatibility condition for these implies

(51) $$\bar{\theta} = p(u, \theta) - \sin^{-1}[q(u, \theta)\bar{r}^{-1}],$$

(52) $$\bar{u} = (\bar{r}^2 - q^2)^{\frac{1}{2}} - \bar{r} + s(u, \theta),$$

with $$s_{\theta} = q p_{\theta}.$$

The functions p, q, s are otherwise arbitrary, except for the usual regularity conditions and the need to make the old and new axes of symmetry coincide, i.e.

(53) $$p \to 0, \quad q \to 0 \quad \text{as} \quad \theta \to 0; \quad p \to \pi, \quad q \to 0 \quad \text{as} \quad \theta \to \pi.$$

After evaluating the coefficient of $d\theta^2$ we can find r and γ:

(54) $$r^2 \sin\theta = p_{\theta} \sin p(\bar{r}^2 - q^2) - (q \sin p)_{\theta} (\bar{r}^2 - q^2)^{\frac{1}{2}} + q q_{\theta} \cos p,$$

(55) $$e^{2\gamma} = \frac{[(\bar{r}^2 - q^2)^{\frac{1}{2}} p_{\theta} - q_{\theta}] \sin\theta}{(\bar{r}^2 - q^2)^{\frac{1}{2}} \sin p - q \cos p} \to \frac{p_{\theta} \sin\theta}{\sin p} \quad \text{as} \quad \bar{r} \to \infty.$$

Equation (54) shows that $\bar{r} \to \infty$ implies $r \to \infty$ unless $p_{\theta} \sin p$ is not positive, a case that must be excluded since otherwise the mapping is incomplete. The requirement that $\gamma \to 0$ as $r \to \infty$ then shows that $p_{\theta} \sin\theta = \sin p$, i.e. $\tan \frac{1}{2}p = \exp \nu(u) \tan \frac{1}{2}\theta$. The condition that g_{00} remain positive as $r \to \infty$ implies $p_u = 0$ and therefore $\nu = \text{const.}$

The r equation (54) can be solved for $\bar{r}$ and it is immediately clear that $\bar{r}$ can be expanded in powers of r as required provided r is sufficiently large if q_{θ}, q/θ and $q/(\pi - \theta)$ are bounded which is a consequence of the conditions imposed above. Moreover, $\bar{r} \to \infty$ uniformly as $r \to \infty$, and so the expansion of all functions is guaranteed, as required.

The work can readily be continued without resorting to the expansions. Next, $\beta \to 0$ as $r \to \infty$ implies $s_u = 1$, and so the transformation has been reduced to the generality found above by series expansion.

To interpret the α transformation ($p = \theta$, $s = -\alpha(\theta)$, $q = -\alpha'(\theta)$) in flat space we use the exact equations which are valid everywhere. (In the general case only the series expansion is available and this precludes an approach to regions of small r.) The co-ordinates $\bar{r}$, $\bar{\theta}$

are ordinary spherical polars and thus the light ray $\theta = \text{const.}$ has equation

$$\bar{r} \sin(\bar{\theta}-\theta) = \alpha'(\theta). \tag{56}$$

Thus the ray is parallel to $\bar{\theta} = \theta$. If $\alpha'(\theta)/\cos\theta > 0$ it intersects the equatorial plane $\bar{\theta} = \frac{1}{2}\pi$ at distance $\alpha'(\theta)/\cos\theta$ from the origin, and equally if $\alpha'(\theta)/\cos\theta < 0$ it intersects the axis of symmetry at distance $-\alpha'(\theta)/\sin\theta$ from the origin.

The interpretation of the K transformation is identical in the general and flat space cases as a Lorentz transformation.

PART D. THE NATURE OF THE SOLUTIONS

H. Bondi, F.R.S.

1. News and Mass Loss

Consider an axi-symmetric system that is static for $u \leqslant 0$. As has been shown, the Weyl metric for this system may be transformed to our form, with all the coefficients $c, M, N, C, \dots$ independent of u. By (35) the vanishing of c_0 guarantees the vanishing of M_0, but by (36) M_2 has to vanish also to secure the constancy of N. Similarly, by (34), N has to be of the form (42) to assure the vanishing of C_0, and so on. If $c_0 = 0$ for $u > 0$, the system must remain static, but if c_0 begins to deviate from zero the other coefficients in turn begin to vary. Thus if anything happens at all at the source leading to changes in the field, it can only do so by affecting c_0, and vice versa. Thus all the news in the field is contained in c_0, which there fore merits the name *news function*. In general the structure of our equations indicates that if γ, M and N are known for $u = a$, and the news function c_0 is known for all u in $a \leqslant u \leqslant b$, then the system is fully determined in the interval $a \leqslant u \leqslant b$.

Next we define the mass $m(u)$ of the system as the mean value of $M(u, \theta)$ over the sphere

$$m(u) = \frac{1}{2}\int_0^{\pi} M(u, \theta) \sin\theta \, d\theta. \tag{57}$$

We note that in the static case $m(u) = m$. Now we integrate (35) and obtain

$$m_0 = -\frac{1}{2}\int_0^{\pi} c_0^2 \sin\theta \, d\theta, \tag{58}$$

since the last term in (35) goes out on integration because of conditions (15). Thus we have the central result of this paper:

The mass of a system is constant if and only if there is no news. If there is news, the mass decreases monotonically as long as the news continues.

This result may appear to depend on the definition of mass given above, but this can be avoided if we confine our attention to systems initially and finally static, in which the physical significance of m as mass is clear and unambiguous. Thus a dynamic period interposed between two static periods is bound to imply a loss of mass. We can ascribe this in the only physically reasonable way to the emission of waves by the system. Note from (45) that the physical components of the Riemann tensor have an r^{-1} term if and only if $c_{00} \neq 0$.[7]

Clear-cut and precise though our result is for initially and eventually static systems, it depends for its validity on the possibility of return to a static state. At first sight the conditions for such a return look rather forbidding. If the series expansion of equation (31) were taken beyond the terms given there and in equation (34), then it is easily seen that all the equations for the higher terms have the same general structure as equation (34). The derivative with respect to u of the coefficient of every term is determined by an expression involving the previous coefficients but not involving any differentiation with respect to u. Equations (35) and (36) must also be considered in this connexion. The problem is essentially this—suppose the system is static before $u = 0$, and then b is allowed to vary for a finite period of time at the end of which it becomes constant. Can we so

[7] The reason for the appearance of c_{00} in the Riemann tensor against c_0 in (58) will be discussed in § 5; very much the same situation occurs for electromagnetic waves.

choose the variation of b that at the end of this period of excitation we return to a situation that (but possibly for a tail) will eventually become static? Thus, when b stops varying, by equation (35) M stops varying. Equation (36) then implies that unless M_2 vanishes at this stage, N will go on varying linearly with time. The first condition on c is therefore that the final function M should be independent of θ. Next, consider the variation of N through the period of wave motion. Once again, the change in N must be such that when N ceases to vary, equation (34) implies that C ceases to vary. Continuing like this one obtains an infinity of conditions on c. To understand this we return to the last section of part A. There it was shown that if we are analyzing the wave equation by means of the first term, that is by the equivalent of c_0, then we first have to sort out the various angular dependences, and the part corresponding to P_n is then such that n conditions have to be applied to this part in order to ensure that the system goes from a static to a static situation. Owing to the non-linearity of our equations, the sorting out into the different types of angular dependence cannot be done in any exact form. What is, however, reasonably plausible is that when we deal with these different forms of angular dependence, the part involving P_n need only be pursued as far as the nth coefficient. For it does not matter if anything of this is left over for the higher coefficients. In accordance with the last equation of part A all this might do is to generate a tail to the wave.

To put it differently, if the ideas of the linear equation are applied to our case, then the structure of the equations of condition becomes reasonably clear. In part, these equations of condition make sure that we return from a static to a static case and do not embark on a non-radiative motion. For the rest, our equations may lead to the production of tails. These tails, it is true, will upset the convergence of our series but, once again following the first part, it appears that this difficulty need only arise when u is allowed to exceed $2r$. Therefore the method can be applied with reasonable confidence as long as we are dealing with a sandwich wave, that is, with c constant outside a period of finite length. If we want to investigate this situation, we have merely got to go sufficiently far out for r to exceed substantially one-half of the period of fluctuation. Thus the structure of our equations does not in this respect imply any essential difference from the scalar wave equation considered in part A.

2. Linearized Form of the Equations

For some purposes it is useful to consider the linear form of our equations. Since our co-ordinates have been chosen in order to simplify the non-linear problem, the relation of our metric to the usual linear ones requires elucidation. Assuming all our variables to be small and to satisfy the boundary conditions at infinity previously imposed, equation (22) shows that β is negligible, and equation (23) becomes

$$(r^4 U_1)_1 = -2r^2(\partial/\partial\theta + 2\cot\theta)\gamma_1, \tag{59}$$

showing immediately that power series expansion of U implies the vanishing of the r^{-2} term in γ. The integration of (24) turns out not to be required beyond the zero order approximation $V = r$, and then (25) becomes

$$2\left(\gamma_{01}+\frac{\gamma_0}{r}\right)-\gamma_{11}-\frac{2\gamma_1}{r}+\left(\frac{\partial}{\partial\theta}-\cot\theta\right)\left(\frac{1}{2}\frac{\partial}{\partial r}+\frac{1}{r}\right)U = 0. \tag{60}$$

Since all the terms in this equation tend to zero at infinity at least like r^{-2}, no information is lost by multiplying it by r and then differentiating with respect to z. Similarly multiplication by

$$\partial/\partial\theta + 2\cot\theta$$

does not involve any loss of information owing to the regularity conditions on the axis. It is now possible to substitute for γ from (59), and after a little simplification we obtain

$$2\left(\frac{\partial}{\partial r}+\frac{5}{r}\right)\left(\frac{\partial}{\partial r}+\frac{3}{r}\right)U_{01} = \left(\frac{\partial}{\partial r}+\frac{5}{r}\right)\left(\frac{\partial}{\partial r}+\frac{3}{r}\right)U_1+\frac{1}{r^2}$$

$$\times\left(\frac{\partial^2}{\partial\theta^2}+\cot\theta\frac{\partial}{\partial\theta}-\frac{1}{\sin^2\theta}\right)\left(\frac{\partial}{\partial r}+\frac{3}{r}\right)U_1. \tag{61}$$

Inspection of (61) shows that it does not serve to determine the r^{-2} part of U. Multiplying the equation by r^5, integrating with respect to r and dividing by r^5 we obtain

$$2\left(\frac{\partial}{\partial r}+\frac{3}{r}\right)U_{01} = \left(\frac{\partial}{\partial r}+\frac{3}{r}\right)^2 U_1+\frac{1}{r^2}\left(\frac{\partial^2}{\partial\theta^2}+\cot\theta\frac{\partial}{\partial\theta}-\frac{1}{\sin^2\theta}\right) \times U_1+\frac{l(u,\theta)}{r^5}, \tag{62}$$

where $l(u, \theta)$ is a function of integration which serves to make the r^{-2} part of U as indeterminate in (62) as it is in (61). If $\overline{U}(u, r, \theta)$ differs from U only by a term in r^{-2} and satisfies (62) with $l = 0$, put $r^2\overline{U}_1 = Q_2(u, r, \theta)$, substitute in (62), and integrate with respect to θ. Assuming the part of Q independent of θ (which is irrelevant for our purposes) to satisfy the same equation as the rest of Q this is immediately seen to be equivalent to equation (2). Accordingly Q satisfies the scalar wave equation

$$\Box Q = 0. \tag{63}$$

The procedure for constructing a linear approximation to solutions of our main equations is therefore to take an axially symmetric solution of (63) representing outgoing waves tending to zero at infinity. By forming

$$\overline{U} = -\int_r^\infty \frac{Q_2}{r^2}\,dr, \tag{64}$$

U is found except for the leading term and, by (59), γ is determined, again except for the leading term. Thus all the higher coefficients have been found, and N is known as the coefficient of r^{-3} in U, but c and M remain to be determined from the supplementary conditions which, in linearized form, are

$$-3N_0 = M_2, \tag{65}$$

$$M_0 = \tfrac{1}{2}(c_{22}+3c_2\cot\theta-2c)_0. \tag{66}$$

Since Q can be expanded as a series of $P_n(\cos\theta)$, (64) shows that U

(and with it F) is a series in $P_n^{(1)}$ and, by (59), $\gamma - r^{-1}c$ correspondingly a series in $P_n^{(2)}$, so that the boundary conditions (15) on the axis are automatically satisfied except of course for the as yet undetermined M and c. Next, (65) determines M but for a function $h(u)$ which may be defined as the coefficient of P_0 in the expansion of M, the factors of $P_n (n \geqslant 1)$ being found from (65). The left-hand side of (66) thus being known, the news function c_0 is readily obtained by a double integration of (66), which yields no admissible complementary functions. Moreover, it can be seen (e.g. by expansion of c_0 in $P_n^{(2)}$) that the c so obtained will not satisfy the boundary conditions on the axis unless $dh/du = 0$ and $d^2 l(u)/du^2 = 0$, where $l(u)$ is the coefficient of $P_n^{(1)}$ in N. Thus the mass must be constant in time and the dipole moment must vary linearly in time. Both these are direct consequences of the conservation laws in the linear approximation.

The time-independent part of c cannot be determined from Q and is clearly irrelevant in the linear approximation. Putting, as before

$$Q = \sum_{k=0}^{\infty} \frac{\overset{k}{L}(u, \theta)}{r^{k+1}}, \tag{67}$$

we find in detail

$$6N = -\overset{1}{L}_2, \tag{68}$$

$$-4M = \overset{0}{L}_{22} + \overset{0}{L}_2 \cot\theta + \text{const.}, \tag{69}$$

$$2(c_0 \sin^2\theta)_2 = -\sin^2\theta \overset{0}{L}_{02}, \tag{70}$$

$$6C_2 + 12C\cot\theta = -\overset{0}{L}_2. \tag{71}$$

3. Non-radiative Motions

Consider a system in which c_0 vanishes but M is not independent of θ, though, owing to the constancy of c, it is independent of u. A system in this state is clearly non-radiative both because of the vanishing of the leading terms of the curvature tensor and also because there is no loss of mass. On the other hand, it follows from equation

(36) and from the basic equations that the system is not at rest. This is, therefore, a case of non-radiative motion. How can this be interpreted?

One very simple case is clearly illustrated by the work in part C. If a K transformation is applied to the Schwarzschild metric ($\gamma = \beta = U = 0$, $V = r - 2\bar{m}$) we find

$$M = \frac{\bar{m}}{(\cosh v + \cos\theta \sinh v)^3}. \tag{72}$$

Thus a mass $\bar{m}$ moving along the axis of symmetry with constant velocity $\tanh v$ produces a field with M given by (72) and therefore depending on θ. Also note that $m = \bar{m} \cosh v$ contains the correct contribution for the kinetic energy.

This interesting phenomenon represents a kind of Doppler shift of the mass aspect. It is plausible to suggest that this is not confined to the simple case displayed in equation (72). If we imagine that our material system consists of several masses moving in various directions, then there would be some form of a superposition of these Doppler shifts, leading to M being a function with complicated angular dependence but independent of u. We therefore get the notion that there is a class of non-radiative motions in which different parts of the material system move with constant velocity in various directions. This case is easy to visualize when the different particles are sufficiently far away from each other for their own gravitational effects on each other to be negligible. Otherwise we know that their motions will not be uniform and that, in particular, oscillations are likely to occur. How would these show themselves in our equations? It follows from (72) applied to (36) that N is a linear function of u, that C is a quadratic function of u and so on. Of course this fits in perfectly with the picture of the moving mass. The dipole moment represented by N will increase linearly with time, the quadrupole moment represented by C will increase quadratically with time and so on. In the case of the moving single mass all these statements are strictly correct. In the more complicated cases where M, though independent of u, depends on angle in a more complex fashion, we will still have it that N increases linearly with u, C quadratically and so on. However, the more complicated angular dependence now implies that N no longer just represents the dipole moment and C no longer represents the quadrupole moment. If we could sort out the terms and put together all the terms representing the quadrupole moment and so on (which of course because of the non-linearity are not independent of each other) then each of them would be represented by a whole power series in u. The accelerations which are known to occur may well be expressed by such power series converging for all values of u, or convergence may cease for some finite value of u. In either case, we see that the future behaviour of a system in this class is entirely determined by the present, i.e. there is no news. This might suggest identification of such behaviour with purely gravitational motions. This identification is strongly supported by the work of Infeld (1960), who found that a system of particles in motion need not radiate. On this basis, we may tentatively identify non-radiative motions on the one hand with the class of solutions in which M depends on angle but not on time, and on the other with motion under purely gravitational forces. The relation between the non-radiative motions discussed here and those discussed in part A is clear.

4. Construction of Solutions

In this section we give a method for the construction of sandwich wave solutions. We shall suppose that the system is static except during an interval extending from $u = -1$ to $u = +1$. Thus outside this interval c_0 vanishes and within the interval it may be given in terms of the expansion

$$c_0 = \sum_{n=0}^{\infty} f_n(\mu)\, P_n(u), \quad -1 \leqslant u \leqslant 1, \quad \mu = \cos\theta. \tag{73}$$

In order to ensure continuity at the beginning and end of the interval, we must have

$$\sum_{n=0}^{\infty} f_{2n} = \sum_{n=0}^{\infty} f_{2n+1} = 0. \tag{74}$$

In order to fit the boundary conditions on the axis we have

$$f_n(1-\mu^2)^{-1} \quad \text{bounded as} \quad \mu \to \pm 1. \tag{75}$$

Now substitute in equation (35) and integrate throughout the interval of emission. We obtain for the change in M

$$[M] = -2 \sum_{n=0}^{\infty} \frac{f_n^2}{2n+1} + \frac{d^2}{d\mu^2}[(1-\mu^2]f_0]. \tag{76}$$

If we wish to go from a static to a static solution then the change in M must be independent of angle. It can readily be established that in order to allow for the divisibility of the first term on the right-hand side of (76) by $(1-\mu^2)^2$, as implied by (75), we must have

$$f_0 = -l(1-\mu^2)(3-\mu^2)+g(\mu), \quad l = \text{const.}, \quad g(\mu) \sim (1-\mu^2)^3. \tag{77}$$

Then by substitution

$$[M] = -16l, \tag{78}$$

$$4[g(1-\mu^2)^{-1}-l(3-\mu^2)]^2+4\sum_{u=1}^{\infty}\frac{1}{2n+1}\left(\frac{f_n}{1-\mu^2}\right)^2$$
$$-60l - \frac{2}{(1-\mu^2)^2}\frac{d^2}{d\mu^2}[(1-\mu^2)g(\mu)] = 0. \tag{79}$$

Equation (79) is a condition on the higher terms of the series, which together with (74) and (75) can be satisfied provided l is positive, which by (78) implies a loss of mass.

A simple example may illustrate this part of the method. Suppose

$$g = 0, \quad f_n = 0 \quad \text{unless} \quad n = 0, 2, 4, \quad f_0+f_2+f_4 = 0, \tag{80}$$

$$4(3-\mu^2)^2l^2-60l+\frac{4}{5}\left(\frac{f_2}{1-\mu^2}\right)^2+\frac{4}{9}\left[\frac{f_2}{1-\mu^2}+l(3-\mu^2)\right]^2 = 0, \tag{81}$$

$$f_2 = \tfrac{5}{14}l\{3-\mu^2 \pm 3^{\frac{3}{2}}[14l^{-1}-(3-\mu^2)^2]^{\frac{1}{2}}\}(1-\mu^2), \tag{82}$$

which is real provided $0 \leqslant l \leqslant \frac{14}{9}$.

The method may be continued to evaluate the changes in N, C and so on. With sufficient terms in the series we can make N_0, C_0, etc., all vanish initially and finally. Though this method of ensuring change from a static system to a static system is laborious, it is very much like the method applicable to the scalar wave equation when this is treated as in part A. Also the degree of freedom left appears to be the same. Accordingly, it seems likely that the change from radiative to static system is not only possible, but immediate, i.e. that Huygens's principle applies to gravitational waves. It may be worth mentioning that the u derivative of each coefficient in the expansion of γ is given by an expression analogous to (34) in which the only coefficient to enter linearly is the immediate predecessor which enters through the associated Legendre operator for $P_n^{(2)}$. There is little doubt that further development of this method would lead to an improved understanding of the equations.

5. The Reception of Gravitational Waves

In order to clarify the energy concept for gravitational waves the problem of their reception is now considered. The simplest receiver to discuss is freely falling, for otherwise one would not know whether some of the energy obtained from the field was not derived from the framework holding the receiver. In order that the gravitational terms entering should be easy to express, the receiver should be small and it should be as simple as possible. One wants, therefore, a device that can make use of the curvature tensor, that is, of the relative acceleration of neighbouring particles. Our ideal receiver of the simplest type is then a quadrupole receiver. It consists of two massive particles which are arranged with a motor between them, such that their distance from each other can be varied at will. Then the machinery in the receiver will absorb energy whenever the motion of the particles is such that the relative gravitational force, as given by the curvature tensor, does work in the motion in question. On the other hand, the receiver loses energy because, being a quadrupole of variable moment, it will itself radiate gravitational waves and so energy to space. The crucial question then is of how the particles should be moved in order to maximize the gain of energy, that is, the difference between energy received and energy re-radiated. In this respect the gravitational receiver is completely analogous to the electromagnetic receiver studied by Bondi (1961). Let the mass of each of the particles be M and let the distance of each from the common centre of mass be x. Then the quadrupole moment Q is given by

$$Q = 2Mx^2. \tag{83}$$

We now suppose the receiver to fall freely, moving so that its time axis coincides with the T axis of the tetrad used at the end of part B. With the quadrupole lined up along the P axis of the tetrad, the gravitational acceleration of each particle with respect to the mid-point is the product $xR_{(TTPP)}$, and so we can regard the particles as being acted upon by a Newtonian force

$$\pm Mc_{00}x/r. \tag{84}$$

Multiplying this by the velocity of the particles with respect to the mid-point and adding for the two particles, we obtain for the rate of doing work of the field on the receiver

$$\frac{2Mc_{00}x\dot{x}}{r} = \frac{1}{2}\frac{c_{00}}{r}Q_0. \tag{85}$$

Next we have to consider the re-radiation from the receiver. For this purpose the receiver has to be considered as a transmitter and all our previous work can be applied. To avoid confusion, the corresponding symbols will now be barred. Here, however, we have to proceed with caution, making a certain number of approximations in order to obtain manageable expressions and in order to be able to apply our previous work. In as far as this work was exact it never identified the precise nature of the source. The quadrupole moment was not identified in the moving but only in the static case. We shall now suppose that we can regard our receiver as quasistatic, so that the identification of the quadrupole moment contained in equation (42) can be applied and, moreover, we shall suppose the entire radiation from the receiver to be so small that, until we come to the final stages, we may work in the linear approximation. Then, from equation (42), we have

$$\bar{C} = \tfrac{1}{2}Q(\bar{u}) \sin^2 \bar{\theta}. \tag{86}$$

From (34)

$$(87)\qquad 4\bar{C}_0 = \bar{N}\cot\bar{\theta} - \bar{N}_2 \quad \text{so that } \bar{N} = 8Q_0 \sin\bar{\theta}\cos\bar{\theta}.$$

Then (36) gives

$$(88)\qquad \bar{M}_2 = -3\bar{N}_0. \quad \text{Thus } \bar{M} = -3Q_{00}\sin^2\bar{\theta} + p(\bar{u}).$$

Substitute into (35) and obtain

$$(89)\qquad \bar{c} = \tfrac{1}{2}Q_{00}\sin^2\bar{\theta}.$$

Still using the quasi-static approach and combining (35) and (57) we have

$$-4\bar{m}_0 = -2\int_0^\pi \bar{M}_0 \sin\bar{\theta}\, d\bar{\theta} = 2\int_0^\pi \bar{c}_0^2 \sin\bar{\theta}\, d\bar{\theta} = \tfrac{1}{2}Q_{000}^2\int_0^\pi \sin^5\bar{\theta}\, d\bar{\theta}$$

$$(90)\qquad = \tfrac{8}{15}Q_{000}^2.$$

Thus the rate of radiation of energy is

$$(91)\qquad -\bar{m}_0 = \tfrac{2}{15}Q_{000}^2.$$

This is a well-known result.

The limitations of our approach are evident. We have supposed a linear superposition of incident radiation and re-radiation and the whole treatment of the re-radiation has been distinctly crude. Nevertheless, it seems unlikely that when the incident wave is weak there should be any major mistake in this calculation. We arrive, therefore, at the answer that the total amount of energy received in the interval of reception is given by

$$(92)\qquad \int du\left[\frac{1}{2}\,\frac{c_{00}}{r}\,Q_0 - \frac{2}{15}\,Q_{000}^2\right].$$

This expression is completely equivalent to the corresponding expression for the reception of electromagnetic waves (equation (3) of Bondi, 1961) and the entire analysis given in that paper can be applied here. The method in brief is to use the usual approach of the calculus of variations to find that variation of quadrupole moment Q with time that will maximize expression (92). The result is easily obtained and shows that if initially c was constant then the maximum possible rate of absorption of energy is given by

$$\frac{15}{16}\frac{(c-c_{\text{initial}})^2}{r^2}. \tag{93}$$

Comparing this with the electromagnetic case, where we suppose both receiving and transmitting aerials to be magnetic dipoles, one sees that this expression is identical (apart from a numerical factor) with the electromagnetic one, provided c is replaced by the time derivative of the current in the transmitter coil. We are now faced with precisely the same problem as arises in electromagnetism in the case, there very unusual, in which after the period of transmission the time derivative of the transmitting current does not equal its value before the period of transmission. It will be recalled that in that case no unambiguous treatment of energy reception can be carried out unless the near field (induction) is considered as well as wave field. However, in the gravitational case this is not the unusual but the only possible situation. For we cannot have c returning to its initial value after the period of transmission, except, possibly, in a few isolated directions. This can immediately be verified by referring to equation (35) and considering its significance along the axis. Since c vanishes on the axis, the first term on the right-hand side vanishes there. Accordingly a change in M on the axis is entirely due to a change in c there. Since M, in going from a static to a static situation, must change independently of angle, as was previously shown, it follows that c, at least near the axis, must differ from its previous value. This result can also be established in a more comprehensive way by looking at equation (73). The change in c is given by $2f_0$. Equation (77) shows that in no circumstances can f_0 vanish, which establishes the same result. Thus, as far as energy is concerned we must be very careful in the consideration of reception of energy from a wave. Energy can only be taken from the whole field, including the non-wave parts. Serious though this consideration is, one must remember that the change in c following upon the change in M is of the second order. Therefore the rate of energy reception can be approximated to by expression (93), particularly for an oscillatory type of wave. One must, however, be careful not to take this too far for otherwise the receiver will appear to be able to continue to absorb energy *ad infinitum* after the wave has passed, that is, after c has taken on a fixed value different from its initial value. Looked at in a different way, expression (93) underlines what has previously been said about the desirability of considering only sandwich waves. The curvature tensor is proportional to c_{00}, the rate of loss of mass to c_0^2, and the rate of energy absorption to the change in c itself. If one did not stick to sandwich waves, then one might be faced with the absurd situation of c varying linearly with time. This would have the effect that there was no wave term in the curvature tensor but that, nevertheless, there was a constant loss of mass and a constant ability to adsorb energy. But all these difficulties arise only from the consideration of rather unnatural situations. As long as one is determined to work only with c constant initially and finally, that is with a sandwich wave, the internal connexions between the change in c, its first and its second derivative are enough to ensure that one obtains only sensible results.

A problem that is as little solved in the general case in gravitational theory as in electromagnetic theory is the maximum permissible proximity of receivers. In order to reconcile equation (93) with an energy loss proportional to c_0^2 requires one to make a statement about receivers, in order that they may not interfere with each other, having to be a distance of the order of the wavelength apart. Artificial as the restriction to harmonic waves and perfectly definite wavelength is in the electromagnetic case, there the linearity enables one to get over the worst consequences. In the gravitational case no such escape is in sight and the question of the maximum permissible number of receivers remains in a rather unsatisfactory state. One can, however, proceed rather differently to show that the energy transmitted can indeed eventually be absorbed. Suppose the transmitting region is enclosed by a large empty region, which, in turn, is bounded by a material shell of matter beyond which again space is empty. Within the outer shell of matter the situation is exactly as described by our equations, provided we suppose the shell not to send out any radiation inwards. Suppose now, moreover, that outside the outer shell we have a spherically symmetrical solution, that is, a Schwarzschild solution with a necessarily fixed value of the mass. Then it will be possible to conceive of a transition from the metric inside the shell to the metric outside the shell involving certain pressures and densities and stresses in the shell. If the mass of the exterior solution is made large enough one can always ensure that the density within the shell is everywhere positive and large compared with the stresses, that is, that one has a physical situation. We can hence speak of the outermost solution as defining the mass of the entire system of central transmitter together with shell, and this is constant in time. If then we have an initially static situation in the interior and equally an eventually static situation which means a final mass of the transmitter necessarily less than its initial mass, then the mass of the shell must increase by exactly the amount by which the mass of the transmitter has diminished. Hence such a shell constitutes a perfectly matched absorbing receiver.

6. The Classification of Time Variation

The work of this paper allows one to discriminate between various types of time variation of empty space fields surrounding isolated material system. Although only axially symmetric cases have been considered here, the generalization of the work of this paper to arbitrary systems by Sachs (1962) enables one to apply this classification without restriction.

(i) *Radiative Class*

This is characterized by the existence of news, and the non-vanishing of the news function c_0 defines the class. A mass loss necessarily occurs, and in general the physical components of the Riemann tensor $\sim r^{-1}$. An exception occurs only if $c_{00} = 0$ although $c_0 \neq 0$. Though this seems to be a case of little physical significance one should perhaps put into *subclass* (i*): *mass loss without radiative Riemann tensor*.

(ii) *Time-dependent Systems Without News*

This occurs whenever some terms in the field are time dependent but there is no news ($c_0 = 0$) and accordingly no mass loss ($M_0 = 0$). The physical interpretation of this class is perhaps the biggest outstanding problem in the subject. What distinguishes locally those source motions that do not give rise to a radiation field from those that do? The methods of this paper do not lend themselves to answering this question but perhaps the alternatives might be stated.

It may be that there is no locally significant way of distinguishing between these types of motion, unsatisfactory as this would be. If there is such a locally significant distinction it seems likely that it will be related to whether one is dealing with free gravitational motion. The lack of radiation for freely falling particles emerges from Infeld's work, but one would like to generalize this to non-singular equations of state. The most clear-cut case then would seem to be pressure-free dust ($T^{\mu\nu} = \varrho v^\mu v^\nu$), but beyond this it is tempting to suggest that perfectly elastic equations of state do not lead to radiation. Pursuing this line of thought one is driven to the following conclusion:

If the distinction between radiative and non-radiative motions is locally significant then the clearest self-consistent distinction appears to be between cases where the equations of state do not involve the time explicitly and are time reversible (no dissipation), and others. A system of the second kind clearly contains news, for either the time enters explicitly into the equation of state (time bomb), or, through the action of dissipation, the system continually reaches *new* states in which its behaviour is not a consequence of its previous behaviour ("fatigue"). A system of the first kind does not contain news in this sense. Its future is a clear consequence of its past, and it would seem difficult to draw a distinguishing line between different systems of this kind though conceivably the pressure-free might be only nonradiative material, all others radiating if in motion.

The distant field of time-dependent systems without news could be divided into two subclasses:

(*a*) $M_2 \neq 0$ (natural non-radiative moving system),
(*b*) $M_2 = 0$ (non-natural non-radiative moving system).

The presence of moving masses in (ii*a*) is a clear consequence of the case discussed in § 3, but the time dependence may only enter the coefficient through N or C or later in the series, corresponding to (ii*b*). The necessarily very peculiar cancellation of mass motion terms leads to the name suggested.

(iii) *Stationary Systems*

Time-independent metrics not reflexion symmetric and

(iv) *Static Systems*

are well known.

The author is deeply indebted to many colleagues for help in understanding and presenting the subject of this paper, particularly to Dr F. A. E. Pirani and to Dr R. K. Sachs, who has recently been able to extend this theory to the general case without any symmetries in a paper now in the press.

References

BONDI, H. 1957 *Nature, Lond.* **179,** 1072.
BONDI, H. 1960 *Nature, Lond.* **186,** 535.
BONDI, H. 1961 *Proc. Roy. Soc.* A, **261,** 1.
FRIEDLANDER, F. G. 1962 *Proc. Roy. Soc.* A, **269,** 53.
HOGARTH, J. E. 1952 London University Ph. D. thesis.
HOGARTH, J. E. 1961 *Proc. Roy. Soc.* A, **267,** 365.
INFELD, L. 1960 Cf. L. Infeld and Plebanski, *Motion and Relativity*. London: Pergamon Press.
SACHS, R. K. 1962 *Proc. Roy. Soc.* A (in the press).
WHEELER, J. A. and FEYNMANN, R. P. 1949 *Rev. Mod. Phys.* **21,** 425.

Appendix

1. *List of 3-Index Symbols*

$$\Gamma^0_{00} = 2\beta_0+\frac{V}{2r^2}-\frac{V_1}{2r}-\frac{\beta_1 V}{r}+r^2 e^{2(\beta-\gamma)}U\left(U_1+\frac{U}{r}+\gamma_1 U\right)$$

$$= -\frac{M+cc_0}{r^2}+\ \dots,$$

$$\Gamma^0_{01} = \Gamma^0_{11} = \Gamma^0_{12} = 0,$$

$$\Gamma^0_{02} = \beta_2 - r^2 e^{2(\gamma-\beta)}\left(\frac{1}{2}U_1+\frac{U}{r}+\gamma_1 U\right) = \frac{N}{r^2}+\ \dots,$$

$$\Gamma^0_{22} = re^{2(\gamma-\beta)}(1+r\gamma_1) = r+c+\frac{1}{2}\frac{c^2}{r}+\ \dots,$$

$$\Gamma^0_{33} = re^{-2(\gamma+\beta)}\sin^2\theta(1-r\gamma_1) = \sin^2\theta\left(r-c+\frac{1}{2}\frac{c^2}{r}+\ \dots\right),$$

$$\Gamma^1_{00} = \frac{V_0}{2r}-\frac{\beta_0 V}{r}-\frac{V^2}{2r^3}+\frac{VV_1}{2r^2}+\frac{\beta_1 V^2}{r^2}-\frac{UV_2}{2r}-\frac{\beta_2 UV}{r}$$

$$+r^2 e^{2(\gamma-\beta)}\left[U^2\left(U_2+\gamma_2 U-\frac{V}{r^2}-\frac{\gamma_1 V}{r}+\gamma_0\right)-\frac{UU_1 V}{r}\right]$$

$$= -\frac{M_0}{r}+\ \dots,$$

$$\Gamma^1_{01} = \frac{V_1}{2r}-\frac{V}{2r^2}+\frac{\beta_1 V}{r}-\beta_2 U-\frac{1}{2}r^2 e^{2(\gamma-\beta)}UU_1 = \frac{M}{r^2}\ \dots\ +,$$

$$\Gamma^1_{11} = 2\beta_1 = \frac{c^2}{r^3}+\ \dots,$$

$$\Gamma^1_{02} = \frac{V_2}{2r}+r^2 e^{2(\gamma-\beta)}\left[U\left(\frac{V}{r^2}+\frac{\gamma_1 V}{r}-\gamma_0-U_2-\gamma_2 U\right)+\frac{U_1 V}{2r}\right]$$

$$= \frac{-M_2+c_0(c_2+2c\cot\theta)}{r}+\ \dots,$$

$$\Gamma^1_{12} = \beta_2+\frac{1}{2}r^2 e^{2(\gamma-\beta)}U_1 = \frac{c_2+2c\cot\theta}{r}-\frac{6N+5cc_2+8c^2\cot\theta}{2r^2}+\dots,$$

$$\Gamma^1_{22} = r^2 e^{2(\gamma-\beta)}\left(\gamma_0 + U_2 + \gamma_2 U - \frac{V}{r^2} - \frac{\gamma_1 V}{r}\right) = -r(1-c_0) + \ \ldots,$$

$$\Gamma^1_{33} = r^2 \sin^2\theta e^{-2(\gamma+\beta)}\left(-\gamma_0 + U\cot\theta - \gamma_2 U - \frac{V}{r^2} + \frac{\gamma_1 V}{r}\right)$$

$$= -r\sin^2\theta\,(1+c_0) + \ \ldots,$$

$$\Gamma^2_{00} = -U_0 + U\left(2\beta_0 - 2\gamma_0 - U_2 - \gamma_2 U + \frac{V}{2r^2} - \frac{V_1}{2r} - \frac{\beta_1 V}{r}\right) + e^{2(\beta-\gamma)}$$

$$\times \frac{V_2 + 2\beta_2 V}{2r^3} + r^2 e^{2(\gamma-\beta)} U^2\left(U_1 + \frac{U}{r} + \gamma_1 U\right)$$

$$= \frac{c_{02} + 2c_0\cot\theta}{r^2} - \frac{2N_0 + c_0 c_2 + 3cc_{02} + 4cc_0\cot\theta + M_2}{r^3},$$

$$\Gamma^2_{01} = -\frac{1}{2}U_1 - \frac{U}{r} - \gamma_1 U + \frac{\beta_2}{r^2} e^{2(\beta-\gamma)} = \frac{N}{r^4} + \ \ldots,$$

$$\Gamma^2_{11} = 0,$$

$$\Gamma^2_{02} = \gamma_0 + \beta_2 U - r^2 e^{2(\gamma-\beta)} U\left(\frac{1}{2}U_1 + \frac{U}{r} + \gamma_1 U\right) = \frac{c_0}{r} + O(r^{-3}),$$

$$\Gamma^2_{12} = \frac{1}{r} + \gamma_1 = \frac{1}{r} - \frac{c}{r^2} - \frac{3C}{r^4} + \ \ldots,$$

$$\Gamma^2_{22} = \gamma_2 + r^2 e^{2(\gamma-\beta)} U\left(\frac{1}{r} + \gamma_1\right) = -\frac{2c\cot\theta}{r} + \ \ldots,$$

$$\Gamma^2_{33} = r^2\sin^2\theta e^{-2(\gamma-\beta)} U\left(\frac{1}{r} - \gamma_1\right) - e^{-4\gamma}\sin^2\theta(\cot\theta - \gamma_2)$$

$$= -\sin\theta\cos\theta\left(1 - \frac{2c}{r}\right) + \ \ldots,$$

$$\Gamma^3_{03} = -\gamma_0 = -\frac{c_0}{r} - \frac{C_0}{r^3} + \ \ldots,$$

$$\Gamma^3_{13} = \frac{1}{r} - \gamma_1 = \frac{1}{r} + \frac{c}{r^2} + \frac{3C}{r^4} + \ \ldots,$$

$$\Gamma^3_{23} = \cot\theta - \gamma_2 = \cot\theta - \frac{c^2}{r} - \frac{C_2}{r^3} - \ \ldots$$

2. *The Supplementary Conditions*

$$R_{02} = \beta_{02}-\gamma_{02}+2\gamma_0\gamma_2-2\gamma_0\cot\theta-U(\beta_{22}+2\beta_2^2-2\beta_2\gamma_2+\beta_2\cot\theta)$$
$$-\frac{V_{12}}{2r}+\frac{V_2}{2r^2}+(\gamma_1-\beta_1)\frac{V_2}{r}+r^2e^{2(\gamma-\beta)}\left[\frac{3}{2}UU_{12}+\frac{3UU_2}{r}+2U\right.$$
$$\times\left(\gamma_{01}+\frac{\gamma_0}{r}\right)+\frac{1}{2}U_{01}+2\gamma_{12}U^2+(\gamma_0-\beta_0)U_1+\gamma_1UU_2+(2\gamma_2-\beta_2)$$
$$\times UU_1+U_1U_2-\frac{U_{11}V}{2r}-\frac{UV_1+2U_1V}{r^2}$$
$$-\frac{\gamma_{11}UV+(\gamma_1-\beta_1)U_1V+\gamma_1UV_1}{r}-\frac{\gamma_1UV}{r^2}+\frac{2\gamma_2U^2}{r}+U$$
$$\left.\times\left(\frac{1}{2}U_1+\frac{U}{r}+\gamma_1U\right)\cot\theta\right]-\frac{1}{2}r^4e^{4(\gamma-\beta)}UU_1^2,$$

$$R_{00} = \frac{2\beta_{01}V}{r}-\frac{VV_{11}}{2r^2}-\frac{\beta_{11}V^2}{r^2}-\frac{\beta_1V^2}{r^3}-\frac{\beta_1VV_1}{r^2}-\frac{V_0-2\beta_0V}{r^2}$$
$$+\frac{2\beta_{12}UV+\beta_2UV_1+\beta_1U_2V+2\beta_1UV_2}{r}+\frac{2\beta_2UV}{r^2}-\frac{U_2V}{2r^2}$$
$$+\frac{U_2V_1}{2r}-\frac{2UV_2}{r^2}-\frac{U_1V_2}{2r}$$
$$-\frac{2\gamma_1UV_2}{r}-2\beta_{02}U-2\beta_0U_2+2\gamma_{02}U+2\gamma_0U_2+U_{02}+UU_{22}+U_2^2$$
$$+2(\gamma_2-\beta_2)UU_2+\frac{UV_{12}}{r}+(2\beta_2^2-2\beta_2\gamma_2+\gamma_{22})U^2+2\gamma_0^2$$
$$-\cot\theta\left(2\beta_0U-2\gamma_0U-U_0-UU_2-\gamma_2U^2+\frac{UV}{2r^2}-\frac{UV_1}{2r}-\frac{\beta_1UV}{r}\right)$$
$$+r^2e^{2(\gamma-\beta)}\left[-UU_{01}-2\left(\gamma_{01}+\frac{\gamma_0}{r}\right)U^2-2(\gamma_0-\beta_0)UU_1-2U^2U_{12}\right.$$
$$-2UU_1U_2-2\gamma_{12}U^3-\frac{2}{r}\gamma_2U^3-3\gamma_2U^2U_1+2\beta_2U^2U_1$$
$$+\frac{UU_{11}V}{r}+\frac{4UU_1V}{r^2}+2(\gamma_1-\beta_1)\frac{UU_1V}{r}+\frac{\gamma_{11}U^2V}{r}+\frac{\gamma_1U^2V_1}{r}$$

$$+\frac{\gamma_1 U^2 V}{r^2}-\frac{3U^2U_2}{r}-\gamma_1 U^2 U_2+\frac{U^2V_1}{r^2}+\frac{U_1^2 V}{2r}$$

$$-U^2\left(U_1+\frac{U}{r}+\gamma_1 U\right)\cot\theta\Big]+\frac{1}{2}r^4 e^{4(\gamma-\beta)}U^2U_1^2$$

$$-\frac{1}{2r^3}e^{2(\beta-\gamma)}[V_{22}+2\beta_{22}V+(2\beta_2-2\gamma_2+\cot\theta)(V_2+2\beta_2 V)].$$

3. *Transformation Formulae* (*Part C*)

$$K(\bar\theta)=\cosh\nu+\sinh\nu\cos\bar\theta,\quad \alpha=\alpha(\bar\theta),$$

$$u=\left(\frac{\bar u}{K}+\alpha\right)-\frac{K}{2\bar r}\left(\frac{K'}{K^2}\bar u-\alpha'\right)^2+O(\bar r^{-2}).$$

$$r=K\bar r+\frac{1}{K}\left[\bar u(1-K\cosh\nu)+\frac{1}{2}K^2(K''\alpha'+K\alpha'')\right]+O(\bar r^{-1}),$$

$$\theta=2\tan^{-1}\left(e^{-\nu}\tan\frac{1}{2}\bar\theta\right)+\frac{1}{\bar r}\left[\frac{K'}{K^2}\bar u-\alpha'\right]+O(\bar r^{-1}),$$

$$\bar c=\frac{c}{K}+\frac{1}{2}K\left[\alpha'\cot\bar\theta-\alpha''-2\alpha'\frac{K'}{K}\right],$$

$$\bar c_0=\frac{c_0}{K^2},\qquad \bar c_{00}=\frac{c_{00}}{K^3},$$

$$\bar M=\frac{M}{K^3}-\frac{c_0}{K^2}\left[2\bar u\,\frac{1-K\cosh\nu}{K^2}-\frac{1}{2}K\alpha''+\alpha'K'-\frac{3}{2}K\alpha'\cot\bar\theta\right]$$

$$-\frac{1}{K^2}c_{02}\left(\frac{K'}{K}\bar u-\alpha'\right)+\frac{1}{2K}c_{00}\left(\frac{K'}{K^2}\bar u-\alpha'\right)^2.$$

$$\bar N=\frac{N}{K^3}+\frac{1}{K^2}(M+cc_0)\left(\frac{K'}{K^2}\bar u-\alpha'\right)-\frac{1}{2K}[2c_0(K\cot\bar\theta-K')+c_{02}]$$

$$\times\left(\frac{K'}{K}\bar u-\alpha'\right)^2+\frac{1}{6}c_{00}\left(\frac{K'}{K^2}\bar u-\alpha'\right)^3.$$

$$\bar C=\frac{C}{K^3}+\frac{N}{K^2}\left(\frac{K'}{K^2}\bar u-\alpha'\right)+\frac{M+cc_0}{2K}\left(\frac{K'}{K^2}\bar u-\alpha'\right)^2$$

$$-\frac{1}{6}[2c_0(K\cot\bar\theta-K')+c_{02}]\left(\frac{K'}{K^2}\bar u-\alpha'\right)^2+\frac{1}{24}c_{00}K\left(\frac{K'}{K^2}\bar u-\alpha'\right)^4.$$

4. *Transformation of the Weyl Metric*

Write the metric (37) in the form

$$ds^2 = e^{2\psi}\,dt^2 - e^{2\sigma-2\psi}(dR^2 + R^2\,d\Theta^2) - e^{-2\psi}R^2\sin^2\Theta\,d\phi^2$$

and put

$$t = u + f(R, \theta), \quad \Theta = \Theta(R, \theta),$$

in order to make g_{11} and g_{12} vanish. Hence

$$e^{2\psi}f_R^2 = e^{2\sigma-2\psi}[1 + R^2\Theta_R^2]; \qquad e^{2\psi}f_R f_\theta = e^{2\sigma-2\psi}R^2\Theta_R\Theta_\theta.$$

Eliminate f: $\quad R^2\Theta_R(e^{2\sigma-4\psi})_\theta = \Theta_\theta\left[e^{2\sigma-4\psi}\dfrac{R^4\Theta_R^2}{1+R^2\Theta_R^2}\right]_R.$

Let

$$e^{2\sigma-4\psi} = 1 + \frac{4m}{R} + \frac{p(\Theta)}{R^2} + \frac{q(\Theta)}{R^3} + \ldots .$$

The case $c = 0$ corresponds to $\lim\limits_{R\to\infty} \Theta = 1$.

Then

$$\Theta = \theta + \frac{p'}{4R^2} + \frac{q' - 6mp'}{12R^3} + \ldots,$$

$$f = R + 2m\log R - \frac{\frac{1}{2}p - 2m^2}{R} - \frac{\frac{1}{2}q - mp + 4m^3}{2R^2} + \ldots .$$

Introduce r in the usual way (13)

$$r^4\sin^2\theta = e^{-2\psi}R^2\sin^2\Theta[e^{2\sigma-2\psi}R^2\Theta_\theta^2 - e^{2\psi}f_\theta^2].$$

Solve for R

$$R = r - m - \tfrac{1}{2}[\tfrac{1}{4}p'\cot\theta + \tfrac{1}{2}p - 3m^2 + \tfrac{1}{4}p'']r^{-1}$$
$$- \tfrac{1}{2}[\tfrac{1}{12}q'' + \tfrac{1}{12}q'\cot\theta + \tfrac{1}{2}q - mp + 4m^3]r^{-2} + \ldots .$$

We now express p and q by (39) and (40) in terms of the dipole moment D and the quadrupole moment Q.

$$p = 4D\cos\Theta + m^2(7 + \cos^2\Theta),$$
$$q = 2Q(3\cos^2\Theta - 1) + 4mD(3 + \cos^2\Theta) + 6m^3(1 + \cos^2\Theta).$$

Hence

$$R = r-m-\tfrac{1}{2}m^2(1-\cos^2\theta)r^{-1}-\tfrac{1}{2}m\sin^2\theta[2D\cos\theta+m^2]\,r^{-2}+\ \ldots .$$

Finally
$$\frac{V}{r}e^{2\psi}f_R R_r = e^{2\psi}+e^{2\psi}\frac{(f_R R_\theta+f_\theta)^2}{R^2e^{2\sigma-4\psi}\Theta_\theta^1-f_\theta^2}\,.$$

Substitute and obtain

$$\frac{V}{r} = 1-\frac{2m}{r}-\frac{2D\cos\theta}{r^2}-\frac{Q(3\cos^2\theta-1)}{r^3}+\ \ldots .$$

Thus from (33) $M = m,\quad N = D\sin\theta,\quad C = \tfrac{1}{2}Q\sin^2\theta.$

With the transformation equations of Appendix 3 this gives equations (42).

NOTES ON EXTRACT 9

LIKE Extract 7, this paper has had great influence on the development of the general theory in recent years, but (probably because of the war) the lead it gave was not taken up for a much longer time.

EXTRACT 9†

On Continued Gravitational Contraction

J. R. OPPENHEIMER AND H. SNYDER

University of California, Berkeley, California
[Received July 10, 1939]

When all thermonuclear sources of energy are exhausted a sufficiently heavy star will collapse. Unless fission due to rotation, the radia'ion of mass, or the blowing off of mass by radiation, reduce the star's mass to the order of that of the sun, this contraction will continue indefinitely. In the present paper we study the solutions of the gravitational field equations which describe this process. In I, general and qualitative arguments are given on the behavior of the metrical tensor as the contraction progresses: the radius of the star approaches asymptotically its gravitational radius; light from the surface of the star is progressively reddened, and can escape over a progressively narrower range of angles. In II, an analytic solution of the field equations confirming these general arguments is obtained for the case that the pressure within the star can be neglected. The total time of collapse for an observer comoving with the stellar matter is finite, and for this idealized case and typical stellar masses, of the order of a day; an external observer sees the star asymptotically shrinking to its gravitational radius.

I

Recently it has been shown[1] that the general relativistic field equations do not possess any static solutions for a spherical distribution of cold neutrons if the total mass of the neutrons is greater than $\sim 0{\cdot}7\odot$. It seems of interest to investigate the behavior of nonstatic solutions of the field equations.

In this work we will be concerned with stars which have large masses, $> 0{\cdot}7\odot$, and which have used up their nuclear sources of

† *Phys. Rev.* **56,** 455 (1939).

[1] J. R. Oppenheimer and G. M. Volkoff, *Phys. Rev.* **55,** 374 (1939).

energy. A star under these circumstances would collapse under the influence of its gravitational field and release energy. This energy could be divided into four parts: (1) kinetic energy of motion of the particles in the star, (2) radiation, (3) potential and kinetic energy of the outer layers of the star which could be blown away by the radiation, (4) rotational energy which could divide the star into two or more parts. If the mass of the original star were sufficiently small, or if enough of the star could be blown from the surface by radiation, or lost directly in radiation, or if the angular momentum of the star were great enough to split it into small fragments, then the remaining matter could form a stable static distribution, a white dwarf star. We consider the case where this cannot happen.

If then, for the late stages of contraction, we can neglect the gravitational effect of any escaping radiation or matter, and may still neglect the deviations from spherical symmetry produced by rotation, the line element outside the boundary r_b of the stellar matter must take the form

(1) $$ds^2 = e^{\nu}\, dt^2 - e^{\lambda}\, dr^2 - r^2(d\theta + \sin^2\theta\, d\varphi^2)$$

with $$e^{\nu} = (1 - r_0/r)$$

and $$e^{\lambda} = (1 - r_0/r)^{-1}.$$

Here r_0 is the gravitational radius, connected with the gravitational mass m of the star by $r_0 = 2mg/c^2$, and constant. We should now expect that since the pressure of the stellar matter is insufficient to support it against its own gravitational attraction, the star will contract, and its boundary r_b will necessarily approach the gravitational radius r_0. Near the surface of the star, where the pressure must in any case be low, we should expect to have a local observer see matter falling inward with a velocity very close to that of light; to a distant observer this motion will be slowed up by a factor $(1 - r_0/r_b)$. All energy emitted outward from the surface of the star will be reduced very much in escaping, by the Doppler effect from the receding source, by the large gravitational red-shift, $(1 - r_0/r)^{\frac{1}{2}}$, and by the gravitational deflection of light which will prevent the escape of radiation except through a cone about the outward normal of progressively shrinking aperture as the star contracts. The star thus tends to close itself off from any communication with a distant observer; only its gravitational field persists. We shall see later that although it takes, from the point of view of a distant observer, an infinite time for this asymptotic isolation to be established, for an observer comoving with the stellar matter this time is finite and may be quite short.

Inside the star we shall still suppose that the matter is spherically distributed. We may then take the line element in the form (1). For this line element the field equations are

$$-8\pi T_1^1 = e^{-\lambda}(\nu'/r+1/r^2)-1/r^2, \tag{2}$$

$$8\pi T_4^4 = e^{-\lambda}(\lambda'/r-1/r^2)+1/r^2, \tag{3}$$

$$\begin{aligned} -8\pi T_2^2 &= -8\pi T_3^3 \\ &= e^{-\lambda}\left(\frac{\nu''}{2}+\frac{\nu'^2}{4}-\frac{\nu'\lambda'}{4}+\frac{\nu'-\lambda'}{2r}\right) \\ &\quad -e^{-\nu}(\ddot{\lambda}/2+\dot{\lambda}^2/4-\dot{\lambda}\dot{\nu}/4), \end{aligned} \tag{4}$$

$$8\pi T_4^1 = -8\pi e^{\nu-\lambda}T_1^4 = -e^{-\lambda}\dot{\lambda}/r; \tag{5}$$

in which primes represent differentiation with respect to r and dots differentiation with respect to t.

The energy-momentum tensor T_ν^μ is composed of two parts: (1) a material part due to electrons, protons, neutrons and other nuclei, (2) radiation. The material part may be thought of as that of a fluid which is moving in a radial direction, and which in moving coordinates would have a definite relation between the pressure, density, and temperature. The radiation may be considered to be in equilibrium with the matter at this temperature, except for a flow of radiation due to a temperature gradient.

We have been unable to integrate these equations except when we place the pressure equal to zero. However, one can obtain some information about the solutions from inequalities implied by the differential equations and from conditions for regularity of the solutions.

From Eqs. (2) and (3) one can see that unless λ vanishes at least as rapidly as r^2 when $r \to 0$, T_4^4 will become singular and that either or both T_1^1 and ν' will become singular. Physically such a singularity would mean that the expression used for the energy-momentum tensor does not take account of some essential physical fact which would really smooth the singularity out. Further, a star in its early stage of development would not possess a singular density or pressure; it is impossible for a singularity to develop in a finite time.

If, therefore, $\lambda(r = 0) = 0$, we can express λ in terms of T_4^4, for, integrating Eq. (3)

$$\lambda = -\ln\left\{1 - \frac{8\pi}{r}\int_0^r T_4^4 r^2\, dr\right\}. \tag{6}$$

Therefore $\lambda \geqslant 0$ for all r since $T_4^4 \geqslant 0$.

Now that we know $\lambda \geqslant 0$, it is easy to obtain some information about ν' from Eq. (2);

$$\nu' \geqslant 0, \tag{7}$$

since λ and $-T_1^1$ are equal to or greater than zero.

If we use clock time at $r = \infty$, we may take $\nu(r = \infty) = 0$. From this boundary condition and Eq. (7) we deduce

$$\nu \leqslant 0. \tag{8}$$

The condition that space be flat for large r is $\lambda(r = \infty) = 0$. Adding Eqs. (2) and (3) we obtain:

$$8\pi(T_4^4 - T_1^1) = e^{\varkappa}(\lambda' + \nu')/r. \tag{9}$$

Since T_4^4 is greater than zero and T_1^1 is less than zero we conclude

$$\lambda' + \nu' \geqslant 0. \tag{10}$$

Because of the boundary conditions on λ and ν we have

$$\lambda + \nu \leqslant 0. \tag{11}$$

For those parts of the star which are collapsing, i.e., all parts of the star except those being blown away by the radiation, Eq. (5) tells us that λ is greater than zero. Since λ increases with time, it may (a) approach an asymptotic value uniformly as a function of r; or (b) increase indefinitely, although certainly not uniformly as a function of r, since $\lambda(r = 0) = 0$. If λ were to approach a limiting value the star would be approaching a stationary state. However, we are supposing that the relationships between the T_μ^ν do not admit any stationary solutions, and therefore exclude this possibility. Under case (b) we might expect that for any value of r greater than zero, λ will become greater than any preassigned value if t is sufficiently large. If this were so the volume of the star

$$V = 4\pi \int_0^{r_b} e^{\lambda/2} r^2 \, dr \tag{12}$$

would increase indefinitely with time; since the mass is constant, the mean density in the star would tend to zero. We shall see, however, that for all values of r except r_0, λ approaches a finite limiting value; only for $r = r_0'$ does it increase indefinitely.

II

To investigate this question we will solve the field equations with the limiting form of the energy-momentum tensor in which the pressure is zero. When the pressure vanishes there are no static solutions to the field equations except when all components of T_μ^ν vanish. With $p = 0$ we have the free gravitational collapse of the matter. We believe that the general features of the solution obtained this way give a valid indication even for the case that the pressure is not zero, provided that the mass is great enough to cause collapse.

For the solution of this problem, we have found it convenient to follow the earlier work of Tolman[2] and use another system of coordinates, which are comoving with the matter. After finding a solution, we will introduce a coordinate transformation to put the line element in form (1).

[2] R. C. Tolman, *Proc. Nat. Acad. Sci.* **20**, 3 (1934).

We take a line element of the form:

(13) $$ds^2 = d\tau^2 - e^{\bar{\omega}}\, dR^2 - e^{\omega}(d\theta^2 + \sin^2\theta\, d\varphi^2).$$

Because the coordinates are comoving with the matter and the pressure is zero,

(14) $$T_4^4 = \varrho$$

and all other components of the energy momentum tensor vanish.

The field equations are:

(15) $$8\pi T_1^1 = 0 = e^{-\omega} - e^{-\bar{\omega}}\frac{\omega'^2}{4} + \ddot{\omega} + \frac{3}{4}\dot{\omega}^2 = 0,$$

$$8\pi T_2^2 = 8\pi T_3^3 = 0 = -e^{-\bar{\omega}}\left(\frac{\omega''}{2} + \frac{\omega'^2}{4} - \frac{\bar{\omega}'\omega'}{4}\right)$$

(16) $$+\frac{\ddot{\bar{\omega}}}{2} + \frac{\dot{\bar{\omega}}}{4} + \frac{\ddot{\omega}}{2} + \frac{\dot{\omega}^2}{4} + \frac{\dot{\bar{\omega}}\dot{\omega}}{4},$$

(17) $$8\pi T_4^4 = 8\pi\varrho = e^{-\omega} - e^{-\bar{\omega}}\left(\omega'' + \frac{3}{4}\omega'^2 - \frac{\bar{\omega}'\omega'}{2}\right) + \frac{\omega^2}{4} + \frac{\dot{\bar{\omega}}\dot{\omega}}{2},$$

(18) $$8\pi e^{\bar{\omega}} e T_4^1 = -8\pi T_1^4 = 0 = \frac{\omega'\dot{\omega}}{2} - \frac{\dot{\bar{\omega}}\omega'}{2} + \dot{\omega}'$$

with primes and dots here and in the following representing differentiation with respect to R and v, respectively. The integral of Eq. (18) is given by Tolman:[3]

(19) $$e^{\bar{\omega}} = e^{\omega}\omega'^2/4f^2(R)$$

with $f^2(R)$ a positive but otherwise arbitrary function of R. We find a sufficiently wide class of solutions if we put $f^2(R) = 1$.

Substituting (19) in (15) with $f^2(R) = 1$ we obtain

(20) $$\ddot{\omega} + \tfrac{3}{4}\dot{\omega}^2 = 0.$$

[3] We wish to thank Professor R. C. Tolman and Mr. G. Omer for making this portion of the development available to us, and for helpful discussion.

The solution of this equation is:

$$e^{\omega} = (F\tau + G)^{4/3}, \tag{21}$$

in which F and G are arbitrary functions of R.

The substitution of (19) in (16) gives a result equivalent to (20). Therefore the solution of the field equations is (21).

For the density we obtain from (17), (19), and (21)

$$8\pi\varrho = 4/3(\tau + G/F)^{-1}(\tau + G'/F')^{-1}. \tag{22}$$

There is less real freedom in (21) than is apparent from the two arbitrary functions F and G; for taking R a function of a new variable R^* the differential equations (15), (17) and (18) will remain of the same form. We may therefore choose

$$G = R^{3/2}. \tag{23}$$

At a particular time, say τ equal zero, we may assign the density as a function of R. Eq. (22) then becomes a first-order differential equation for F,

$$FF' = 9\pi R^2 \varrho_0(R). \tag{24}$$

The solution of this equation contains only one arbitrary constant. We now see that the effect of setting $f^2(R)$ equal to one allows us to assign only a one-parameter family of functions for the initial values of $\dot{\varrho}_0$, whereas in general one should be able to assign the initial values of $\dot{\varrho}_0$ arbitrarily.

We now take, as a particular case of (24):

$$FF' = \begin{cases} \text{const.} \times R^2; & \text{const.} > 0; \quad R < R_b \\ 0 & ; \quad R > R_b. \end{cases} \tag{25}$$

A particular solution of this equation is:

$$F = \begin{cases} -\frac{3}{2} r_0^{\frac{1}{2}} (R/R)^{\frac{3}{2}}; & R < R_b \\ -\frac{3}{2} r_0^{\frac{1}{2}} & ; \quad R > R_b \end{cases} \tag{26}$$

in which the constant r_0 is introduced for convenience, and is the gravitational radius of the star.

We wish to find a coordinate transformation which will change the line element into form (1). It is clear, by comparison of (1) and (13), that we must take

$$e^{\omega/2} = (F\tau+G)^{\frac{2}{3}} = r. \tag{27}$$

A new variable t which is a function of τ and R must be introduced so that the $g_{\mu\nu}$ are of the same form as those in Eq. (1). Using the contravariant form of the metric tensor, we find that

$$g^{44} = e^{-\nu} = \dot{t}^2 - t'^2/r'^2 = \dot{t}^2(1-\dot{r}^2), \tag{28}$$

$$g^{11} = -e^{-\lambda} = -(1-\dot{r}^2), \tag{29}$$

$$g^{14} = 0 = \dot{t}\dot{r} - t'/r'. \tag{30}$$

Here (30) is a first-order partial differential equation for t. Using the values of r given by (27) and the values of F and G given by (26) and (23) we find:

$$t'/\dot{t} = \dot{r}r' = \begin{cases} -(r_0R)^{\frac{1}{2}}[R^{\frac{3}{2}} - \frac{3}{2}r_0^{\frac{1}{2}}\tau]^{-\frac{2}{3}} & ; \quad R > R_b \\ -r_0^{\frac{1}{2}}RR_b{}^{-\frac{3}{2}}[1-\frac{3}{2}r_0^{\frac{1}{2}}\tau R_b^{-\frac{3}{2}}]^{\frac{1}{3}}; & \quad R < R_b. \end{cases} \tag{31}$$

The general solution of (31) is:

$$t = L(x) \quad \text{for} \quad R > R_b, \quad \text{with} \quad x = \frac{2}{3r_0^{\frac{1}{2}}}(R^{\frac{3}{2}} - r^{\frac{3}{2}}) - 2(rr_0)^{\frac{1}{2}} + r_0 \ln \frac{r^{\frac{1}{2}}+r_0^{\frac{1}{2}}}{r^{\frac{1}{2}}-r_0^{\frac{1}{2}}}$$

$$t = M(y) \quad \text{for} \quad R < R_b, \quad \text{with} \quad y = \tfrac{1}{2}[(R/R_b)^2-1]+R_br/r_0R \tag{32}$$

where L and M are completely arbitrary functions of their arguments.

Outside the star, where R is greater than R_b, we wish the line element to be of the Schwarzschild form, since we are again neglecting the gravitational effect of any escaping radiation; thus

$$e^{\lambda} = (1-r_0/r)^{-1}, \tag{33}$$

$$e^{\nu} = (1-r_0/r). \tag{34}$$

This requirement fixes the form of L; from (28) we can show that we must take $L(x) = x$, or

$$t = x. \tag{35}$$

At the surface of the star, R equals R_b, we must have L equal to M for all τ. The form of M is determined by this condition to be:

$$t = M(y) = \tfrac{3}{2} r_0^{-\frac{1}{2}}\left(R_b^{\frac{3}{2}} - r_0^{\frac{3}{2}} y^{\frac{3}{2}}\right) - 2r_0 y^{\frac{1}{2}} + r_0 \ln \frac{y^{\frac{1}{2}}+1}{y^{\frac{1}{2}}-1}. \tag{36}$$

Eq. (36), together with (27), defines the transformation from R, τ to r and t, and implicitly, from (28) and (29), the metrical tensor.

We now wish to find the asymptotic behavior of e^{λ}, e^{ν}, and τ for large values of t. When t is large we obtain the approximate relation from Eqs. (36) and (27):

$$t \sim -r_0 \ln \left\{\tfrac{1}{2}[(R/R_b)^2 - 3] + R_b/r_0 \left(1 - 3 r_0^{\frac{1}{2}} \tau / 2R_b^{\frac{3}{2}}\right)^{\frac{2}{3}}\right\}. \tag{37}$$

From this relation we see that for a fixed value of R as t tends toward infinity, τ tends to a finite limit, which increases with R. After this time τ_0 an observer comoving with the matter would not be able to send a light signal from the star; the cone within which a signal can escape has closed entirely. For a star which has an initial density of one gram per cubic centimeter and a mass of 10^{33} grams this time τ_0 is about a day.

Substituting (27) and (37) into (28) and (29) we find

$$e^{-\lambda} \simeq 1 - (R/R_b)^2 \{e^{-t/r_0} + \tfrac{1}{2}[3 - (R/R_b)^2]\}^{-1}, \tag{38}$$

$$e^{\nu} \simeq e^{\lambda - 2t/r_0} \{e^{-t/r_0} + \tfrac{1}{2}[3 - (R/R_b)^2]\}. \tag{39}$$

For R less than R_b, e^{λ} tends to a finite limit as t tends to infinity. For R equal to R_b, e^{λ} tends to infinity like e^{t/r_0} as t approaches infinity. Where R is less than R_b, e^{ν} tends to zero like e^{-2t/r_0} and where R is equal to R_b, e^{ν} tends to zero like e^{-t/r_0}.

This quantitative account of the behavior of e^{λ} and e^{ν} can supplement the qualitative discussion given in I. For λ tends to a finite limit for $r < r_0$ as t approaches infinity, and for $r = r_0$ tends to infinity. Also for $r \leqslant r_0$, ν tends to minus infinity. We expect that this behavior will be realized by all collapsing stars which cannot end in a stable stationary state. Of course, actual stars would collapse more slowly than the example which we studied analytically because of the effect of the pressure of matter, of radiation, and of rotation.

NOTES ON EXTRACT 10

IF Extracts 7 and 9 provided new physical insights into the further developments of the theory, a new mathematical calculus to carry them out was provided by the present one. Although the idea of a spinor had been known for a long time, Penrose's method of using them is strikingly different in emphasis, and provides a great simplification.

EXTRACT 10†

A Spinor Approach to General Relativity

ROGER PENROSE

St. John's College, Cambridge, England‡

A calculus for general relativity is developed in which the basic role of tensors is taken over by spinors. The Riemann–Christoffel tensor is written in a spinor form according to a scheme of Witten. It is shown that the curvature of empty space can be uniquely characterized by a totally symmetric four-index spinor which satisfies a first order equation formally identical with one for a zero rest-mass particle of spin two. However, the derivatives used here are covariant, so that on iteration, instead of the usual wave equation, a nonlinear "source" term appears. The case when a source-free electromagnetic field is present is also considered. (No quantization is attempted here.)

The "gravitational density" tensor of Robinson and Bel is obtained in a natural way as a striking analogy with the spinor expression for the Maxwell stress tensor in the electromagnetic case. It is shown that the curvature tensor determines four gravitational principal null directions associated with flow of "gravitational density" which supplement the two electromagnetic null directions of Synge. The invariants and Petrov type of the curvature tensor are analyzed in terms of these, and a natural classification of curvature tensors is given.

An essentially coordinate-free method is outlined, by which any analytic solution of Einstein's field equations may, in principle, be found. As an elementary example the gravitational and gravitational-electromagnetic plane wave solutions are obtained.

1. Introduction

An essentially coordinate-free attitude to general relativity will be adopted here. The tensors and spinors occurring are best thought of not

† *Ann. of Phys.* **10,** 171 (1960).

‡ Present address: Palmer Physical Laboratory, Princeton University, Princeton, New Jersey.

as sets of components, but as geometric objects subject to certain formal rules of manipulation. A spinor formalism will be used instead of the usual tensor one, spinors appearing to fit in with general relativity in a remarkably natural way. This adds to a belief that spinors are basically simpler and perhaps more deep-rooted than tensors.

The usual correspondence between tensors and spinors (*1*, *2*) is obtained by the use of a mixed quantity† $\sigma_\mu^{AB'}$ satisfying the equation

$$\sigma^A_{\mu C'}\sigma^{BC'}_\nu + \sigma^A_{\nu C'}\sigma^{BC'}_\mu = g_{\mu\nu}\varepsilon^{AB}, \tag{1.1}$$

where ε^{AB}, together with ε_{AB}, $\varepsilon^{A'B'}$, and $\varepsilon_{A'B'}$ is a skew-symmetric "metric" spinor for the 2-dimensional complex spin space. The components of the ε's may be taken as ± 1, 0. (To raise or lower a spinor index, one of the ε's must be used, e.g., $\xi^A = \varepsilon^{AB}\xi_B$, $\xi_B = \xi^A\varepsilon_{AB}$.) Primed indices‡ refer to the complex conjugate spin space. Italic capitals are used here for spinor indices and Greek letters for tensor indices. The spinor equivalent of any tensor is a quantity which has an unprimed and a primed spinor index replacing each tensor index. For example, for a tensor $X^{\lambda\mu}{}_\nu$, we have

$$X^{\lambda\mu}{}_\nu \leftrightarrow X^{AB'CD'}{}_{EF'},$$

where

$$X^{AB'CD'}{}_{EF'} = \sigma_\lambda{}^{AB'}\sigma_\mu{}^{CD'}X^{\lambda\mu}{}_\nu\sigma^\nu{}_{EF'}$$

and

$$X^{\lambda\mu}{}_\nu = \sigma^\lambda{}_{AB'}\sigma^\mu{}_{CD'}X^{AB'CD'}{}_{EF'}\sigma_\nu{}^{EF'},$$

(with $\sigma^\mu{}_{AB'} = g^{\mu\nu}\sigma_\nu{}^{CD'}\varepsilon_{CA'}\varepsilon_{D'B'}$). We have

$$g_{AB'CD'} = \varepsilon_{AC}\varepsilon_{B'D'}, \quad \delta^{AB'}_{CD'} = \delta^A_C\delta^{B'}_{D'}, \quad g^{AB'CD'} = \varepsilon^{AC}\varepsilon^{B'D'}. \tag{1.2}$$

The algebraic tensor operations can now all be interpreted as spinor operations. Also the notions of reality of tensors, and of complex conjugate, are interpreted in spinor form with

$$\overline{X}^{\lambda\mu}{}_\nu \leftrightarrow \overline{X}^{A'BC'D}{}_{E'F}$$

† For each of the four values of μ, $\sigma_\mu^{AB'}$ is a (2×2) Hermitian matrix.

‡ Primed indices are used here rather than the more usual dotted indices, for typographical reasons.

so that the roles of primed and unprimed indices are interchanged.[1] Thus reality of tensors is expressed as a Hermitian property of the corresponding spinors.

In addition to the usual correspondence between tensors and spinors given above, there is also a well-known correspondence between real skew-symmetric second rank tensors and symmetric second rank spinors (*2*). Thus if $F_{\mu\nu}$ is real and skew-symmetric, we have

$$F_{AB'CD'} = \tfrac{1}{2}\{\varphi_{AC}\varepsilon_{B'D'} + \varepsilon_{AC}\bar{\varphi}_{B'D'}\}, \tag{1.3}$$

where φ_{AB} is a uniquely defined symmetric spinor. The right-hand side of (1.3) expresses $F_{AB'CD'}$ as the sum of the part symmetric in A, C (and therefore skew in B', D') and the part skew in A, C (and symmetric in B', D'). (Any skew pair of spinor indices may be split off as an ε-spinor.) A corresponding procedure can be applied to *any* skew-symmetric pair of tensor indices. A tensor with r skew-symmetric pairs of indices thus gives rise to 2^r spinors each with r symmetric pairs of indices in a decomposition similar to (1.3). If the tensor is real, these spinors are paired off as complex conjugates. For an example, see (2.2).

If the tensor $H_{\mu\nu}$ "dual" to $F_{\mu\nu}$ is defined by

$$H_{\mu\nu} = \tfrac{1}{2}\sqrt{(-g)}F^{\varrho\sigma}\varepsilon_{\mu\nu\varrho\sigma}, \tag{1.4}$$

we have

$$H_{AB'CD'} = \tfrac{1}{2}\{-i\varphi_{AC}\varepsilon_{B'D'} + i\varepsilon_{AC}\varphi_{B'D'}\}, \tag{1.5}$$

since if

$$\mathscr{E}^{\alpha\beta}_{\mu\nu} = \sqrt{(-g)}\varepsilon_{\varrho\sigma\mu\nu}g^{\alpha\varrho}g^{\beta\sigma},$$

then

$$\mathscr{E}^{AB'CD'}_{EF'GH'} = i\delta^A_E\delta^C_G\delta^{B'}_{H'}\delta^{D'}_{F'} - i\delta^A_G\delta^C_E\delta^{B'}_{F'}\delta^{D'}_{H'}. \tag{1.6}$$

(Actually, formulas (1.5) and (1.6) are only correct for one class of choices of $\sigma_\mu{}^{AB'}$ satisfying (1.1). If $\sigma_\mu{}^{AB'}$ had been chosen from the other class of solutions, the signs of the right-hand sides of (1.5) and

[1] Many authors would omit the bar on the right-hand side. The coihce here here is made for reasons of clarity.

(1.6) would be reversed. It will be supposed that the $\sigma_{\mu}{}^{AB'}$ have, in fact, been selected from the appropriate class.) In a similar way any tensor possessing a pair of skew-symmetric indices may be "dualized" with respect to that pair of indices. The spinor decomposition of the "dualized" tensor then differs from that of the original tensor in that the relevant ε_{AC} and $\varepsilon_{B'D'}$ are, respectively, multiplied by i and by $-i$. This again follows from (1.6). For an example, see (2.6).

General relativity requires, in addition to algebraic properties of tensors, the notion of covariant derivative. The symbol ∂_{μ}, or correspondingly $\partial_{AB'}$, will be used here to denote *covariant* differentiation. The covariant derivatives of $g_{\mu\nu}$ and of $\sigma_{\mu}{}^{AB'}$ are both required to be zero.[2] This implies that

$$\partial_{\mu}\{\varepsilon_{AB}\varepsilon_{C'D'}\} = 0$$

(see *2*). The stronger conditions

$$\partial_{\mu}\,\varepsilon_{AB} = 0 \quad \text{and} \quad \partial_{\mu}\varepsilon_{A'C'} = 0 \tag{1.7}$$

will be adopted here (*3*). This enables one to raise and lower spinor indices under the derivative symbol, but it precludes the use of phase transformations of the spinors to generate the electromagnetic field. However, the electromagnetic field will appear here as being associated with spinor transformations in a different way (see 3.13). These two procedures do not appear to combine in an altogether natural way. The simplest formalism, when charges are not present, seems to be obtained when such phase transformations are not permitted.

The point of view adopted here is nearer to that of Rainich (*4*) and of Misner and Wheeler (*5*) in which the electromagnetic field is obtained from the curvature of space–time alone. These phase transformations would not be related in any way to the geometry of the space–time.

[2] "Spin affinities" $\Gamma^{A}{}_{B\mu}$, $\overline{\Gamma}^{A'}{}_{B'\mu}$ are introduced to deal with the spinor indices. The conditions (1.7) imply that these spin affinities can be expressed explicitly in terms of $\sigma_{\mu}{}^{AB'}$ and its coordinate derivatives (see Ruse (*3*)).

2. The Curvature Spinors

Since the symbol ∂_μ here stands for covariant differentiation, we have

$$\partial_\mu \partial_\nu \neq \partial_\nu \partial_\mu,$$

the commutation of two ∂'s giving rise to the Riemann–Christoffel tensor $R_{\mu\nu\varrho\sigma}$. In fact, we have

$$\{\partial_\mu \partial_\nu - \partial_\nu \partial_\mu\} X_\varrho = R_{\mu\nu\varrho\sigma} X^\sigma. \tag{2.1}$$

The tensor $R_{\mu\nu\varrho\sigma}$ is skew-symmetric in μ, ν and in ϱ, σ. Thus, following Witten (*6*), we can apply the procedure outlined in Section 1 and obtain

$$\begin{aligned} R_{AE'BF'CG'DH'} = \tfrac{1}{2}\{&\chi_{ABCD}\varepsilon_{E'F'}\varepsilon_{G'H'} + \varepsilon_{CD}\varphi_{ABG'H'}\varepsilon_{E'F'} \\ &+ \varepsilon_{AB}\bar{\varphi}_{E'F'CD}\varepsilon_{G'H'} + \varepsilon_{AB}\varepsilon_{CD}\bar{\chi}_{E'F'G'H'}\}. \end{aligned} \tag{2.2}$$

The spinors χ_{ABCD} and φ_{ABGH} are the uniquely defined *curvature spinors*. However, this differs from Witten's form by a factor $\frac{1}{2}$ which is included here for reasons of convenience. From the symmetries of $R_{\mu\nu\varrho\sigma}$, it follows that

$$\chi_{ABCD} = \chi_{BACD} = \chi_{ABDC} = \chi_{CDAB} \tag{2.3}$$

and

$$\varphi_{ABC'D'} = \varphi_{BAC'D'} = \varphi_{ABD'C'} = \bar{\varphi}_{C'D'AB}. \tag{2.4}$$

Let the right dual $S_{\mu\nu\varrho\sigma}$ of $R_{\mu\nu\varrho\sigma}$ be defined by

$$S_{\mu\nu\varrho\sigma} = \tfrac{1}{2}\sqrt{(-g)}R_{\mu\nu}{}^{\alpha\beta}\varepsilon_{\alpha\beta\varrho\sigma}. \tag{2.5}$$

Then from (1.6), we have

$$\begin{aligned} S_{AE'BF'CG'DH'} = \frac{i}{2}\{&-\chi_{ABCD}\varepsilon_{E'F'}\varepsilon_{G'H'} + \varepsilon_{CD}\varphi_{ABG'H'}\varepsilon_{E'F'} \\ &- \varepsilon_{AB}\bar{\varphi}_{E'F'CD}\varepsilon_{G'H'} + \varepsilon_{AB}\varepsilon_{CD}\bar{\chi}_{E'F'G'H'}\}. \end{aligned} \tag{2.6}$$

Now, the symmetry relation $R_{\mu\nu\varrho\sigma} + R_{\mu\varrho\sigma\nu} + R_{\mu\sigma\nu\varrho} = 0$ is equivalent to

$$S_{\mu\nu\varrho}{}^{\nu} = 0,$$

so that multiplying (2.6) by $\varepsilon^{BD}\varepsilon^{F'H'}$ should give zero (see 1.2). Hence,

$$-\chi_{ABC}{}^{B}\varepsilon_{E'G'}-\varphi_{ACG'E'}+\bar{\varphi}_{E'G'CA}+\varepsilon_{AC}\bar{\chi}_{E'F'G'}{}^{F'} = 0.$$

The φ terms cancel by (2.4), so we have

$$\chi_{ABC}{}^{B} = \lambda\varepsilon_{AC}, \tag{2.7}$$

where λ is *real* and given by

$$\lambda = \tfrac{1}{2}\chi_{AB}{}^{AB} = \tfrac{1}{2}\bar{\chi}_{E'F'}{}^{E'F'}. \tag{2.8}$$

The reality of λ is, in fact, the only thing new we get out of this identity since (2.7) is implied by (2.3) in any case.

The relations (2.3), (2.4), and (2.8) are the only algebraic relations necessarily satisfied by χ_{ABCD} and $\varphi_{ABC'D'}$ for a general Riemannian space, since they imply that an $R_{\mu\nu\varrho\sigma}$ given by (2.2) has the required symmetry properties. These relations are all to be found in Witten's paper. However, χ_{ABCD} and $\varphi_{ABC'D'}$ also satisfy a differential relation obtained from the Bianchi identity

$$\partial_\tau R_{\mu\nu\varrho\sigma}+\partial_\varrho R_{\mu\nu\sigma\tau}+\partial_\sigma R_{\mu\nu\tau\varrho} = 0.$$

This is equivalent to

$$\partial^\sigma S_{\mu\nu\varrho\sigma} = 0,$$

i.e., (by 2.6)

$$\partial^{D}{}_{G'}\chi_{ABCD}\varepsilon_{E'F'}-\partial_{C}{}^{H'}\varphi_{ABG'H'}\varepsilon_{E'F'}+\varepsilon_{AB}\,\partial^{D}{}_{G'}\bar{\varphi}_{E'F'CD}-\varepsilon_{AB}\,\partial_{C}{}^{H'}\bar{\chi}_{E'F'G'H'} = 0.$$

Separating this into the two equations obtained by, respectively, symmetrizing and skew-symmetrizing with respect to A, B, we get

$$\partial^{D}{}_{G}\chi_{ABCD} = \partial_{C}{}^{H'}\varphi_{ABG'H'} \tag{2.9}$$

and its complex conjugate. The Bianchi identity is therefore equivalent to (2.9).

There are also relations connecting χ_{ABCD} and $\varphi_{ABG'H'}$ with covariant second derivatives of spinors, corresponding to the vector relation

(2.1). Let ξ_A be an arbitrary spinor field and define

$$X_{PR'QS'} = \xi_P \xi_Q \varepsilon_{R'S'}. \tag{2.10}$$

Now (2.1) generalizes to (and in fact implies)

$$\{\partial_\mu \partial_\nu - \partial_\nu \partial_\mu\} X_{\varrho\sigma} = R_{\mu\nu\varrho\alpha} X^{\alpha}{}_{\sigma} + R_{\mu\nu\sigma\alpha} X_{\varrho}{}^{\alpha}. \tag{2.11}$$

But $\partial_\mu \partial_\nu - \partial_\nu \partial_\mu$ is skew-symmetric in μ, ν so that the decomposition (1.3) can be applied:

$$\begin{aligned} \partial_{AC'} \partial_{BD'} - \partial_{BD'} \partial_{AC'} \equiv {} & \tfrac{1}{2} \varepsilon_{C'D'} \{\partial_{AF'} \partial_B{}^{F'} + \partial_{BF'} \partial_A{}^{F'}\} \\ & + \tfrac{1}{2} \varepsilon_{AB} \{\partial_{EC'} \partial^E{}_{D'} + \partial_{ED'} \partial^E{}_{C'}\}. \end{aligned} \tag{2.12}$$

Thus, (2.11) can be split into two equations each of which must hold separately, one symmetric in A, B (and skew in C', D') and the other skew in A, B (and symmetric in C', D'). Also, any skew pair of indices can be split off as an ε-spinor and these may be cancelled throughout the equation. Hence by (2.2), the equation symmetric in A, B is

$$\begin{aligned} \{\partial_{AF'} \partial_B{}^{F'} + \partial_{BF'} \partial_A{}^{F'}\} \xi_P \xi_Q \varepsilon_{R'S'} = {} & \chi_{ABPC} \xi^C \xi_Q \varepsilon_{R'S'} \\ & + \chi_{ABQC} \xi_P \xi^C \varepsilon_{R'S'} - \phi_{ABR'C'} \xi_P \xi_Q \delta^{C'}_{S'} + \phi_{ABS'C'} \xi_P \xi_Q \delta^{C'}_{R'}. \end{aligned} \tag{2.13}$$

The ϕ terms cancel and, because of (1.7), the $\varepsilon_{R'S'}$ term may be divided out. Also,

$$\partial_\mu \partial_\nu (\xi_P \xi_Q) = \xi_P\, \partial_\mu \partial_\nu \xi_Q + \xi_Q\, \partial_\mu \partial_\nu \xi_P + (\partial_\nu \xi_P)(\partial_\mu \xi_Q) + (\partial_\mu \xi_Q)(\partial_\nu \xi_P),$$

whence

$$\{\partial_\mu \partial_\nu - \partial_\nu \partial_\mu\}(\xi_P \xi_Q) = \xi_P \{\partial_\mu \partial_\nu - \partial_\nu \partial_\mu\} \xi_Q + \xi_Q \{\partial_\mu \partial_\nu - \partial_\nu \partial_\mu\} \xi_P.$$

It follows that

$$\begin{aligned} \{\partial_{AF'} \partial_B{}^{F'} + \partial_{BF'} \partial_A{}^{F'}\}(\xi_P \xi_Q) &= \xi_P \{\partial_{AF'} \partial_B{}^{F'} + \partial_{BF'} \partial_A{}^{F'}\} \xi_Q \\ &\quad + \xi_Q \{\partial_{AF'} \partial_B{}^{F'} + \partial_{BF'} \partial_A{}^{F'}\} \xi_P \\ &= \xi_P \chi_{ABQC} \xi^C + \xi_Q \chi_{ABPC} \xi^C \end{aligned}$$

by (2.13). Multiplying this equation by $\eta^P \eta^Q$ where η^A is chosen arbitrarily, we get

$$2(\eta^P\xi_P)(\eta^Q\{\partial_{AF'}\partial_B{}^{F'}+\partial_{BF'}\partial_A{}^{F'}\}\xi_Q) = 2(\eta^P\xi_P)(\eta^Q\chi_{ABQC}\xi^C).$$

Since η^A is arbitrary, we may divide by $2(\eta^P\xi_P)$ and obtain

$$\{\partial_{AF'}\partial_B{}^{F'}+\partial_{BF'}\partial_A{}^{F'}\}\xi_Q = \chi_{ABQC}\xi^C. \tag{2.14}$$

Also the equation obtained from (2.11) which is skew in A, B and symmetric in C', D' gives rise to

$$\{\partial_{EC'}\partial^E{}_{D'}+\partial_{ED'}\partial^E{}_{C'}\}\xi_Q = \phi_{QAC'D'}\xi^A \tag{2.15}$$

in an exactly similar way. The corresponding results for a primed spinor $\zeta_{A'}$ are obtained by taking the complex conjugates of (2.14) and (2.15). Thus,

$$\{\partial_{EC'}\partial^E{}_{D'}+\partial_{ED'}\partial^E{}_{C'}\}\zeta_{A'} = \bar{\chi}_{C'D'A'B'}\zeta^{B'} \tag{2.16}$$

and

$$\{\partial_{AF'}\partial_B{}^{F'}+\partial_{BF'}\partial_A{}^{F'}\}\zeta_{C'} = \phi_{ABC'D'}\zeta^{D'}. \tag{2.17}$$

The corresponding relations for spinors with more than one index can be obtained from (2.14), . . ., (2.17) since any spinor can be expressed as a linear combination of products of one-index spinors. Spinors with upper indices present no extra problem because the derivative of an ε-spinor is zero. As an example, we have

$$\{\partial_{BAF'}{}^{F'}+\partial_{BF'}\partial_A{}^{F'}\}\beta_C{}^{DE'} = \chi_{ABCP}\beta^{PDE'}+\chi_{AB}{}^D{}_P\beta_C{}^{PE'}+\phi_{AB}{}^{E'}{}_{Q'}\beta_C{}^{DQ'}.$$

In particular, by applying this to a "vector" $X^{DE'}$, and using (2.12) and (2.2), we can get back to (2.1). (It is not so easy to obtain (2.14), . . ., (2.17) directly from (2.1) rather than from (2.11), since the fact that the ε-spinors are constant must be used somewhere in the argument.)

The geometry of a Riemannian space (with signature $+---$) can thus be described entirely in spinor terms, with the role of the curvature tensor being taken over by spinors χ_{ABCD}, $\phi_{ABE'F'}$ satisfying (2.3), (2.4), (2.8), (2.9), (2.14), (2.15), (2.16), and (2.17).

3. The Einstein Conditions

The theory of Section 2 will now be specialized to two cases of particular note, namely empty space–time and source-free electromagnetic field.

The Ricci tensor $R_{\mu\nu} = R^{\sigma}_{\mu\sigma\nu}$ has the spinor form

$$R_{AC'BD'} = \tfrac{1}{2}\{\chi_{EA}{}^{E}{}_{B}\varepsilon_{C'D'} - 2\phi_{ABC'D'} + \varepsilon_{AB}\bar{\chi}_{F'C'}{}^{F'}{}_{D'}\}$$
$$= \lambda\varepsilon_{AB}\varepsilon_{CD} - \phi_{ABC'D'}$$

by (2.2), (2.7, (2.8). The scalar curvature $R = R^{\sigma}_{\sigma}$ is given by

$$R = 4\lambda \tag{3.1}$$

because of the symmetry of $\phi_{ABC'D'}$. The Einstein tensor $G_{\mu\nu} = R_{\mu\nu} - \frac{1}{2}g_{\mu\nu}R$ takes the form

$$G_{AC'BD'} = -\lambda\varepsilon_{AB}\varepsilon_{C'D'} - \phi_{ABC'D'}. \tag{3.2}$$

Einstein's equations $G_{\mu\nu} = 0$ for empty space clearly give

$$\phi_{ABC'D'} = 0 \tag{3.3}$$

and

$$\lambda = 0.$$

On the other hand, if it is required to include a cosmological term in Einstein's equations, we have only $\phi_{ABC'D'} = 0$, the cosmological constant being equal to λ by (3.1).

Supposing for the moment that the cosmological constant is zero, (2.7) gives

$$\chi_{ABC}{}^{B} = 0,$$

that is, χ_{ABCD} is symmetric in B and D. But by (2.3), it is also symmetric A, B and in C, D. It is therefore *completely symmetric* in all its indices.

It is a remarkable and perhaps significant fact, that only for a manifold with the apparently arbitrary $+---$ signature of our space–time, and which satisfies the Enstein equations for empty space, can its curvature be characterized by so natural an object as a totally symmetric four-index spinor. The geometry of this spinor will be dealt with in Section 3.

If a cosmological term (or matter) is present, we can write

$$\chi_{ABCD} = \psi_{ABCD} + \frac{\lambda}{3}\{\varepsilon_{AC}\varepsilon_{BD} + \varepsilon_{AD}\varepsilon_{BC}\} \tag{3.4}$$

and then ψ_{ABCD} will be totally symmetric even if $\lambda \neq 0$. The spinor ψ_{ABCD} defined by (3.4) will be called here the *gravitational* spinor (even in cases where $\phi_{ABC'D'} \neq 0$). It corresponds uniquely to Weyl's conformal tensor $C_{\mu\nu\varrho\sigma}$.

Relation (2.9) gives (with $\phi_{ABC'D'} = 0$)

$$\partial^{DE'}\psi_{ABCD} = 0 \tag{3.5}$$

and of course $\partial^{DE'}\lambda = 0$ also. Equation (3.5) has the suggestive appearance of being formally identical with a spinor equation for a zero rest-mass particle of spin two. (See Dirac (7) and compare (3.10).) However, the differentiation used here is covariant, so that derivatives do not commute. Hence, new features arise with second and higher derivatives. In particular, it is not true that Eq. (3.5) leads to the covariant wave equation upon iteration with $\partial_{FE'}$. We have

$$\partial_{FE'}\,\partial_D{}^{E'} \equiv \tfrac{1}{2}\{\partial_{FE'}\partial_D{}^{E'} + \partial_{DE'}\partial_F{}^{E'}\} + \tfrac{1}{2}\varepsilon_{FD}\Box, \tag{3.6}$$

where

$$\Box \equiv \partial_\mu\partial^\mu \equiv \partial_{FE'}\partial^{FE'}.$$

Also,

$$\{\partial_{FE'}\partial_D{}^{E'} + \partial_{DE'}\partial_F{}^{E'}\}\xi_A = \psi_{FDAB}\xi^B - \frac{\lambda}{3}\{\xi_D\varepsilon_{FA} + \xi_F\varepsilon_{DA}\} \tag{3.7}$$

by (2.14) and (3.4). Now (3.6) gives

$$0 = \partial_{FE'}\partial_D{}^{E'}\psi_{ABC}{}^D = \tfrac{1}{2}\{\partial_{FE'}\partial_D{}^{E'} + \partial_{DE'}\partial_F{}^{E'}\}\psi_{ABC}{}^D - \tfrac{1}{2}\Box\psi_{ABCF}.$$

By (3.7), this leads to

$$\begin{aligned}\Box\psi_{ABCD} &= \psi_{ABEF}\psi_{CD}{}^{EF} + \psi_{ACEF}\psi + \psi_{DB}{}^{EF} + \psi_{ADEF}\psi_{BC}{}^{EF} - 2\lambda\psi_{ABCD}\\ &= 3\psi_{(AB}{}^{EF}\psi_{CD)EF} - 2\lambda\psi_{ABCD} \qquad (3.8)\end{aligned}$$

where the indices between the brackets are to be symmetrized.[3] Thus, even when $\lambda = 0$ there is the nonlinear term on the right. This shows that the ψ-field can perhaps be thought of as acting as its own source to a certain extent. If ψ_{ABCD} is small we have

$$\Box \psi_{ABCD} \triangleq 0$$

since λ is small in any case. Equation (3.8) indicates that we can only expect to have *exact* solutions for plane gravitational waves moving with the velocity of light when $\lambda = 0$ and $\psi_{(AB}{}^{EF}\psi_{CD)EF} = 0$. This question will be returned to in Section 4 where this condition will be interpreted geometrically and in Section 5 where such an exact solution will be given.

The tensor $T_{\mu\nu\varrho\sigma}$ whose spinor equivalent is given by

$$T_{AE'BF'CG'DH'} = \psi_{ABCD}\bar{\psi}_{E'F'G'H'} \tag{3.9}$$

is of considerable interest. It has the properties of complete symmetry in its tensor indices, vanishing traces (as easily follows from (3.9)) and vanishing covariant divergence (with or without λ-term in Einstein's equations), since by (3.5)

$$\partial^{DH'}\{\psi_{ABCD}\bar{\psi}_{E'F'G'H'}\} = 0.$$

It would therefore appear that $T_{\mu\nu\varrho\sigma}$ is a multiple of the "gravitational density" (or "super-energy") tensor due independently to I. Robinson (unpublished seminars) and to Bel (*8, 9*).[4] As is easily verified, $T_{\mu\nu\varrho\sigma}$ is, in fact, proportional to the Robinson–Bel tensor. Equation (3.9) bears a striking resemblance to the corresponding Eq. (3.11) for the electromagnetic case.

[3] The tensor form of this relation is

$$\Box R_{\mu\nu\varrho\sigma} = R_{\mu\nu}{}^{\alpha\beta}R_{\alpha\beta\varrho\sigma} + 4R^{\alpha}{}_{\mu\beta[\varrho}R^{\beta}{}_{\sigma]\alpha\nu} - 2\lambda R_{\mu\nu\varrho\sigma}.$$

[4] This tensor was also found by R. Sachs working with the group at Hamburg, and by A. Komar. However, only Robinson noticed the *total* symmetry of the tensor expression. It is not hard to see in the spinor formalism that, if $R_{\mu\nu} = 0$, *any* four-index tensor, quadratic and homogeneous in $R^{\mu}{}_{\nu\varrho\sigma}$ and with vanishing divergence, must be totally symmetric. Robinson's tensor expression (with $S_{\mu\nu\varrho\sigma}$ as in (2.5)) is

$$T_{\mu\nu\varrho\sigma} = R_{\mu\alpha\nu\beta}R_{\varrho}{}^{\alpha}{}_{\sigma}{}^{\beta} + S_{\mu\alpha\nu\beta}S_{\varrho}{}^{\alpha}{}_{\sigma}{}^{\beta}.$$

Let us now suppose that there is a source-free electromagnetic field present. The field tensor $F_{\mu\nu}$ can be expressed according to (1.3) in terms of a symmetric spinor ϕ_{AB}. This spinor can be used instead of $F_{\mu\nu}$ to represent the electromagnetic field (*2*), and the Maxwell field equations (in covariant form) become

$$\partial^{AC'}\phi_{AB} = 0. \tag{3.10}$$

The energy-momentum tensor $T_{\mu\nu}$ for the electromagnetic field is given by

$$T_{AC'BD'} = \tfrac{1}{2}\phi_{AB}\bar{\phi}_{C'D'} \tag{3.11}$$

(see *10*). Now Einstein's equations with cosmological term are

$$G_{\mu\nu}+\lambda g_{\mu\nu} = -\varkappa T_{\mu\nu}\,.$$

The λ defined by (2.8) is still the cosmological constant, because $T^{\nu}{}_{\nu} = 0$ and by (3.1), $4\lambda = R = -G^{\nu}{}_{\nu}$. Choosing units suitably so that $\varkappa = 2$ (or absorbing the constant into the definition of ϕ_{AB}) we have, from (3.2), (3.11)

$$\phi_{ABC'D'} = \phi_{AB}\bar{\phi}_{C'D'}\,.$$

Equation (2.9) now gives

$$\partial^{D}{}_{G'}\psi_{ABCD} = \bar{\phi}_{G'H'}\partial_{C}{}^{H'}\phi_{AB} \tag{3.12}$$

by (3.4) and (3.10) since λ is necessarily constant. Thus the ϕ-field appears as a kind of source term to the ψ-field (here in the first-order equation).

From (2.17) we have

$$\{\partial_{FE'}\partial_{D}{}^{E'}+\partial_{DE'}\partial_{F}{}^{E'}\}\zeta_{G'} = \phi_{DF}\bar{\phi}_{G'H'}\zeta^{H'} \tag{3.13}$$

and (3.7) still holds. The second order equation arising from (3.10) turns out to be

$$\Box\phi_{AB} = \psi_{ABCD}\phi^{CD}-\tfrac{4}{3}\lambda\phi_{AB}$$

so that even the Maxwell field does not exactly satisfy the covariant wave equation (compare Eddington, *11*, Section 74). Also, (3.5) leads to

$$\Box\psi_{ABCD} = 3\psi_{(AB}{}^{EF}\psi_{CD)EF}-2\lambda\psi_{ABCD}-2\bar{\phi}_{G'H'}\partial_{A}{}^{G'}\partial_{B}{}^{H'}\phi_{CD}.$$

4. The Geometry and Invariants of ψ_{ABCD}

It is known that a general electromagnetic field determines two real principal null directions at each point (*12*). These are given in the general case by the real eigenvectors of the field tensor $F^{\mu}{}_{\nu}$ considered as a matrix. (There are also two complex null directions given by the complex eigenvectors, but these add nothing to the geometry as they are determined by their orthogonality with the real ones.) An alternative method of obtaining these principal null directions is to use a spinor approach. Any null vector x^{μ} corresponds to the product of a dotted with an undotted spinor

$$x^{AB'} = \eta^{A}\theta^{B'}.$$

If x^{μ} is real, $\theta^{B'}$ is a multiple of $\bar{\eta}^{B'}$, positive if x^{μ} points to the future. Any direction along the light cone therefore corresponds uniquely to a one-index spinor ray (set of spinors proportional to a given spinor). Now $F_{\mu\nu}$ corresponds uniquely to ϕ_{AB} (by 1.3) and we have

$$F^{AC'}{}_{BD'} = -\tfrac{1}{2}\{\phi^{A}{}_{B}\delta^{C'}_{D'} + \delta^{A}_{B}\bar{\phi}^{C'}{}_{D'}\}.$$

It is easily verified from this that the eigenvectors of $F^{AC'}{}_{BD'}$ are $\eta^{A}\bar{\eta}^{B'}$, $\zeta^{A}\bar{\zeta}^{B'}$ (corresponding to the real null vectors) and $\eta^{A}\bar{\zeta}^{B'}$, $\zeta^{A}\bar{\eta}^{B'}$ (corresponding to the complex null vectors) where

$$\phi_{AB} = \tfrac{1}{2}\{\eta_{A}\zeta_{B} + \eta_{B}\zeta_{A}\} = \eta_{(A}\zeta_{B)}. \tag{4.1}$$

See also Witten *(13)*. A decomposition exactly analogous to (4.1) exists for the gravitational spinor. We have

$$\psi_{ABCD} = \alpha_{(A}\beta_{B}\gamma_{C}\delta_{D)}, \tag{4.2}$$

which expresses the gravitational spinor uniquely (except for scale factors) as a symmetrized product of one-index spinors. The bracket here denotes symmetrization as before, so that written out in full, there would be 24 terms on the right-hand side. The existence and uniqueness of (4.2) follows from the fundamental theorem of algebra:

$$\psi_{ABCD}\xi^{A}\xi^{B}\xi^{C}\xi^{D} = (\alpha_{A}\xi^{A})(\beta_{B}\xi^{B})(\gamma_{C}\xi^{C})(\delta_{D}\xi^{D}) \tag{4.3}$$

expresses the general binary quartic form as a product of linear factors.

These factors are essentially unique, and equating coefficients gives (4.2).

Now the spinors $\alpha_A, \beta_B, \gamma_C, \delta_D$ determine four directions along the light cone. These are uniquely determined by ψ_{ABCD} and will be called the *gravitational principal null directions*.[5] They supplement the two electromagnetic principal null directions corresponding to η_A and ζ_A. The gravitational principal null directions are only undefined if $\psi_{ABCD} = 0$ but they may coincide in special cases. In particular, for the case of the Schwarzschild solution, it follows from the symmetry that they must coincide in pairs at every point, one pair pointing towards the origin along the light cone and the other pair pointing away from it. (Time reversal symmetry shows that they cannot all four coincide or coincide three and one.) The coincidence of the two electromagnetic null directions is the condition for the electromagnetic field to be null. (The electromagnetic directions are, of course, only undetermined if $\phi_{AB} = 0$.) Thus, for an electromagnetic plane wave, the principal null directions coincide and, naturally enough, point in the direction of motion of the wave. Similarly, it turns out that for a gravitational plane wave, all the gravitational null directions coincide. This question will be returned to later. Gravitational radiation is sometimes analysed in terms of the invariants of the Riemann tensor *(17)* and it will be useful first to relate these invariants to the null directions defined above.

The number of independent invariants of the Riemann tensor in empty space is well known to be four. These may be interpreted as the real and imaginary parts of two independent complex invariants of ψ_{ABCD}, e.g.,

$$(4.4) \qquad I = \psi_{ABCD}\psi^{ABCD}, \quad J = \psi^{AB}{}_{CD}\psi^{CD}{}_{EF}\psi^{EF}{}_{AB}$$

(see Witten, *6*, p. 359). These may be thought of as invariants of the binary quartic form (4.3). According to the theory of invariants of binary forms, *I* and *J* are independent and any invariant of the quartic

[5] These four null directions are implicit in the work of Ruse (*14*). They correspond to the self-conjugate lines of the Riemannian complex. *Note added in proof:* They have been further exploited by Debever (*15*, *16*).

form (4.3) is a function of them (see Grace and Young, *18*). Thus the real and imaginary parts of I and J are a complete set of curvature invariants for empty space. The invariants I and J take the following tensor form if $R_{\mu\nu} = \lambda g_{\mu\nu}$:

$$I = \frac{1}{2} R_{\mu\nu\varrho\sigma}R^{\mu\nu\varrho\sigma} + \frac{i}{4}\sqrt{-g}R_{\mu\nu}{}^{\alpha\beta}\varepsilon_{\alpha\beta\varrho\sigma}R^{\mu\nu\varrho\sigma} - \frac{4}{3}\lambda^2,$$

$$J = \frac{1}{2}\left\{R^{\mu\nu}{}_{\varrho\sigma} + \frac{i}{2}\sqrt{-g}R^{\mu\nu\alpha\beta}\varepsilon_{\alpha\beta\varrho\sigma}\right\}R^{\varrho\sigma}{}_{\gamma\delta}R^{\gamma\delta}{}_{\mu\nu} - 2\lambda I - \frac{8}{9}\lambda^3$$

with $\lambda = \frac{1}{4}R$. These relations are obtained from (2.2), (2.5), (2.6), and (3.4). For a general curvature tensor,[6] the tensor $R_{\mu\nu\varrho\sigma}$ in the above expressions must be replaced by

$$\tfrac{1}{2}R_{\mu\nu\varrho\sigma} + \tfrac{1}{8}gR^{\alpha\beta\gamma\delta}\varepsilon_{\alpha\beta\mu\nu}\varepsilon_{\gamma\delta\varrho\sigma}.$$

Binary forms have a geometrical interpretation as sets of points on a complex projective line. The equation

$$\psi_{ABCD}\xi^A\xi^B\xi^C\xi^D = 0$$

is satisfied if and only if at least one of the factors $\alpha_A\xi^A, \beta_B\xi^B, \gamma_C\xi^C, \delta_D\xi^D$ vanishes, each of the conditions $\alpha_A\xi^A = 0, \ldots, \delta_D\xi^D = 0$ representing a point on the line. Thus ψ_{ABCD} corresponds to four points A, B, C, D on a complex projective line, the coordinates of these points being the components of $\alpha_A, \beta_A, \gamma_A, \delta_A$, respectively. Now any four collinear points have a projective invariant, namely, their cross-ratio

$$\mu = \frac{(\alpha_A\beta^A)(\gamma_B\delta^B)}{(\alpha_C\delta^C)(\gamma_D\beta^D)}.$$

[6] It is perhaps worth remarking that a general method of converting expressions involving ψ_{ABCD} into the corresponding expressions for $R_{\mu\nu\varrho\sigma}$ would be to use the formula

$$\psi_{ABCD} = \tfrac{1}{2}R_{AE'BF'CG'DH'}\varepsilon^{E'F'}\varepsilon^{G'H'} - \tfrac{1}{12}R\{\varepsilon_{AC}\varepsilon_{BD} + \varepsilon_{AD}\varepsilon_{BC}\}$$

but the conversion of spinor contractions to an equivalent tensor form is sometimes complicated.

This cross-ratio is the only independent invariant of the four points and is therefore the only independent invariant of ψ_{ABCD} which is unchanged if ψ_{ABCD} is multiplied by a non-zero complex number. Thus, the four real invariants of the curvature of empty space can be interpreted as a complex cross-ratio,[7] and a phase and a magnitude[8] for ψ_{ABCD}.

This phase is associated with duality rotations of the curvature tensor (suggested to me first by I. Robinson) which are exactly analogous to electromagnetic duality rotations (*5*). In each case the duality rotation invariance of the first-order equation (3.5), (3.10) is broken only when sources are present. Letting

$$\psi_{ABCD} \to e^{i\theta}\psi_{ABCD},$$

where θ is a real constant, we have, assuming for simplicity that $\phi_{ABC'D'}$ and λ both vanish,

$$R_{\mu\nu\varrho\sigma} \to \cos\theta R_{\mu\nu\varrho\sigma} - \sin\theta S_{\mu\nu\varrho\sigma}$$

by (2.2) and (2.6), $S_{\mu\nu\varrho\sigma}$ being the right (or equivalently the left) dual of $R_{\mu\nu\varrho\sigma}$ defined by (2.5). This is exactly analogous to

$$\phi_{AB} \to e^{i\theta}\phi_{AB}$$

giving

$$F_{\mu\nu} \to \cos\theta F_{\mu\nu} - \sin\theta H_{\mu\nu}$$

where the dual $H_{\mu\nu}$ of $F_{\mu\nu}$ is given by (1.4). Unlike the electromagnetic case, however, duality rotations of the ψ-field of an empty space solution do not in general give rise to new exact solutions of the field equations. (See, for example, **Eq.** (3.8).)

It will be observed that the Robinson–Bel tensor $\psi_{ABCD}\bar{\psi}_{E'F'G'H}$ determines ψ_{ABCD} up to a duality rotation in the same way that

[7] The idea of using a complex cross-ratio as an invariant defined by four nu rays has also been independently suggested by I. Robinson (unpublished).

[8] This phase and magnitude of ψ_{ABCD} can be interpreted in an invariant way as the argument and modulus of, say, $\sqrt{I}$. This not really satisfactory, however, since I may vanish. It might be better to use the argument and modulus of the $\varkappa$ which is defined by the relations (4.5). This only need vanish if $I = J = 0$, the condition for three of the null directions to coincide. Its definition depends on an arbitrary ordering of the null directions, however, as does the definition of μ.

$\phi_{AB}\bar{\phi}_{C'D'}$ determines ϕ_{AB} up to a duality rotation. The principal null directions are therefore associated even more closely with these "energy" expressions than with the field quantities themselves. These expressions are completely characterized by the principal null directions, apart from their actual magnitude. It might be expected that the gravitational null directions are in some way associated with flow of "gravitational density". There does appear to be such a connection, as may be seen from the following argument.

$$x_{AB'} = \xi_A \bar{\xi}_{B'}.$$

Then by (3.9) and (4.3)

$$\begin{aligned} T_{\mu\nu\varrho\sigma} x^\mu x^\nu x^\varrho x^\sigma &= (\psi_{ABCD}\xi^A\xi^B\xi^C\xi^D)(\bar{\psi}_{E'F'G'H'}\bar{\xi}^{E'}\bar{\xi}^{F'}\bar{\xi}^{G'}\bar{\xi}^{H'}) \\ &= (a_\mu x^\mu)(b_\nu x^\nu)(c_\varrho x^\varrho)(d_\sigma x^\sigma), \end{aligned}$$

where

$$a_{AB'} = \alpha_A \bar{\alpha}_{B'}, \quad b_{AB'} = \beta_A \bar{\beta}_{B'}, \quad c_{AB'} = \gamma_A \bar{\gamma}_{B'}, \quad d_{AB'} = \delta_A \delta_{B'}.$$

The vectors a_μ, b_μ, c_μ, d_μ are null vectors, pointing into the future, corresponding to the gravitational principal null directions. Thus $T_{\mu\nu\varrho\sigma} x^\mu x^\nu x^\varrho x^\sigma$ only vanishes for null vectors x^μ which point in one of the gravitational principal null directions. Otherwise it is positive. But for any time-like vector t^μ, the expression

$$\frac{T_{\mu\nu\varrho\sigma} t^\mu t^\nu t^\varrho t^\sigma}{(t_\alpha t^\alpha)^2} \tag{4.5}$$

measures the gravitational density for an observer whose time axis is t^μ (see Bel (*8, 9*)). It is positive (for empty space) unless $R_{\mu\nu\varrho\sigma} = 0$. Thus the gravitational principal null directions are characterized by the fact that for observers travelling with a given velocity infinitesimally less than c, the gravitational density will be a minimum for those observers who travel approximately along a principal null direction.

It is convenient, from a geometrical point of view, to represent null directions as points on a sphere. This sphere may be thought of as the field of vision of some observer. It may also be interpreted as a realization of the complex projective line mentioned above. (A complex projective line is, topologically, a real 2-sphere.) This sphere is the Argand sphere of the ratio of the two components of a one-index spinor (see Penrose, *19*, p. 138). Any Lorentz transformation corresponds to a bilinear transformation of this ratio and therefore to a projective (conformal) transformation of the sphere, which sends circles into circles.

Four points on the sphere are concyclic if and only if their cross-ratio is real. A particular case of this are harmonic points for which the cross-ratio is $-1, 2$, or $\frac{1}{2}$ according to the order in which the points are taken. The symmetry of a harmonic set is best exhibited when the points are equally spaced around a great circle. The symmetries are then just the symmetries of a square. Any harmonic set can be brought into this form by a suitable projective (Lorentz) transformation, since any three points on the sphere can be transformed into any three others. Harmonic sets are of interest here because they have a greater symmetry than a general set of four points. They correspond to the vanishing of the invariant J (see Grace and Young, *18*, p. 206). Also of interest is the equianharmonic set which has an even greater symmetry. The cross-ratio here is $-\omega$ or $-\omega^2$ where $\omega = e^{i2\pi/3}$. By a suitable projective transformation these four points can be made the vertices of a regular tetrahedron. Equianharmonic points correspond to the vanishing of the invariant I (*18*, p. 206).

In the case of a general cross-ratio μ, the symmetry is given by the Klein 4-group (except that there are also some reflectional symmetries if μ is real or has modulus unity). There is a unique projective transformation (involution) which interchanges any pair of the points with the remaining pair. These and the identity constitute the complete projective symmetry group provided that μ is different from $-1, 2, \frac{1}{2}$, $-\omega$, $-\omega^2$, 0, 1, or ∞, the cases 0, 1, and ∞ occurring when a pair of points coincide. The value of μ can be obtained from I^3/J^2 since it can be shown that

$$(4.6) \qquad I = 6\varkappa^2(\mu+\omega)(\mu+\omega^2), \quad J = 6\varkappa^3(\mu+1)(\mu-\tfrac{1}{2})(\mu-2)$$

for some $\varkappa$ (see (*16*) p. 205). There are in general six values of μ for a given value of I^3/J^2. They correspond to different orders in which the four points can be taken. The values are μ, $1-\mu$, $1/\mu$, $1-(1/\mu)$, $1/(1-\mu)$, $\mu/(\mu-1)$. The symmetries in the general case can also be realized as rotational symmetries of the sphere similarly to the two cases considered above. By a suitable projective transformation the four points, A, B, C, D can be transformed into the vertices of a tetrahedron which has opposite edges equal in pairs (a disphenoid). Such a tetrahedron has three orthogonal dyad axes of symmetry. These axes are the joins of the midpoint of opposite edges. If the cross-ratio is real, the tetrahedron is flattened into a rectangle but the three symmetry axes remain.

To see that such a transformation exists consider the three pairs (E, F,), (G, H), (K, L) of united points for the three involutions which send (A, B, C, D) into (B, A, D, C) (C, D, A, B) and (D, C, B, A), respectively. Now the involution which sends. (A, B, C, D) into (B, A, D, C) transforms the other involutions into themselves. It therefore sends G into H and K into L. Hence, (E, F) is harmonic with respect to (G, H) and also with respect to (K, L). Similarly (G, H) is harmonic with respect to (K, L). Now, E, G, F, H can be transformed (as above) into four points equally spaced, in that order, around the equator. K and L will then be the north and south poles, so that the six points form the vertices of a regular octahedron. The three involutions are then represented as rotations through π about the three axes EF, GH, KL. The point A is rotated into B, C, D by means of these involutions giving the symmetrical tetrahedron described above.

This symmetrical representation of the points A, B, C, D is of interest because it is related to Petrov's canonical representation of the Riemann tensor with $R_{\mu\nu} = 0$ (*17*, *20*). The rest frame in which the gravitational principal null directions appear to have this symmetrical form determines the canonical time axis, the three canonical space axes arising from the three axes of symmetry. These four axes are orthogonal to each other and are called the Riemann principal directions. They are uniquely defined provided that A, B, C, D are all distinct. If A, B, C, D coincide in pairs they can still be considered to exist but they are not uniquely defined.

The rotational symmetries of the tetrahedron $A\,B\,C\,D$ in the general case give rise to the corresponding symmetries for $R_{\mu\nu\varrho\sigma}$, since being dyad axes the only other possibility would be $R_{\mu\nu\varrho\sigma} \to -R_{\mu\nu\varrho\sigma}$ (a duality rotation of π).[9] Such an alternative is easily ruled out as impossible. It follows that, for the canonical choice of axes,

$$R_{ijkl} = 0 \quad \text{whenever} \quad i = k \quad \text{and} \quad j \neq l$$

as is required in Petrov's canonical form. Conversely, the above condition is sufficient for the Riemann principal directions to be the axes.

The usual definition of the Riemann principal directions is in terms of the intersections of certain planes which are determined by the "eigenbivectors" of $R_{\mu\nu\varrho\sigma}$, i.e., from the nonzero (complex) skew tensors $x^{\mu\nu}$ which satisfy a relation

$$R^{\mu\nu}{}_{\varrho\sigma} x^{\varrho\sigma} = \alpha x^{\mu\nu}. \tag{4.7}$$

Writing this in a spinor form with

$$x^{AC'BD'} = \tfrac{1}{2}\{\eta^{AB}\varepsilon^{C'D'} + \varepsilon^{AB}\bar{\zeta}^{C'D'}\}$$

(see 1.3) η^{AB} and ζ^{AB} being symmetric, (4.7) becomes

$$\psi^{AB}{}_{EF}\eta^{EF}\varepsilon^{C'D'} + \varepsilon^{AB}\bar{\psi}^{C'D'}{}_{E'F'}\bar{\zeta}^{E'F'} = \alpha\{\eta^{AB}\varepsilon^{C'D'} + \varepsilon^{AB}\bar{\zeta}^{C'D'}\}$$

(since $\phi_{ABC'D'} = 0$, $\lambda = 0$) so that

$$\psi^{AB}{}_{EF}\eta^{EF} = \alpha\eta^{AB}, \quad \psi^{AB}{}_{EF}\zeta^{EF} = \bar{\alpha}\zeta^{AB}.$$

One or the other of η^{AB}, ζ^{AB} may be zero. The eigenbivectors of $R^{\mu\nu}{}_{\varrho\sigma}$ are thus expressible in terms of "eigenspinors" of $\psi^{AB}{}_{CD}$, the eigenvalues of $R^{\mu\nu}{}_{\varrho\sigma}$ being those of $\psi^{AB}{}_{CD}$ and their complex conjugates. Witten (6) also considers these eigenspinors.

[9] In the special cases where the set of points A, B, C, D has an additional rotational symmetry, this does not always lead to a corresponding symmetry of $R_{\mu\nu\varrho\sigma}$, although it does for the case when A, B, C, D coincide in pairs. In particular, in the equianharmonic case, the triad axes of symmetry give rise to duality rotations through angles $2\pi/3$, $4\pi/3$.

Now if the eigenvalues of $\psi^{AB}{}_{CD}$ are $\alpha_1, \alpha_2, \alpha_3$ (the space of symmetric ξ^{AB} being three-dimensional) we have

$$\alpha_1+\alpha_2+\alpha_3 = \psi^{AB}{}_{AB} = 0,$$
$$\alpha_1{}^2+\alpha_2{}^2+\alpha_3{}^2 = \psi^{AB}{}_{CD}\psi^{CD}{}_{AB} = I,$$
$$\alpha_1{}^3+\alpha_2{}^3+\alpha_3{}^3 = \psi^{AB}{}_{CD}\psi^{CD}{}_{EF}\psi^{EF}{}_{AB} = J.$$

With the expressions for I and J given in (4.6), it is easily verified that these relations are satisfied by

$$(4.8) \qquad \alpha_1 = \varkappa(2\mu-1), \quad \alpha_2 = \varkappa(2-\mu), \quad \alpha_3 = \varkappa(-1-\mu).$$

The six eigenvalues of $R^{\mu\nu}_{\varrho\sigma}$ are therefore these three numbers and their complex conjugates. It will be seen that the vanishing of just one of the eigenvalues (4.8) is the condition for the principal null directions to form a harmonic set. If two of them vanish they must all vanish and $I = J = 0$. This is the condition for at least three of the principal null directions to coincide (since they form both a harmonic and an equianharmonic set). If two of the eigenvalues (4.8) coincide, this is the condition $F = 0$, 1, or ∞ for a pair of principal null directions to coincide. This is the case $I^3 = 6J^2$ (*18*, p. 198).

The three eigenspinors η^{AB}, ζ^{AB}, θ^{AB} of $\psi^{AB}{}_{CD}$ will next be considered. They are symmetric and therefore each is expressible as a symmetrized product of a pair of one-index spinors (see 4.1). Each of η^{AB}, ζ^{AB}, θ^{AB} corresponds to a pair of points on the projective line considered earlier, so in the general case we have six points on this line determined by A, B, C, D. These can only be E, F, G, H, K, L since a general quartic form has only one sextic covariant (*18*, pp. 92, 94). This sextic covariant is

$$\psi_{PQRA}\psi^{PQ}{}_{BC}\psi^{R}{}_{DEF}\xi^A\xi^B\xi^C\xi^D\xi^E\xi^F,$$

whence

$$\psi_{PQR(A}\psi^{PQ}{}_{BC}\psi^{R}{}_{DEF)} = \eta_{(AB}\zeta_{CD}\theta_{EF)},$$

choosing the scale factor suitably. The vanishing of this expression is the condition for A, B, C, D to coincide in pairs, since E, F, G, H, K, L are not then defined uniquely. It does not vanish if just two of A, B, C, D coincide, or if they coincide three and one.

The planes determined by the eigenbivectors of $R^{\mu\nu}{}_{\varrho\sigma}$ are those determined by η^{AB}, ζ^{AB}, θ^{AB}. They are therefore the three planes of the pairs of null directions corresponding to EF, GH, KL and the three orthogonal complements of these planes. Their intersections give the Riemann principal directions defined here, as is required. This is easily seen from the symmetrical representation of A, B, C, D given above.

These considerations have so far been essentially only concerned with Petrov's tensors $R_{\mu\nu\varrho\sigma}$ of Type I. This is the case when the eigenbivectors of $R^{\mu\nu}{}_{\varrho\sigma}$ span the six-dimensional space of bivectors. In special cases these eigenbivectors span only a four-dimensional space (Type II) and in very special cases, a two-dimensional space (Type III). In spinor terms, this means that Type I occurs when the eigenspinors of $\psi^{AB}{}_{CD}$ span a three-dimensional space, Type II when they span a two-dimensional space and Type III when they span only a one-dimensional space. Thus, Type II can only occur when at least two of the eigenvalues (4.8) are equal and Type III when they are all equal (and therefore all zero). We have seen that equality of eigenvalues implies coincidences among A, B, C, D so the cases where such coincidences occur must now be considered.

There are six different cases to be distinguished including the general case [1111] where the null directions are all distinct. There is the case [211] where exactly two of them coincide, [22] where they coincide in pairs, [31] where they coincide three and one, and [4] where all four directions are the same. Finally, there is the case [—] when $\psi_{ABCD} = 0$ and the null directions are undefined. This gives us a natural classification of Riemann tensors in empty space into six types (see also Géhéniau (*21*) for a closely related procedure[10]). In each case, the eigenspinors can be obtained by observing what happens to E, F, G, H, K, L when A, B, C, D are specialized. However, this must be done with care so that possible limiting positions of E, F, G, H, K, L are not omitted. Figure 1 shows how the different special cases arise from one another. The vertical specializations can be carried out keeping

[10] *Note added in proof:* See also, more explicitly, Debever (*15, 16*).

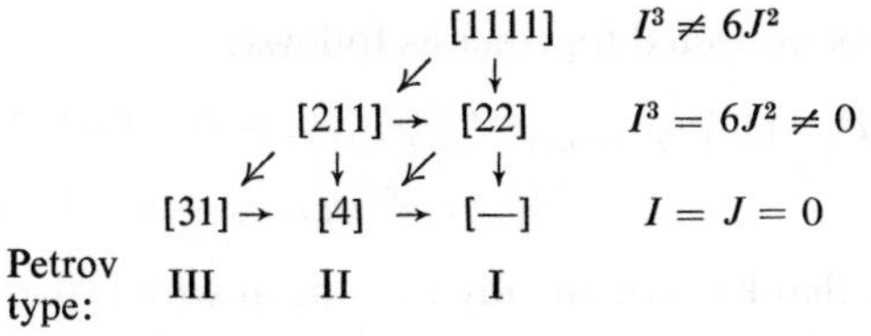

$$\begin{array}{lcccccl}
 & & & & & [1111] & I^3 \neq 6J^2 \\
 & & & & \swarrow & \downarrow & \\
 & & & [211] & \rightarrow & [22] & I^3 = 6J^2 \neq 0 \\
 & & \swarrow & \downarrow & \swarrow & \downarrow & \\
 & [31] & \rightarrow & [4] & \rightarrow & [-] & I = J = 0 \\
\text{Petrov type:} & \text{III} & & \text{II} & & \text{I} &
\end{array}$$

FIG.1. Classification scheme for ψ_{ABCD} in terms of coincidences between principal null directions.

the positions of E, F, G, H, K, L fixed, but in the diagonal specializations, further pairs of them are forced to coincide. (For example, in the case [1111] $\rightarrow$ [211] if $B \rightarrow X$ and $A \rightarrow X$, we have $(G, H, \rightarrow \rightarrow (X, X), (K, L) \rightarrow (X, X)$ and $(E, F) \rightarrow (X, Y)$ where Y is the harmonic conjugate of X with respect to the limiting positions of C and D.) The Petrov type for each case may be obtained in this way and the results are shown in Fig. 1. Each column corresponds to a particular type. Thus, [1111], [22], and [—] are Type I, [211] and [4] are Type II, while [31] is Type III. The different rows can be distinguished by the invariants I and J (or by the eigenvalues). Hence the invariants and Petrov type together serve to characterize ψ_{ABCD}.

It is of interest to see how this classification is in accord with that given by the classical canonical form of $\psi^{AB}{}_{CD}$ considered as a (3×3) matrix. These corresponding canonical forms are given in Fig. 2.

The various algebraic conditions for each case (or one of its specializations)

$$\begin{array}{ccccc}
 & & & & \begin{pmatrix} \varkappa(2\mu-1) & & \\ & \varkappa(2-\mu) & \\ & & \varkappa(-1-\mu) \end{pmatrix} \\
 & & & \swarrow & \downarrow \\
 & & \begin{pmatrix} \varkappa & 1 & \\ & \varkappa & \\ & & -2\varkappa \end{pmatrix} & \rightarrow & \begin{pmatrix} \varkappa & & \\ & \varkappa & \\ & & -2\varkappa \end{pmatrix} \\
 & \swarrow & \downarrow & \swarrow & \downarrow \\
\begin{pmatrix} 0 & 1 & \\ & 0 & 1 \\ & & 0 \end{pmatrix} & \rightarrow & \begin{pmatrix} 0 & 1 & \\ & 0 & \\ & & 0 \end{pmatrix} & \rightarrow & \begin{pmatrix} 0 & & \\ & 0 & \\ & & 0 \end{pmatrix}
\end{array}$$

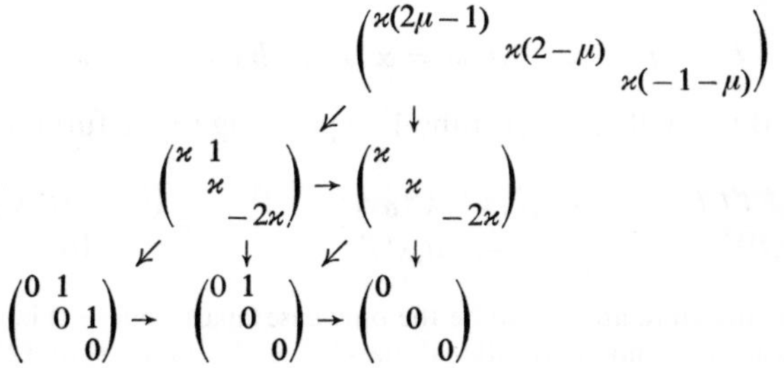

FIG. 2. Classification in terms of matrix canonical form of $\psi^{AB}{}_{CD}$.

to occur may be collected together as follows:

$$[211]\colon I^3 = 6J^2,\quad [22]\colon \psi_{PQR(A}\psi^{PQ}{}_{BC}\psi^{R}{}_{DEF)} = 0,\quad [31]\colon I = J = 0,$$
$$[4]\colon \psi_{(AB}{}^{EF}\psi_{CD)EF} = 0,\quad [-]\colon \psi_{ABCD} = 0.$$

The only case that has not already been dealt with is the condition for [4] to occur. The quartic form $\psi_{AB}{}^{EF}\psi_{CDEF}\xi^A\xi^B\xi^C\xi^D$ is the Hessian of the form

$$\psi_{ABCD}\xi^A\xi^B\xi^C\xi^D$$

and its vanishing is known to be the condition for the latter form to be a perfect fourth power (*18*, p. 235). The interest of this condition lies in the fact that $\psi_{(AB}{}^{EF}\psi_{CD)EF}$ is precisely the term (in the case $\lambda = 0$) which prevents Eq. (3.8) from being a covariant wave equation[11] for ψ_{ABCD}. Thus, plane wave solutions can only reasonably be expected in case [4].[12] This is Petrov's Type II with vanishing invariants and is apparently the case characteristic of a "*pure*" gravitational radiation field (*8*, *22*, and *23*). The other cases which might conceivably also be considered as "pure gravitational radiation" are [211] and [31] (see *17*). Case [211] would seem to be wrong since [22], which is a special case of it, would also have to be considered as pure gravitational radiation. But we have seen that the Schwarzschild solution is [22].

Case [31] is, however, worthy of consideration in this respect since it shares with Case [4] the property that the gravitational density (4.5) can be made as small as we please by a suitable choice of time axis ("following the wave"). If

$$\psi_{ABCD} = \alpha_{(A}\alpha_B\alpha_C\beta_{D)}$$

and

$$t_\mu = a_\mu + \varepsilon x_\mu,\quad a_{AB'} = \alpha_A\bar{\alpha}_{B'},\quad b_{AB'} = \beta_A\bar{\beta}_{B'},$$

where $\varepsilon > 0$ is small and x_μ is time-like pointing to the future, we have

$$\frac{T_{\mu\nu\varrho\sigma}t^\mu t^\nu t^\varrho t^\sigma}{(t_\tau t^\tau)^2} \simeq \frac{\frac{1}{4}\varepsilon^3|\beta_A\alpha^A|^2(\alpha_B\bar{\alpha}_{C'}x^{BC'})^3}{4\varepsilon^2(a_\tau x^\tau)^2} = \varepsilon\,\frac{(b_\mu a^\mu)\,(a_\nu x^\nu)}{16}.$$

[11] Case [4] therefore appears to be the only case (apart from [−]) in which the gravitational field has no "gravitational mass". See also Bondi *et al.* (*23*), p. 532.

[12] However, a point perhaps worth mentioning is that in case [22], ψ_{ABCD} and $\psi_{(AB}{}^{EF}\psi_{CD)EF}$ are proportional.

If $\beta_A = \alpha_A$, the right-hand side would be of order ε^2 instead of ε. Thus, the gravitational density tends to zero for observers, whose velocity approaches the multiple principal null direction, both in Case [31] and in Case [4], but it tends to zero more rapidly in Case [4]. It would appear to be correct to call Case [4] "pure" radiation field[13] but not Case [31]. Case [4] is like a null electromagnetic field ("pure" electromagnetic radiation field) in that it determines only one null direction, and in that it is the general limiting case obtained as a result of a high velocity Lorentz transformation (see also *24*). (However, it is worth remarking that for a null electromagnetic field, the energy $T_{\mu\nu}t^\mu t^\nu/(t_\tau t^\tau)$ can only be made to tend to zero to order ε by "following the wave", like Case [31] above.)

The invariants of ψ_{ABCD} have been treated in considerable detail above. It now remains to give a brief discussion of the combined system ψ_{ABCD}, ϕ_{AB} for the case when electromagnetic field is present. We expect to find just three more complex invariants, since ϕ_{AB} is determined by its phase and magnitude, and by the positions relative to A, B, C, D of the two complex points Y and Z on the argand sphere, corresponding to the electromagnetic principal null directions. There is the obvious invariant

$$K = \phi_{AB}\phi^{AB}$$

of ϕ_{AB} alone. This is the discriminant of the binary form $\phi_{AB}\xi^A\xi^B$, the condition $K = 0$ being necessary and sufficient for the points X and Y to coincide, that is, for the field to be null.[13a] The list is completed by the two independent invariants

$$L = \phi_{AB}\psi^{AB}{}_{CD}\phi^{CD}, \quad M = \phi_{AB}\psi^{AB}{}_{CD}\psi^{CD}{}_{EF}\phi^{EF}.$$

The fact that I, J, K, L, M are in general independent is most easily seen if $\psi^{AB}{}_{CD}$ is thought of as a matrix and ϕ^{AB} as a "vector" which may then be expanded in terms of the eigenspinors of $\psi^{AB}{}_{CD}$ with

[13] It is probably preferable, however, to call case [4] simply a *null* gravitationa field (as suggested by Robinson) analogously to the electromagnetic case.

[13a] The real and imaginary parts of K are the usual invariants $F_{\mu\nu}F^{\mu\nu}$ and $\frac{1}{2}\sqrt{(-g)}\, F^{\mu\nu}F^{\sigma\varrho}\varepsilon_{\mu\nu\varrho\sigma}$ respectively, of $F_{\mu\nu}$.

arbitrary coefficients. K, L, and M then become independent linear functions of the squares of these coefficients.

However, I, J, K, L, and M do not form a complete system of invariants in the sense of invariant theory (*18*). That is, not every algebraic invariant of ψ_{ABCD} and ϕ_{AB} can be expressed as a polynomial in $I, \ldots, M$. The invariant

$$N = \phi_{AB}\psi^{AB}{}_{CD}\psi^{CD}{}_{EF}\phi^{E}{}_{G}\psi^{FG}{}_{PQ}\phi^{PQ}$$

clearly is not even a rational function of $I, \ldots, M$ since every such function is of even order in ϕ_{AB}. Also N does not vanish identically. On the other hand, N is *algebraically* dependent on $I, \ldots, M$, there being the *syzygy*

$$N^2 = \tfrac{1}{2}JKLM - \tfrac{1}{6}JL^3 - \tfrac{1}{2}M^3 - \tfrac{1}{8}I^2KL^2 - \tfrac{1}{6}IJK^2L - \tfrac{1}{18}J^2K^3 + \tfrac{1}{4}IKM^2 + \tfrac{1}{4}IL^2M.$$

The system I, J, K, L, M, N does, in fact, form a complete system of invariants for ψ_{ABCD} and ϕ_{AB}.

The condition for an electromagnetic principal null direction to coincide with a gravitational principal null direction is that the resultant of the quartic and quadratic forms should vanish. Expressed in terms of invariants this condition turns out to be

$$2K^2I - 4KM + L^2 = 0.$$

The condition for both electromagnetic null directions to lie along a gravitational null direction is therefore

$$K = 0, \quad L = 0.$$

The electromagnetic and gravitational fields together have ten independent real invariants, namely the real and imaginary parts of I, J, K, L, M. However, only nine of these are determined by the curvature $R_{\mu\nu\varrho\sigma}$ since it is unaffected by duality rotations of the electromagnetic field. These are the nine independent real invariants of ψ_{ABCD} and $\phi_{ABC'D'} = \phi_{AB}\bar{\phi}_{C'D'}$. The phase of ϕ_{AB} is undetermined by $\phi_{ABC'D'}$, so we can take for these invariants[14]

$$I, J, |K|, |L|, |M|$$

and the arguments of the two ratios

$$K{:}L{:}M.$$

(The invariants $|K|^2$, $|L|^2$, $|M|^2$, $K\bar{L}$, $L\bar{M}$, $M\bar{K}$ are easily expressible in terms of ψ_{ABCD} and $\phi_{ABC'D'}$.)

5. Analytic Solutions of Einstein's Equations

Let $\mathfrak{M}$ be an analytic (connected) Riemannian manifold. Then starting from any point O on $\mathfrak{M}$ at which the curvature tensor $R_{\mu\nu\varrho\sigma}$ and all its covariant derivatives are known, it is possible to calculate the curvature tensor (and its derivatives) at any other point by means of a power series:

$$(5.1)\quad (R_{\mu\nu\varrho\sigma})_x = (R_{\mu\nu\varrho\sigma})_0 + x^\alpha(\partial_\alpha R_{\mu\nu\varrho\sigma})_0 + \frac{1}{2!}\, x^\alpha x^\beta(\partial_\alpha\partial_\beta R_{\mu\nu\varrho\sigma})_0 + \ldots .$$

The point x is that point on $\mathfrak{M}$ whose geodesic distance from O is $\sqrt{(x_\alpha x^\alpha)}$ and which lies on the geodesic through O which starts off in the direction of x^α (Riemannian coordinates). (If x^α is null this has to be interpreted suitably.) The $R_{\mu\nu\varrho\sigma}$ at the point x is referred to axes which are those at O transferred in parallel along this geodesic. If the power series does not converge, the point x may be reached in several steps, using intermediate points, in the manner of analytic continuation. This power series expression and its convergence is considered by Thomas (*25*, p. 234).

Equation (5.1) is a special case of the more general situation, whereby any analytic tensor field may be calculated from a knowledge

[14] When the electromagnetic field is null there still remain the seven real invariants given by $I, J, |L|, |M|$ and the argument of L/M. Thus Witten (*6*) is mistaken when he claims that there remain only the four real invariants of ψ_{ABCD} in this case. For example, the invariant $\phi_{ABE'F'}\psi^{AB}{}_{CD}\bar{\psi}^{E'F'}{}_{G'H'}\phi^{CDG'H'} = |L|^2$, need not vanish when $K = 0$. Such an invariant could appear as a quotient of invariants built up from Witten's list.

of the tensor and all its covariant derivatives at the point O alone:

$$(f\ldots)_x = (f\ldots)_0 + x^\alpha(\partial_\alpha f\ldots)_0 + \frac{1}{2!}x^\alpha x^\beta(\partial_\alpha\partial_\beta f\ldots)_0 + \ldots$$

$$= [\exp(x^\alpha\partial_\alpha)f\ldots]_0 = \left[\lim_{n\to\infty}\left(1+\frac{1}{n}x^\alpha\partial_\alpha\right)^n f\ldots\right]_0. \tag{5.2}$$

The ∂_α's are to be taken as acting only on $f\ldots$ and not on x^α. (This last expression can be used to obtain the power series expression since

$$(f\ldots)_{\varepsilon x} = [(1+\varepsilon x^\alpha\partial_\alpha)f\ldots]_0 + O(\varepsilon^2),$$

which may be applied n times with $\varepsilon = 1/n$, giving $(f\ldots)_x$ correct to order $1/n$.)

These power series can be used as the basis for a coordinate-free approach to Riemannian geometry. Instead of specifying a space by giving the metric tensor $g_{\mu\nu}$ as a function of some coordinates, the space may be determined (except possibly for some of its topological properties in the large) by specifying $R_{\mu\nu\varrho\sigma}$, $\partial_\alpha R_{\mu\nu\varrho\sigma}$, $\partial_\alpha\partial_\beta R_{\mu\nu\varrho\sigma}$, ... at a point O. To specify a set of tensors at a point does not require coordinates since their algebraic tensorial properties need only be given. The metric tensors $g_{\mu\nu}$, $g^{\mu\nu}$ and the alternating tensor $\sqrt{(\pm g)}\varepsilon_{\mu\ldots\sigma}$ are also supposed to be specified at the point O. They are an essential part of the tensor algebra at O.

A difficulty about specifying a space in this way is that $R_{\mu\nu\varrho\sigma}$, $\partial_\alpha R_{\mu\nu\varrho\sigma}$, $\partial_\alpha\partial_\beta R_{\mu\nu\varrho\sigma}$, ... are not algebraically (tensorially) independent of one another. Relation (2.1) implies identities (Ricci) connecting second derivatives with the curvature tensor, and also there is the Bianchi identity which is the consistency condition for (2.1). The Bianchi identity is in fact the only consistency condition required (*25*, pp. 131, 132). Applying these two types of identity to the higher derivatives of $R_{\mu\nu\varrho\sigma}$ a host of relations is obtained. It is therefore of importance to be able to single out a set of tensors which are algebraically independent (in the general case) and from which $R_{\mu\nu\varrho\sigma}$ and all its derivatives are obtainable by algebraic operations. It is possible to show that the following set of tensors, in fact, has all these properties:

$$Q^{\mu\nu}{}_{\varrho\sigma} = R^{\mu}{}_{(\varrho}{}^{\nu}{}_{\sigma)}, \quad Q^{\mu\nu}{}_{\varrho\sigma\alpha} = \partial_{(\alpha}R^{\mu}{}_{\varrho}{}^{\nu}{}_{\sigma)}, \quad Q^{\mu\nu}{}_{\varrho\sigma\alpha\beta} = \partial_{(\alpha}\partial_{\beta}R^{\mu}{}_{\varrho}{}^{\nu}{}_{\sigma)} \quad \text{etc.}$$

Each $Q\ldots$ has the symmetry given by a Young tableau operator corresponding to a partition $(r-2, 2)$. That is to say, we have

$$Q_{\mu\nu\varrho\sigma\ldots\beta} = Q_{(\mu\nu)(\varrho\sigma\ldots\beta)} \quad \text{and} \quad Q_{\mu(\nu\varrho\sigma\ldots\beta)} = 0.$$

Apart from these symmetries and from certain considerations of convergence, the Q's may be chosen arbitrarily.[15] Unfortunately, however, if it is required to impose a condition such as Einstein's $R^{\mu}{}_{\nu\mu\sigma} = 0$ (or $\lambda g_{\sigma\nu}$) on the space, this implies a condition not only on $Q^{\mu\nu}{}_{\varrho\sigma}$, but also on $Q^{\mu\nu}{}_{\varrho\sigma\alpha}$, $Q^{\mu\nu}{}_{\sigma\varrho\alpha\beta}$, etc. These conditions are not linear, and they appear to be somewhat complicated. It seems for this reason that an approach based explicitly on these Q's would not be usually very convenient for general relativity. (However, in a later paper it is proposed to give a class of special solutions using this method.) On the other hand, if a spinor approach is used, these conditions take on a particularly simple form. This approach will now be described in more detail.

Suppose that $\mathfrak{M}$ has four dimensions and signature $(+ - - -)$, and that $R^{\mu}{}_{\nu\mu\sigma} = \lambda g_{\nu\sigma}$. Then we have seen that $R_{\mu\nu\varrho\sigma}$ can be represented uniquely by a totally symmetric spinor ψ_{ABCD} (λ being known). We wish to find a set of algebraically independent spinors from which

$$\psi_{ABCD}, \quad \partial_E{}^{P'}\psi_{ABCD}, \quad \partial_E{}^{P'}\partial_F{}^{Q'}\psi_{ABCD}, \ldots \tag{5.3}$$

(at the point O) can be constructed by means of algebraic spinor operations. The identities relating the spinors (5.3) arise from the equivalent of the Bianchi identity, namely (3.5):

$$\partial^{AP'}\psi_{ABCD} = 0 \quad \text{or} \quad \varepsilon^{EA}\partial_E{}^{P'}\psi_{ABCD} = 0 \tag{5.4}$$

[15] These Q's are somewhat analogous to (but different from) the "normal tensors" (see Thomas *25*, p. 102).

and the equivalent of (2.1), namely, (2.14), (2.15), (2.16), and (2.17):

(5.5)
$$\varepsilon_{R'S'}\{\partial_G{}^{R'}\,\partial_H{}^{S'}+\partial_H{}^{R'}\,\partial_G{}^{S'}\}\xi_A = \psi_{GHAB}\xi^B-\frac{\lambda}{3}\{\xi_G\varepsilon_{HA}+\xi_H\varepsilon_{GA}\},$$
$$\varepsilon_{R'S'}\{\partial_G{}^{R'}\,\partial_H{}^{S'}+\partial_H{}^{R'}\,\partial_G{}^{S'}\}\eta^{P'} = 0$$

(5.6)
$$\varepsilon^{GH}\{\partial_G{}^{R'}\,\partial_H{}^{S'}+\partial_G{}^{S'}\,\partial_H{}^{R'}\}\xi_A = 0,$$
$$\varepsilon^{GH}\{\partial_G{}^{R'}\,\partial_H{}^{S'}+\partial_G{}^{S'}\,\partial_H{}^{R'}\}\eta^{P'} = \bar{\psi}^{R'S'P'}{}_{Q'}\eta^{Q'}-\frac{\lambda}{3}\{\eta^{R'}\varepsilon^{S'P'}+\eta^{S'}\varepsilon^{R'P'}\}$$

(see 3.7 and 3.3) applied to ψ_{ABCD} and its derivatives.

The various derivatives of (5.4) must all hold identically also. Hence the algebraic relations on the spinors (5.3) arising from (5.4) are

(5.7)
$$\varepsilon^{HA}(\partial_E{}^{P'}\ \ldots\ \partial_G{}^{R'}\,\partial_H{}^{S'}\psi_{ABCD}) = 0.$$

This expresses a condition on (namely, the vanishing of) the part of $\partial_E{}^{P'}\ \ldots\ \partial_H{}^{S'}\psi_{ABCD}$ which is skew in H, A and says nothing about the part symmetric in H, A. Moreover the relations (5.5) connect

$$\varepsilon_{R'S'}(\partial_E{}^{P'}\ \ldots\ \partial_G{}^{R'}\,\partial_H{}^{S'}\ \ldots\ \partial_K{}^{V'}\,\psi_{ABCD}) + \varepsilon_{R'S'}(\partial_E{}^{P'}\ \ldots\ \partial_H{}^{R'}\,\partial_G{}^{S'}\ \ldots\ \partial_K{}^{V'}\psi_{ABCD})$$

with lower derivatives of ψ_{ABCD}, while (5.6) connect

$$\varepsilon^{GH}(\partial_E{}^{P'}\ \ldots\ \partial_G{}^{R'}\,\partial_H{}^{S'}\ \ldots\ \partial_K{}^{V'}\psi_{ABCD}) + \varepsilon^{GH}(\partial_E{}^{P'}\ \ldots\ \partial_G{}^{S'}\partial_H{}^{R'}\ \ldots\ \partial_K{}^{V'}\psi_{ABCD'})$$

with lower derivatives of ψ_{ABCD}. These express conditions only on parts of $\partial_E{}^{P'} \ldots \partial_K{}^{V'}\psi_{ABCD}$ which are skew in a pair of primed indices or in a pair of unprimed indices. Thus the algebraic relations arising from (5.4), (5.5), and (5.6) connecting the spinors (5.3) are all concerned with parts of $\partial_E{}^{P'}\ \ldots\ \partial_K{}^{V'}\psi_{ABCD}$ which are skew in at least one pair of indices. They imply no conditions on the parts totally symmetric in all primed indices and in all unprimed indices. (It might, perhaps, be thought that other relations could be obtained by expanding skew parts of $\partial_E{}^{P'} \dots \partial_K{}^{V'}\psi_{ABCD}$ in two different ways. However, these all lead back to (5.7) which is the only consistency condition implied.) Hence the spinors

$$\psi_{ABCD}, \psi_{ABCDE}{}^{P'} = \partial_{(E}{}^{P'}\psi_{ABCD)}, \psi_{ABCDEF}{}^{P'Q'} = \partial_{(E}{}^{(P'}\partial_F{}^{Q')}\psi_{ABCD)}, \dots \tag{5.8}$$

are all algebraically independent and can therefore be specified arbitrarily (apart from convergence considerations) at the point 0.

The problem is now to show, conversely, that all the spinors (5.3) can be obtained algebraically from the spinors (5.8). For then ψ_{ABCD}, $\psi_{ABCDEP'}$, $\psi_{ABCDEFP'Q'}$, ... will be a complete set of algebraically independent spinors at O, which can be used to generate the space $\mathfrak{W}$. In order to show that they form such a complete set, an argument by induction will be used. We wish to express $\partial_E{}^{P'} \dots \partial^K{}_{V'}\psi_{ABCD}$ in terms of $\psi_{ABCDE\dots K}{}^{P'\dots V'}$ and lower order derivatives of ψ_{ABCD} since it may be supposed as the inductive hypothesis that all these lower derivatives have already been expressed algebraically in terms of symmetrized derivatives $\psi_{AB\dots G}{}^{P'\dots R'}$. Now if we add together all the spinors obtained from $\partial_E{}^{P'} \dots \partial_K{}^{V'}\psi_{ABCD}$ by permuting $P', \dots, V'$ in all possible ways and $A, B, C, D, E, \dots K$ in all possible ways, we get a multiple of $\psi_{AB\dots K}{}^{P'\dots V'}$. Thus, if it can be shown that each of the spinors obtained by such permutations differs from $\partial_E{}^{P'} \dots \partial_K{}^{V'}\psi_{ABCD}$ by expressions involving only lower derivatives of ψ_{ABCD} the result will be proved. The spinor $\partial_E{}^{P'} \dots \partial_K{}^{V'}\psi_{ABCD}$ will then be seen to differ from $\psi_{AB\dots K}{}^{P'\dots V'}$ by a spinor built up from lower derivatives of ψ_{ABCD}.

Any two spinors obtained by such a permutation of indices from

$$\partial_E{}^{P'} \dots \partial_K{}^{V'}\psi_{ABCD}$$

will be called *equivalent* (denoted by $\sim$) if they differ from each other by expressions built up from lower order derivatives of ψ_{ABCD}. This is clearly an equivalence relation. It is required to show that all such spinors are, in fact, equivalent to one another. Now since

$$\partial_W{}^{X'}\partial_Y{}^{Z'} - \partial_Y{}^{Z'}\partial_W{}^{X'} \equiv \tfrac{1}{2}\varepsilon^{X'Z'}\varepsilon_{M'N'}\{\partial_W{}^{M'}\partial_Y{}^{N'} + \partial_Y{}^{M'}\partial_W{}^{N'}\}$$
$$+ \tfrac{1}{2}\varepsilon_{WY}\varepsilon^{ST}\{\partial_R{}^{X'}\partial_T{}^{Z'} + \partial_S{}^{Z'}\partial_T{}^{X'}\}$$

(see 2.12), we have, applying (4.5) and (5.6)

$$\ldots \partial_W{}^{X'} \partial_Y{}^{Z'} \ldots \psi_{ABCD} \sim \ldots \partial_Y{}^{Z'} \partial_W{}^{X'} \ldots \psi_{ABCD}.$$

Hence any permutation of the $\partial_M{}^{N'}$ symbols gives rise to an equivalent spinor. (Any permutation can be expressed as a product of transpositions of adjacent elements.) That is, any permutation of $P', \ldots, V'$ can be applied to $\partial_E{}^{P'} \ldots \partial_K{}^{V'}\psi_{ABCD}$ provided that the same permutation is applied to $E, \ldots, K$ and an equivalent spinor is obtained. It remains to show that $E, \ldots, K, A, B, C, D$ can be permuted independently and an equivalent spinor is still obtained. The symmetry of ψ_{ABCD} implies that A, B, C, D can be permuted without change. Furthermore, from 5.7, K and A can be interchanged in $\partial_E{}^{P'} \ldots \ldots \partial_K{}^{V'}\psi_{ABCD}$. Also,

$$\ldots \partial_Y{}^{Z'} \ldots \partial_K{}^{V'}\psi_{ABCD} \sim \ldots \partial_K{}^{V'} \ldots \partial_Y{}^{Z'}\psi_{ABCD},$$

$$\sim \ldots \partial_K{}^{V'} \ldots \partial_A{}^{Z'}\psi_{YBCD} \sim \ldots \partial_A{}^{Z'} \ldots \partial_K{}^{V'}\psi_{YBCD}$$

so that A can be interchanged with any other unprimed index and an equivalent spinor is obtained. It follows that any pair of unprimed indices can be interchanged since

$$\ldots \partial_W{}^{X'} \ldots \partial_Y{}^{Z'} \ldots \psi_{ABCD} \sim \ldots \partial_W{}^{X'} \ldots \partial_A{}^{Z'} \ldots \psi_{YBCD},$$

$$\sim \ldots \partial_Y{}^{X'} \ldots \partial_A{}^{Z'} \ldots \psi_{WBCD} \sim \ldots \partial_Y{}^{X'} \ldots \partial_W{}^{Z'} \ldots \psi_{ABCD}.$$

Hence all the spinors are equivalent and the result is proved.

As examples of the above, we have

$$\partial_E{}^{P'}\psi_{ABCD} = \psi_{ABCDE}{}^{P'},$$

$$\partial_E{}^{P'}\partial_F{}^{Q'}\psi_{ABCD} = \psi_{ABCDEF}{}^{P'Q'} + \varepsilon_{EF}\varepsilon^{P'Q'}\{\tfrac{3}{4}\psi_{(AB}{}^{GH}\psi_{CD)GH} - \tfrac{1}{2}\lambda\psi_{ABCD}\}$$
$$+\varepsilon^{P'Q'}\{\psi_{(ABC}{}^{G}\psi_{D)EFG} + \tfrac{1}{3}\lambda\psi_{E(ABC}\varepsilon_{D)F} + \tfrac{1}{3}\lambda\psi_{F(ABC}\varepsilon_{D)E}\}.$$

Higher derivatives involve $\bar{\psi}_{A'B'C'D'}$, $\bar{\psi}_{A'B'C'D'E'P}$, ... also. We have from (5.1), with $\psi_{ABCD} = (\psi_{ABCD})_0$, etc.,

$$(\psi_{ABCD})_x = \psi_{ABCD} + x^{EP'}\partial_{EP'}\psi_{ABCD} + \frac{1}{2!}x^{EP'}x^{FQ'}\partial_{EP'}\partial_{FQ'}\psi_{ABCD} + \ldots.$$

Hence

$$(\psi_{ABCD})_x = \psi_{ABCD} + x^{EP'}\psi_{AB(\,DEP'} + \tfrac{1}{2}x^{EP'}x^{FQ'}\psi_{ABCDEFP'Q'}$$
$$+\tfrac{1}{2}(x_{EP}x^{EP'})\{\tfrac{3}{4}\psi_{(AB}{}^{GH}\psi_{CD)GH} - \tfrac{1}{2}\lambda\psi_{ABCD}\} + O(x^3).$$

It is possible to obtain a class of exact solutions for gravitational plane waves using this method. Such solutions, obtained using more conventional methods, have been known for some time (for references, see Bondi *et al.*, *23*). Let

$$\psi_{ABCD} = \alpha_0\pi_A\pi_B\pi_C\pi_D, \quad \psi_{ABCDEP'} = \alpha_1\pi_A\pi_B\pi_C\pi_D\pi_E\bar{\pi}_{P'},$$
$$\psi_{ABCDEP'Q'} = \alpha_2\pi_A \ldots \pi_F\bar{\pi}_{P'}\bar{\pi}_{Q'}, \ldots \tag{5.9}$$

at the point O, where π_A is a spinor corresponding to the null direction giving the direction of motion of the wave and $\alpha_0, \alpha_1, \ldots$ are complex numbers. Suppose $\lambda = 0$. It will now be shown that the unsymmetrized derivatives of ψ_{ABCD} are all equal to the symmetrized derivatives, so the situation is much simplified in this case. As an inductive hypothesis we assume that all the derivatives of ψ_{ABCD} of lower order than $\partial_E{}^{P'} \ldots \partial_K{}^{V'}\psi_{ABCD}$ are already symmetric and therefore equal to the corresponding expressions 5.9. The argument given above shows that $\partial_E{}^{P'} \ldots \partial_K{}^{V'}\psi_{ABCD}$ differs from $\psi_{ABCDE\ldots K}{}^{P'\ldots V'}$ by expressions obtained by applying rule (5.5) and (5.6) to ψ_{ABCD} and derivatives of ψ_{ABCD}, and perhaps differentiating further. Since $\lambda = 0$, this leads to terms of the form

$$\psi_{XAG\ldots K}{}^{T'\ldots V'}\psi^X{}_{B\ldots F}{}^{P'\ldots S'} \quad \text{or} \quad \bar{\psi}_{Y'}{}^{P'T'\ldots V'}{}_{G\ldots K}\psi_{A\ldots F}{}^{Y'Q'\ldots S'}$$

only. (By the inductive hypothesis all the derivatives of ψ_{ABCD} which occur are equal to the $\psi_{\ldots}{}^{\ldots}$'s.) These terms all involve contractions between the $\psi_{\ldots}{}^{\ldots}$'s. But with $\psi_{\ldots}{}^{\ldots}$'s given by (5.9), any contraction must clearly vanish (since $\pi_X\pi^X = 0$). Hence

$$(\partial_E{}^{P'} \ldots \partial_K{}^{V'}\psi_{ABCD})_0 = \alpha_r\pi_A\pi_B \ldots \pi_K\bar{\pi}^{P'} \ldots \bar{\pi}^{V'}$$

as required.

The curvature at points other than O can now be calculated:

$$(\psi_{ABCD})_x = \alpha_0\pi_A\pi_B\pi_C\pi_D + \alpha_1 x^{EP'}\pi_A \ldots \pi_E\bar{\pi}_{P'} + \frac{1}{2!}\alpha_2 x^{EP'}x^{FQ'}\pi_A \ldots \pi_F\bar{\pi}_{P'}\bar{\pi}_{Q'} + \ldots$$

$$= f(x^{EP'}\pi_E\bar{\pi}_P)\,\pi_A\pi_B\pi_C\pi_D = f(x^\mu p_\mu)\,\pi_A\pi_B\pi_C\pi_D,$$

where

(5.10) $$f(s) = \alpha_0 + \alpha_1 s + \frac{1}{2!}\alpha_2 s^2 + \frac{1}{3!}a_3 s^3 + \ldots .$$

and $p_{AB'} = \pi_A\bar{\pi}_{B'}$. Thus the curvature is a function of the one parameter $x^\mu p_\mu$ only. It is constant along the (null) 3-spaces $x^\mu p_\mu =$ constant. Furthermore, by (5.2),

$$(\partial_{EP'}\psi_{ABCD})_x = (\partial_{EP'}\psi_{ABCD})_0 + x^{FQ'}(\partial_{FQ'}\{\partial_{EP'}\psi_{ABCD}\})_0 + \frac{1}{2!}x^{FQ'}x^{GR'}(\partial_{FQ'}\partial_{GR'}\{\partial_{EP'}\psi_{ABCD}\})_0 + \ldots$$

$$= f'(x^\mu p_\mu)\pi_A\pi_B\pi_C\pi_D\pi_E\bar{\pi}_{P'},$$

$$(\partial_{EP'}\partial_{FQ'}\psi_{ABCD})_x = f''(x^\mu p_\mu)\pi_A \ldots \pi_F\bar{\pi}_{P'}\bar{\pi}_{Q'},$$

etc. Hence ψ_{ABCD}, $\psi_{ABCDEP'}$, $\psi_{ABCDEFP'Q'}$, ... are all constant along the 3-space $x^\mu p_\mu = 0$. It follows that the whole space $\mathfrak{M}$ admits the three-parameter group of translations[16] in the directions lying in this 3-space. The space $\mathfrak{M}$ thus represents a plane wave which moves uniformly with the velocity of light in the direction represented by p_μ. The intensity and polarization of the wave are determined by the modulus and argument of the function $f(s)$.

Particular cases of interest are:

(i) the constant gravitational field with ψ_{ABCD} constant everywhere.

[16] $\mathfrak{M}$ also admits a two parameter group of rotational (Lorentz) symmetries given by the unimodular matrices $t^A{}_B$ satisfying $t^A{}_B\pi^B = \pm\pi^A$, and disconnected from these, the rotations for which $t^A{}_B\pi^B = \pm\, i\pi^A$. There may also be some reflectional symmetries in special cases. This five parameter group of motions serves to characterize the plane wave solutions (see Bondi *et al.*, (*23*)).

Here $f(s) =$ constant, i.e., $\alpha_1 = \alpha_2 = \ldots = 0$, and $\mathfrak{M}$ admits additional translational motions.

(ii) Sinusoidal waves;

$$f(s) \equiv ae^{ins}+b\bar{e}^{ins}, \quad \text{i.e.,} \quad \alpha_r = a(in)^r+b(-in)^r.$$

In this case $\mathfrak{M}$ admits an additional discrete group of translations.

(iii) Gravitational pulse; for example,

$$f(s) = \begin{cases} b \exp\left(\dfrac{c}{s-a} - \dfrac{c}{s+a}\right) & \text{if} \quad -a < s < a \\ 0 & \text{if} \quad s \leqq -a \quad \text{or} \quad s \geqq a. \end{cases}$$

Case (iii) is not strictly an analytic manifold. $\mathfrak{M}$ has to be constructed from three analytic pieces (two of which are flat). The middle piece fits on smoothly to the other two pieces, the join being C^∞. The space is exactly flat before the pulse arrives and is again exactly flat after the pulse has departed (*23*, p. 523).

An advantage of a method such as this for obtaining spaces satisfying Einstein's equations is that the usual problem of deciding whether an effect is real or merely due to a bad choice of coordinates simply does not arise. The curvature at any point is found directly. However, it will naturally be convenient to be able to introduce coordinates into a space defined in this way, if desired. A coordinate system on $\mathfrak{M}$ may be thought of as a set of four scalar fields $u_{(i)}$ ($i = 0, \ldots 3$). The symmetric derivatives $\partial_{(\alpha} \ldots \partial_{\gamma)}u_{(i)}$ of each $u_{(i)}$ may be specified arbitrarily at the point O. The values of the coordinates $u_{(i)}$ and their derivatives at any other point may then be calculated using (5.2), after some of the unsymmetrized derivatives have been obtained using (2.1). The expression for the metric at each point can be obtained from the first derivatives of the $u_{(i)}$ at that point. This method will be described in detail in a later paper.

The case when an electromagnetic field is present in the space can be treated by an extension of the coordinate-free method for empty space described above. The spinors

$$\psi_{ABCDE\ldots G}{}^{P'\ldots R'} = \partial_{(E}{}^{(P'} \ldots {}_{G}{}^{R')}\psi_{ABCD)}$$

are defined as before and spinors ϕ_{AB}, $\phi_{ABC}{}^{P'}$, $\phi_{ABCD}{}^{P'Q'}$, ... are introduced, defined similarly by

$$\phi_{ABC\ldots E}{}^{P'\ldots R'} = \partial_{(C}{}^{(P'} \ldots \partial_{E}{}^{R')}\phi_{AB)}.$$

By the same kind of argument as before, it follows that ϕ_{AB}, $\phi_{ABC}{}^{P'}$, ..., ψ_{ABCD}, $\psi_{ABCDE}{}^{P'}$, ... are all algebraically independent. Instead of (5.4) we have

$$\varepsilon^{CA}\partial_C{}^{P'}\phi_{AB} = 0 \quad \text{and} \quad -\varepsilon^{EA}\partial_E{}^{P'}\psi_{ABCD} = \bar{\phi}^{P'}{}_{Q'}\partial_D{}^{Q'}\phi_{BC}$$

from (3.10) and (3.12). The first of these states the symmetry of

$$\partial_C{}^{P'} \ldots \partial_E{}^{R'}\phi_{AB}$$

in E, A, while the second expresses the part of

$$\partial_E{}^{P'} \ldots \partial_G{}^{R'}\psi_{ABCD}$$

skew in G, A in terms of derivatives of ϕ_{AB} of at most the same order. They imply no condition on the symmetrized derivatives of ϕ_{AB} or ψ_{ABCD}. Nor do the equivalents of (5.5) and (5.6), which differ from them only in that the second relation (5.5) is replaced by

$$\varepsilon_{R'S'}\{\partial_G{}^{R'}\partial_H{}^{S'} + \partial_H{}^{R'}\partial_G{}^{S'}\}\eta^{P'} = \phi_{GH}\bar{\phi}^{P'}{}_{Q'}\eta^{Q'}$$

(see 3.13) and the first relation (5.6) by

$$\varepsilon^{GH}\{\partial_G{}^{R'}\partial_H{}^{S'} + \partial_G{}^{S'}\partial_H{}^{R'}\}\xi_A = \bar{\phi}_{AB}\xi^B$$

The argument to show that the unsymmetrized derivatives can be expressed algebraically in terms of the symmetrized derivatives is exactly analogous to that for pure gravitational case. The derivative $\partial_C{}^{P'} \ldots \partial_E{}^{R'}\phi_{AB}$ differs from $\phi_{ABC\ldots E}{}^{P'\ldots R'}$ by expressions constructed from lower order derivatives of ϕ_{AB} and ψ_{ABCD}, while $\partial_E{}^{P'} \ldots \partial_G{}^{R'}\psi_{ABCD}$ differs from $\psi_{ABCDE\ldots G}{}^{P'\ldots R'}$ by expressions constructed from derivatives of ϕ_{AB} of the same order or lower and from lower order derivatives of ψ_{ABCD}. Thus, we can construct $\partial_C{}^{P'}\phi_{AB}$, $\partial_E{}^{P'}\psi_{ABCD}$, $\partial_C{}^{P'}\partial_D{}^{Q'}\phi_{AB}$, $\partial_E{}^{P'}\partial_F{}^{Q'}\psi_{ABCD}$, ..., in that order, from the symmetrized derivatives. The symmetric spinors $\phi_{AB}, \phi_{ABC}{}^{P'}, \phi_{ABCD}{}^{P'Q'}, \ldots, \psi_{ABCD}, \psi_{ABCDE}{}^{P'}, \ldots$ can therefore be specified arbitrarily at a point O (apart from convergence considerations) and ϕ_{AB}, ψ_{ABCD} at any other point can be determined from them by (5.2).

A simple example is the case of a combined gravitational-electromagnetic wave (see also *22*). Here $\psi_{ABCD}, \psi_{ABCDEP'}, \ldots$ are given by (5.9) and

$$\phi_{AB} = \beta_0\pi_A\pi_B, \quad \phi_{ABCP'} = \beta_1\pi_A\pi_B\pi_C\bar{\pi}_{P'},$$
$$\phi_{ABCDP'Q'} = \beta_2\pi_A\pi_B\pi_C\pi_D\bar{\pi}_{P'}\bar{\pi}_{Q'}, \ldots$$

at the point O. As was the case, considered earlier, with the pure gravitational wave the unsymmetrized derivatives of ϕ_{AB} and ψ_{ABCD} turn out to be equal to the symmetrized derivatives provided that $\lambda = 0$. Hence

$$(\phi_{AB})_x = g(x^\mu p_\mu)\pi_A\pi_B, \quad (\psi_{ABCD})_x = f(x^\mu p_\mu)\pi_A\pi_B\pi_C\pi_D,$$

where

$$g(s) \equiv \beta_0+\beta_1 s+\frac{1}{2!}\beta_2 s^2+ \ldots$$

and $f(s)$ is given by (5.10) as before. The discussion given in the pure gravitational case applies here also. The function $g(s)$ determines the intensity and polarization of the electromagnetic part of the wave and $f(s)$ the "purely gravitational" part. The electromagnetic field is null everywhere and the gravitational field is [4]. All six principal null directions coincide and point in the direction p_μ giving the motion of the wave.

Table I summarizes some of the many analogies between the electromagnetic and gravitational fields, that are brought out by the spinor formalism.

I should like to offer my thanks to Dr. D. W. Sciama for his early encouragement and for many invaluable discussions.

[*Received: September 16, 1959*]

TABLE I. SUMMARY OF SOME OF THE RESULTS OF THIS PAPER ON THE COMPARISONS BETWEEN ELECTROMAGNETIC AND GRAVITATIONAL FIELDS IN SPINOR FORM

	Maxwell field	Curvature tensor with $R^{\mu}{}_{\gamma\mu\sigma} = 0$
Tensor-spinor correspondence	$F_{\mu\gamma} \leftrightarrow \frac{1}{2}\{\phi_{AB}\varepsilon_{C'D'} + \varepsilon_{AB}\bar{\phi}_{C'D'}\}$	$R_{\mu\nu\varrho\sigma} \leftrightarrow \frac{1}{2}\{\psi_{ABCD}\varepsilon_{E'F'}\varepsilon_{G'H'} + \varepsilon_{AB}\varepsilon_{CD}\bar{\psi}_{E'F'G'H'}\}$
First order equation	Maxwell equations: $\partial^{AC'}\phi_{AB} = 0$	Bianchi identities: $\partial^{AE'}\psi_{ABCD} = 0$
(Super-)energy tensor	Maxwell stress tensor $\leftrightarrow$ $\frac{1}{2}\phi_{AB}\bar{\phi}_{C'D'}$	Robinson–Bel tensor $\leftrightarrow$ $\psi_{ABCD}\bar{\psi}_{E'F'G'H}$
Duality rotations	$\phi_{AB} = e^{i\theta}\phi_{AB}$	$\psi_{ABCD} \rightarrow e^{i\theta}\psi_{ABCD}$
Canonical representation	$\phi_{AB} = \eta_{(A}\zeta_{B)}$	$\psi_{ABCD} = \alpha_{(A}\beta_B\gamma_C\delta_{D)}$
Classification scheme	[11] $K \neq 0$ ↙ ↓ [2]→[—] $K = 0$	[1111] $I^3 \neq 6J^2$ ↙ ↓ [211]→[22] $I^3 = 6J^2 \neq 0$ ↙ ↓ ↙ ↓ [31]→[4]→[—] $I = J = 0$
Plane wave	$\phi_{AB}(x^{\mu}) = g(x^{\mu}p_{\mu})\pi_A\pi_B$	$\psi_{ABCD}(x^{\mu}) = f(x^{\mu}p_{\mu})\,\pi_A\pi_B\pi_C\pi_D$

References

1. L. INFELD AND B. L. VAN DER WAERDEN, *Sitzber. preuss. Akad. Wiss. Physik.-math. Kl.* **9,** 380 (1933).
2. W. L. BADE AND H. JEHLE, *Revs. Modern Phys.* **25,** 714 (1953).
3. H. S. RUSE, *Proc. Roy. Soc. Edinburgh* **57,** 97 (1937).
4. G. Y. RAINICH, *Trans. Am. Math. Soc.* **27,** 106 (1925).
5. C. W. MISNER AND J. A. WHEELER, *Annals of Physics* **2,** 525 (1957).
6. L. WITTEN, *Phys. Rev.* **113,** 357 (1959).
7. P. A. M. DIRAC, *Proc. Roy. Soc. A***155,** 447 (1936).
8. L. BEL, *Compt. rend.* **247,** 1094 (1958).
9. L. BEL, *Compt. rend.* **248,** 1297 (1959).
10. E. M. CORSON, *Introduction to Tensors, Spinors, and Relativistic Wave-equations*, Blackie, London, 1953.
11. A. S. EDDINGTON, *The Mathematical Theory of Relativity*, Cambridge Univ. Press, London and New York, 1923.
12. J. L. SYNGE, *Relativity: The Special Theory*, North-Holland Publ. Co., Amsterdam, 1956.
13. L. WITTEN, *Phys. Rev.* **115,** 206 (1959).

14. H. S. Ruse, *Proc. Roy. Soc. Edinburgh* A**62,** 64 (1944).
15. R. Debever, *Compt. Rend. Acad. Sci.* **249,** 1324 (1959).
16. R. Debever, *Compt. Rend. Acad. Sci.* **249,** 1744 (1959).
17. F. A. E. Pirani, *Phys. Rev.* **105,** 1089 (1957).
18. J. H. Grace and A. Young, *Algebra of Invariants*, Cambridge Univ. Press, London and New York, 1903.
19. R. Penrose, *Proc. Cambridge Phil. Soc.* **55,** 137 (1959).
20. A. Z. Petrov, *Doklady Akad. Nauk SSSR* **105,** 905 (1955).
21. J. Géhéniau, *Compt. rend.* **244,** 723 (1957).
22. A. Lichnerowicz, *Compt. rend.* **246,** 893 (1958).
23. H. Bondi, F. A. E. Pirani and I. Robinson, *Proc. Roy. Soc.* A**251,** 519 (1959).
24. F. A. E. Pirani, *Proc. Roy. Soc.* A**252,** 96 (1959).
25. T. Y. Thomas, *Differential Invariants of Generalized Spaces*, Cambridge Univ. Press, London and New York, 1934.

NOTES ON EXTRACT 11

THE experimental aspect of general relativity still leaves much to be desired. In 1959 it seemed as if a new era had been initiated by the experiment described in this paper; it may still prove to be the case that the theory will be able to be tested experimentally in the laboratory in numerous ways, but how this will come is no longer clear. The momentum generated by the use of the Mössbauer effect has dissipated in the last decade.

EXTRACT 11[†]

Gravitational Red-shift in Nuclear Resonance

R. V. POUND and G. A. REBKA, Jr.

Lyman Laboratory of Physics, Harvard University, Cambridge, Massachusetts
[Received October 15, 1959]

It is widely considered desirable to check experimentally the view that the frequencies of electromagnetic spectral lines are sensitive to the gravitational potential at the position of the emitting system. The several theories of relativity predict the frequency to be proportional to the gravitational potential. Experiments are proposed to observe the timekeeping of a "clock" based on an atomic or molecular transition, when held aloft in a rocket-launched satellite, relative to a similar one kept on the ground. The frequency ν_h and thus the timekeeping at height h are related to that at the earth's surface ν_0 according to

$$\Delta\nu_h = \nu_h - \nu_0 = \nu_0 gh/c^2(1+h/R) \approx \nu_0 h \times (1.09 \times 10^{-18}),$$

where R is the radius of the earth and h is the altitude measure in cm. Very high accuracy is required of the clocks even with the altitudes available with artificial satellites. Although several ways of obtaining the necessary frequency stability look promising, it would be simpler if a way could be found to do the experiment between fixed terrestrial points. In particular, if an accuracy could be obtained allowing the measurement of the shift between points differing as little as one to ten kilometers in altitude, the experiment could be performed between a mountain and a valley, in a mineshaft, or in a borehole.

† *Phys. Rev. Letters*. **3**, 439 (1959).

Recently Mössbauer has discovered[1] a new aspect of the emission and scattering of γ rays by nuclei in solids. A certain fraction f of γ rays of the nuclei of a solid are emitted without individual nuclear recoil. Instead, the recoil momentum is delivered to the crystal lattice as a whole resulting in negligible Doppler shift. Such γ rays are in resonance with nuclei similarly bound in a lattice and a similar fraction f of the electromagnetic resonant cross section

$$\sigma_R = 2\pi\lambda^2\left(\frac{2I_e+1}{2I_g+1}\right)\frac{1}{1+\alpha},$$

where I_e and I_g are the spins of the emitting and the ground states respectively, and α is the internal conversion coefficient, pertains to the scattering. Calculations based on the Debye model of lattice vibrations yield for f at temperatures T much less than the Debye temperature θ_D

$$f = \exp\left\{-\frac{3}{2}\frac{E_\gamma^2}{2Mc^2k\theta_D}\left[1+\frac{2}{3}\left(\frac{\pi T}{\theta_D}\right)^2\right]\right\},$$

where E_γ is the energy of the γ ray, M is the nuclear mass, and k is Boltzmann's constant. The factor $(E_\gamma^2/\alpha Mc^2k\theta_D)$ is the ratio of the recoil energy that would be taken up by the free nucleus to $k\theta_D$. For γ rays much above the 129 keV employed by Mössbauer the factor f becomes very small even at absolute zero.

The most striking evidence for the existence of this effect is the observation that the attenuation of the 129-keV γ rays of Ir^{191} in passing through an iridium absorber is reduced if the source is moved. The speed required to reduce the part of the attenuation caused by resonant scattering to one-half its maximum value was found to be approximately 1.5 cm/sec. From this a half-life of the excited state is derived to be 0.1 mμ sec. Others have repeated this experiment, and extended it to helium temperatures.[2, 3] One other case is reported,[3] that

[1] R. L. Mössbauer, *Z. Physik* **151,** 124 (1958); *Naturwissenschaften* **45,** 538 (1958); *Z. Naturforsch.* **14***a,* 211 (1959).

[2] Craig, Dash, McGuire, Nagle, and Reiswig, *Phys. Rev. Letters* **3,** 221 (1959).

[3] Lee, Meyer-Schutzmeister, Schiffer, and Vincent, *Phys. Rev. Letters* **3,** 223 (1959).

of W^{182} wherein a half-life of 0.6 mμ sec is inferred by the Doppler width of the resonance. This is half the accepted lifetime as measured by delay coincidence techniques. It is not clear whether this discrepancy represents a limit of the technique or whether it is largely an instrumental problem, as the authors suggest, enhanced by the complex array of other γ rays in the Ta^{182} source. Of course, as has been suggested, one should expect to see effects caused by hyperfine structure in these spectra when lifetimes are long enough to allow them to be important. All the effects discussed in connection with the directional correlation of cascade γ rays should have an influence. For example, it would seem desirable to use a source that has a good chance of being in a normal lattice site and electronic state at the time of emission of the final γ ray in question. One could have serious after effects from β decays, from prior emission of high-energy γ rays or from electron captures as well as broadening from imperfections in the crystal lattice or short spin-lattice relaxation.

Even if the further development of the technique does not yield still narrower resonances, those already observed have fractional widths in frequency well below those of all the reference lines yet proposed for "atomic clocks". If the scattering is reduced to one-half its maximum by relative motion of the source and scatterer with velocity v, the Q, the ratio of the frequency to the full width at half-height of the resonance line being observed, is just $c/2v$. In the case of Mössbauer's experiment Q is about 1×10^{10} and in the case of W^{182} it is 7×10^{10}. In general $Q = 1.10 E_\gamma$ (MeV) $\tau_{\frac{1}{2}}$(mμ sec)$\times 10^{12}$.

A measurement of the gravitational red shift could be performed by transmitting γ rays from a source to a scatterer at an altitude different by h and by observing what relative velocity yields maximum scattering. For the predicted shift to be a full half-width of the line, the altitude difference h must be $h_{\frac{1}{2}} = [4.18/E_\gamma$ (MeV) $\tau_{\frac{1}{2}}$ (mμ sec)] km. Thus, for the width reported for W^{182}, 66 km difference of height would be required.

It is exciting to speculate about the possibilities opened up if cases of even less breadth can be found. For example, Fe^{57}, for which $E_\gamma =$

$= 0.0144$ MeV and $\tau_{\frac{1}{2}} = 100$ mμ sec, would require only 2.9 km separation were it to yield its natural breadth. Another example might be Zn^{67} with an excited level at 0.093 MeV, of half-life 9400 mμ sec. For this, if the natural breadth were obtained, $h_{\frac{1}{2}}$ would be 4.74 meters. This possibility represents a considerable extrapolation from present data. We are undertaking to examine these and other isotopes in various environments with the aim of selecting an isotope suitable for a gravitational experiment. Among other things equivalence of or absence of hyperfine structures in the sources and scatterers would be desirable.

Obviously one of the difficulties with large separations between source and scatterer arises from the inverse-square law of intensity. As a consequence of the participation of a large number of identical nuclei in an individual recoil-free scattering process, one anticipates the existence of intense Bragg diffraction from thin crystals. Thus one has the possibility of some degree of focusing with bent crystals. Furthermore one may use the Bragg reflection from thin crystals to separate the γ rays emitted without recoil from all others. In this way irrelevant background γ rays could be eliminated from the detector.

Total external reflection of low-energy γ rays at grazing angles of incidence offers a possibility of a "light-pipe" to increase the effective solid angle that the scatterer subtends at the source. Within the limits set by the small angle of total reflection, this pipe need not be optically straight.[4]

The fixed baseline used for an experiment of this type reduces unwanted Doppler shifts to only those resulting from thermal, seismic, or similar disturbances. To equal the predicted gravitational shift the fractional change required in the height difference is 3.27×10^{-8} per second. Perturbing effects must be kept well below this value but this is also true for the other methods of measuring the red shift. Relative motion could be separated from the red shift by simultaneous observations of beams traveling in both directions.

[4] We wish to thank E. M. Purcell for this suggestion.

Index

Abel 109
Absolute derivative 23
Accelerated frames 7
Accelerated transformations 6, 7
Adams 37
Archimedes 119

Bade 91
Baryons 88
Bass 256
Beisbroek 37
Bel 319, 329, 334
Bianchi 177
Bianchi identities 272
Bianchi identity 324
Birkhoff 248, 260
Black hole 90
Bondi 84, 256, 259
Bonnor 60, 83

Causality 70, 274
Chemical forces 88
Christoffel 27, 71, 141, 233
Clemence 35
Clifford 15, 16, 101, 107, 125
Coefficients of affine connection 18, 27
Collapse 86, 310
Coriolis acceleration 9, 11
Corson 91
Cosmology 37
Covariant derivative 17
Curvature spinors 323
Curvature tensor 23

D'Alembert 263
Debever 76, 332, 340
De Sitter universe 58
Dipole 71
Dipoles 191

Eddington 16, 56, 237, 330
Einstein universe 56
Eisenhart 237
Electromagnetic theory 3
Energy density 65
Energy tensor 64, 167
Equation of geodesic deviation 24, 239
Euler 164
Euler's theorem 104

Faraday 64
Field equations 30, 41, 155
Fock 39, 221
Foucault pendulum 11, 39
Freundlich 170
Friendlander 265

Galileo 4, 119, 142
Gauss 15, 101, 105, 108, 111, 114,141
Gaussian curvature 16
Gauss's theorem 44
Géhéniau 340
Generalized acceleration 18
Generally covariant 148
Geodesics 16, 103
Goldberg 233
Goldberg-Sachs 79
Grace 333
Gravitation 151
Gravitational field 3
Gravitational wave 234
Grommer 63
Grossman 141

Hamiltonian function 156
Harmonic coordinates 227

Harrison 88
Hilbert 163
Hoffman 63, 256
Holton 12
Hubble 60
Huygens's principle 263, 267

Inertial frame 4
Infeld 63, 64, 91, 175, 256

Jacobi 109
Jacobian 40
Jehle 91

Kepler 171
Klein space 237
Komar 329

Lagrange 109
Landau 233
Laplace's equation 20, 24, 176
Laue's scalar 160
Legendre 107
Leverrier 156, 172
Levi-Civita 24, 141
Lewison 175
Lichnerowicz 231, 234, 245
Lifshitz 233
Light 25, 35, 129
Local inertial frames 39
Lorentz condition 68
Lorentz group 13
Lovelock 47, 49

Mach 12, 39
Mach's principle 12, 48
Maxwell 62, 64
McCrea 90
Mercury 35, 156, 172
Metric 27
Metzner 84, 259, 281
Minkowski 141, 149, 283
Misner 322
Mössbauer 37

News 285
News function 85, 259, 285
Newtonian approximation 200

Noether 42
Noether's theorem 45
Normal coordinates 248
Nuclear forces 88

O'Brien 235
Omer 314
Oppenheimer 85, 88, 309

Penrose 91, 95, 319
Perihelion 168
Petrov 76, 79, 81, 94, 229
Pfaff 109
Pirani 24, 72, 76, 79, 229, 301
Poisson's equation 38, 159
Poles 191
Pound 37, 359
Principle of equivalence 3
Projectile 5
Pseudo-tensor 233, 247

Radiation 73
Rainich 322
Rebka 37, 359
Recession 60
Relative velocity 14
Retarded time 69
Ricci 141
Ricci tensor 29
Riemann 15, 16, 101, 107, 125, 141
Riemann–Christoffel tensor 23, 28 151
Riemannian geometry 26
Riemann tensor 72, 155, 230
Robinson 28, 319, 329, 334

Sachs 77, 299, 301, 329
Scalar curvature 41
Schild 178, 239
Schwarzschild 31, 37, 39, 172, 191 268, 279, 298, 342
Schwarzschild singularity 89
Sciama 355
Snyder 85, 88, 309
Special theory 3
Spectral lines 146
Spinors 91, 319

Stevinus 4
Synge 232, 235, 319

Theorema egregium 105
Thomas 248, 345
Tolman 313
Total curvature 114
Trautmann 28
Tulczyjew 66

Van der Burg 84, 259
Van der Waerden 91
Variational principle 40, 44
Volkoff 309
Volume element 40

Weak field 31
Weber 85
Weight 4
Weight of energy 131
Weyl 278, 285
Wheeler 88, 322
White dwarf 310
Wirtz 60
Witten 319, 332, 338

Young 333